大飞机出版工程

总主编　顾诵芬

# 民用飞机
# 适航术语手册

Terminology of
Civil Aircraft Airworthiness

袁 烨 黄 雄 罗 欢 张晓刚 秦 飞 著

上海交通大学出版社
SHANGHAI JIAO TONG UNIVERSITY PRESS

**内容提要**

  本书立足于民用飞机适航术语定义的确定和适航术语的梳理,较全面地考虑了航空法规、国际民航组织文件、各国适航当局文件、航空标准组织出版物、航空器制造商和航空公司出版物中的相关术语,并结合多年的研究成果和工程经验进行提炼,力求内容全面、通俗、简明、准确。

  本书可作为从事民用飞机相关领域研究的工程技术人员的参考用书。

**图书在版编目(CIP)数据**

民用飞机适航术语手册/ 袁烨等著. 一上海:上
海交通大学出版社,2023.12
 ISBN 978 - 7 - 313 - 29623 - 8

 Ⅰ. ①民… Ⅱ. ①袁… Ⅲ. ①民用飞机−适航−术语
−手册 Ⅳ. ①V271.1−62

 中国国家版本馆 CIP 数据核字(2023)第 220066 号

**民用飞机适航术语手册**

**MINYONG FEIJI SHIHANG SHUYU SHOUCE**

| | | | | |
|---|---|---|---|---|
| 著  者:袁 烨 黄 雄 罗 欢 张晓刚 秦 飞 | | | | |
| 出版发行:上海交通大学出版社 | | 地  址:上海市番禺路 951 号 | | |
| 邮政编码:200030 | | 电  话:021 - 64071208 | | |
| 印  制:上海文浩包装科技有限公司 | | 经  销:全国新华书店 | | |
| 开  本:710 mm×1000 mm 1/16 | | 印  张:30 | | |
| 字  数:522 千字 | | | | |
| 版  次:2023 年 12 月第 1 版 | | 印  次:2023 年 12 月第 1 次印刷 | | |
| 书  号:ISBN 978 - 7 - 313 - 29623 - 8 | | | | |
| 定  价:248.00 元 | | | | |

大飞机出版工程

# 丛书编委会

**总主编**

顾诵芬（中国航空工业集团公司科技委原副主任、中国科学院和中国工程院院士）

**副总主编**

贺东风（中国商用飞机有限责任公司董事长）

林忠钦（上海交通大学原校长、中国工程院院士）

**编委会**（按姓氏笔画排序）

王礼恒（中国航天科技集团公司科技委主任、中国工程院院士）

王宗光（上海交通大学原党委书记、教授）

刘　洪（上海交通大学航空航天学院教授）

任　和（中国商用飞机有限责任公司原副总工程师、教授）

李　明（中国航空工业集团沈阳飞机设计研究所研究员、中国工程院院士）

吴光辉（中国商用飞机有限责任公司首席科学家、C919飞机总设计师、中国工程院院士）

汪　海（上海市航空材料与结构检测中心主任、研究员）

张卫红（西北工业大学副校长、教授）

张新国（中国航空工业集团原副总经理、研究员）

陈　勇（中国商用飞机有限责任公司工程总师、ARJ21飞机总设计师、研究员）

陈迎春（中国商用飞机有限责任公司CR929飞机总设计师、研究员）

陈宗基（北京航空航天大学自动化科学与电气工程学院教授）

陈懋章（北京航空航天大学能源与动力工程学院教授、中国工程院院士）

金德琨（中国航空工业集团公司原科技委委员、研究员）

赵越让（中国商用飞机有限责任公司原总经理、研究员）

姜丽萍（中国商用飞机有限责任公司制造总师、研究员）

曹春晓（中国航空工业集团北京航空材料研究院研究员、中国工程院院士）

敬忠良（上海交通大学航空航天学院教授）

傅　山（上海交通大学电子信息与电气工程学院研究员）

# 总　　序

　　国务院在 2007 年 2 月底批准了大型飞机研制重大科技专项正式立项，得到全国上下各方面的关注。"大型飞机"工程项目作为创新型国家的标志工程重新燃起我们国家和人民共同承载着"航空报国梦"的巨大热情。对于所有从事航空事业的工作者，这是历史赋予的使命和挑战。

　　1903 年 12 月 17 日，美国莱特兄弟制作的世界第一架有动力、可操纵、比重大于空气的载人飞行器试飞成功，标志着人类飞行的梦想变成了现实。飞机作为 20 世纪最重大的科技成果之一，是人类科技创新能力与工业化生产形式相结合的产物，也是现代科学技术的集大成者。军事和民生对飞机的需求促进了飞机迅速而不间断的发展和应用，体现了当代科学技术的最新成果；而航空领域的持续探索和不断创新，为诸多学科的发展和相关技术的突破提供了强劲动力。航空工业已经成为知识密集、技术密集、高附加值、低消耗的产业。

　　从大型飞机工程项目开始论证到确定为《国家中长期科学和技术发展规划纲要》的十六个重大专项之一，直至立项通过，不仅使全国上下重视我国自主航空事业，而且使我们的人民、政府理解了我国航空事业半个多世纪发展的艰辛和成绩。大型飞机重大专项正式立项和启动使我们的民用航空进入新纪元。经过 50 多年的风雨历程，当今中国的航空工业已经步入了科学、理性的发展轨道。大型客机项目产业链长、辐射面宽、对国家综合实力带动性强，在国民经济发展和科学技术进步中发挥着重要作用，我国的航空工业迎来了新的发展机遇。

　　大型飞机的研制承载着中国几代航空人的梦想，造出与波音公司波音 737 和空客公司 A320 改进型一样先进的"国产大飞机"已经成为每个航空人心中奋斗的目标。然而，大型飞机覆盖了机械、电子、材料、冶金、仪器仪表、化工等几乎所有工业门类，集成数学、空气动力学、材料学、人机工程学、自动控制学等多种学科，是一个复杂的科技创新系统。为了迎接新形势下理论、技术和工程等方面的严峻挑战，迫切需要引入、借鉴国外的优秀出版物和数据资料，总结、巩固我们的经验和成果，编著一套以"大飞机"为主题的丛书，借以推动服务"大飞机"作为推动服务整个航空科学的切入点，同时对于促进我国航空事业的发展和加快航空紧缺人才的培养，具有十分重要的现实意义和深远的历史意义。

2008 年 5 月，中国商用飞机有限责任公司成立之初，上海交通大学出版社就开始酝酿"大飞机出版工程"，这是一项非常适合"大飞机"研制工作时宜的事业。新中国第一位飞机设计宗师——徐舜寿同志在领导我们研制中国第一架喷气式歼击教练机——歼教 1 时，亲自撰写了《飞机性能及算法》，及时编译了第一部《英汉航空工程名词字典》，翻译出版了《飞机构造学》《飞机强度学》，从理论上保证了我们的飞机研制工作。我本人作为航空事业发展 50 多年的见证人，欣然接受上海交通大学出版社的邀请担任该丛书的总主编，希望为我国的"大飞机"研制发展出一份力。出版社同时也邀请了王礼恒院士、金德琨研究员、吴光辉总设计师、陈迎春总设计师等航空领域专家撰写专著、精选书目，承担翻译、审校等工作，以确保这套"大飞机"丛书具有高品质和重大的社会价值，为我国的大飞机研制以及学科发展提供参考和智力支持。

编著这套丛书，一是总结整理 50 多年来航空科学技术的重要成果及宝贵经验；二是优化航空专业技术教材体系，为飞机设计技术人员的培养提供一套系统、全面的教科书，满足人才培养对教材的迫切需求；三是为大飞机研制提供有力的技术保障；四是将许多专家、教授、学者广博的学识见解和丰富的实践经验总结继承下来，旨在从系统性、完整性和实用性角度出发，把丰富的实践经验进一步理论化、科学化，形成具有我国特色的"大飞机"理论与实践相结合的知识体系。

"大飞机出版工程"丛书主要涵盖了总体气动、航空发动机、结构强度、航电、制造等专业方向，知识领域覆盖我国国产大飞机的关键技术。图书类别分为译著、专著、教材、工具书等几个模块；其内容既包括领域内专家们最先进的理论方法和技术成果,也包括来自飞机设计第一线的理论和实践成果。如:2009 年出版的荷兰原福克飞机公司总师撰写的 *Aerodynamic Design of Transport Aircraft*(《运输类飞机的空气动力设计》);由美国堪萨斯大学 2008 年出版的 *Aircraft Propulsion*(《飞机推进》)等国外最新科技的结晶;国内《民用飞机总体设计》等总体阐述之作和《涡量动力学》《民用飞机气动设计》等专业细分的著作;也有《民机设计 1 000 问》《英汉航空缩略语词典》等工具类图书。

　　该套图书得到国家出版基金资助,体现了国家对"大型飞机"项目和"大飞机出版工程"这套丛书的高度重视。这套丛书承担着记载与弘扬科技成就、积累和传播科技知识的使命,凝结了国内外航空领域专业人士的智慧和成果,具有较强的系统性、完整性、实用性和技术前瞻性,既可作为实际工作指导用书,亦可作为相关专业人员的学习参考用书。期望这套丛书能够有益于航空领域里人才的培养,有益于航空工业的发展,有益于大飞机的成功研制。同时,希望能为大飞机工程吸引更多的读者来关心航空、支持航空和热爱航空,并投身于中国航空事业做出一点贡献。

2009 年 12 月 15 日

# 前　　言

适航是"适航性"的简称，是航空器及其部件和子系统的整体性能和操纵特性在预期运行环境和使用限制下的安全性和物理完整性的一种品质，这种品质要求航空器应始终处于符合其型号设计和安全运行的状态。航空器的适航包括初始适航和持续适航，初始适航是对航空器的设计、制造的控制，持续适航是对航空器使用、维护的控制。适航工作涉及政府适航管理部门、航空器设计制造单位和航空器使用、维修单位。

进入 21 世纪，我国的民用航空器产业发展势头迅猛，各项适航工作都已取得了突破性进展。为了强化适航理念，增加适航知识，推动适航工作，需要在适航知识的普及方面多下功夫。目前尚无对适航术语的定义，但作为设计、制造、使用、维护民用运输类飞机的工作人员，平时会遇到大量术语。基于此，亟须定义与梳理适航术语，《民用飞机适航术语手册》应运而生。

《民用飞机适航术语手册》共收集条目近 2 000 条，主要基于航空法规、国际民航组织文件、各国适航当局文件、航空标准组织出版物、航空器制造厂家和航空公司出版物中的相关内容提炼而成。对《民用飞机适航术语手册》中的每个条目均提供了中、英文名称对照，并提供了中、英文的简要说明或解释。在确定术语的过程中，我们力求做到选准、选好条目，对选定条目的中、英名称力求准确、无误，对条目的解释力求通俗易懂、简明扼要。《民用飞机适航术语手册》中的条目按中文名称的首字母拼音顺序排列。为便于读者查询，《民用飞机适航术语手册》还提供了英文名称索引。

在《民用飞机适航术语手册》的成书过程中，上海飞机设计研究院的颜万亿、陈巴生、冯惠冰、贾少澎、董翠玲等老师给予了精心指导；上海飞机设计研究院的刘文成、王留呆、李大海、谭兆光、周洁等老师也进行了评审；同时，中国民航科学技术研究院的李春生等老师、中国民航大学的冯振宇等老师也提供了大力支持和帮助，在此一并表示衷心感谢。

由于时间仓促且作者水平有限，书中存在谬误在所难免，欢迎广大读者批评指正。

作　者
二○二三年十月

# 适航术语定义

　　术语是在特定学科领域用来表示概念的称谓和专门用语,是通过语音或文字来表达或限定科学概念的约定性语言符号,是思想和认识交流的工具。术语可以是词,也可以是词组,用来正确表达生产技术、科学、艺术、社会生活等各个专门领域中的事物、现象、特性、关系和过程。术语根据其使用范围,可能会有纯术语、一般术语和准术语的区别,其中纯术语专业性最强。术语应具备以下基本特征:

　　(1) 专业性:术语是表达各个专业的特殊概念的,使用范围较小,使用的人有限。

　　(2) 科学性:术语的语义范围准确,与相似的概念相区别。术语应能准确、扼要地表达定义的要旨。进行信息交流时,要求术语尽可能地简明,以提高效率。

　　(3) 单义性:术语在某一特定专业范围内是单义的。有少数术语属于两个或更多专业。在创立新术语之前应先检查有无同义词,并在已有的几个同义词之间,选择能较好满足其他要求的术语。

　　(4) 系统性:在一门科学或技术中,每个术语的地位只有在这一专业的整体概念系统中才能加以规定。术语由某一领域中的一般词汇(包括一些词素)构成,也可来自专名(人名、地名),但一般的专名不是术语。术语也常来自外来语。术语要符合语言习惯,遣词用字务必不引起歧义,不带有褒贬等感情色彩的意蕴。

　　(5) 稳定性:对使用频率较高,范围较广,已经约定俗成的术语,没有重要原因,即使有不理想之处,也不宜轻易变更。

　　在民用航空领域的各种文件中,尚无对"术语"一词的明确规定和具体解释,也无相应的标准。但是,相应的词汇(如名词术语、术语解释、专用术语、定义、缩写、glossary、abbreviation、explanation 等)已在航空领域的各类文件中普遍使

用,例如:

(1) 中国民用航空规章 CCAR - 21 - R4 第一章 第 21.2B 条:定义 民用航空产品:指民用航空器、航空发动机和螺旋桨。

(2) FAA order 1000. 15A GLOSSARY 定义 access taxiway:a taxiway that provides access to a particular location or area(接近滑行道:用于接近到具体位置或区域的滑行道)。

(3) 国际民航组织(ICAO)附件 8 第 Ⅰ 部分 定义 abandoned runway:a runway permanently closed to aircraft operations which may be marked in accordance with current FAA standards for marking and lighting of deceptive, closed, and hazardous area on airports(废弃跑道:对飞机运行永久关闭的跑道,并根据 FAA 现行标准可在机场设置防误解、已关闭、危险区域的标志和灯光)。

值得注意的是,目前尚没有一个文件,对适航管理、设计制造、运行维修等领域的术语进行研究和整理,并成为行业统一的适航术语。考虑到在不同适航文件中对同一术语的描述存在差异,在综合权衡后,本书确定了较为普适性的描述方式,即本书定义的适航术语。

本书定义的适航术语是与民用航空器适航有关的术语,是民用航空器适航管理领域概念称谓的集合,是表达和限定民用航空器适航管理领域的约定性和语言符号,用于正确标记该领域中的事物、现象、特性、关系和过程。适航术语包括词、词组、短句、缩写和名称,具有专业性、科学性、单义性和系统性等特征。

适航术语的研究和制定应符合国家标准文件《标准编写规则 第 1 部分:术语》的相关要求。在对民用航空适航术语进行研究整理和编译,并使之系统化和标准化时考虑了以下几类标准(详见附录):

(1) 国际标准:指国际标准化组织(ISO)、国际电工委员会(IEC)和国际电信联盟(ITU)制定的标准,以及国际标准化组织确认并公布的其他国际组织制定的标准。国际标准在世界范围内统一使用。

(2) 国家标准:指在全国范围内统一的技术要求,随着社会的发展,国家需要制定新的标准来满足人们生产、生活的需要。因此,标准是种动态信息。

(3) 行业标准:指行业的标准化主管部门批准发布的,在行业范围内统一的标准,是在没有国家标准,又需要在全国某个行业范围内统一的技术要求时制定的标准。行业标准由国务院有关行政主管部门制定,并报国务院标准化行政主管部门备案。当同一内容的国家标准公布后,该内容的行业标准即废止。

# 目　录

# A

**AOG 订货；Aircraft on Ground Order**

航空器因航空器材缺件而停场时所实施的订货，为最高等级的紧急订货。

The order implemented for aircraft on ground due to the shortage of the aviation equipment parts，which is the highest levels of emergency order.

**APU 小时；APU Hour**

APU 启动至关停的持续时间。

The duration of APU starting to closing.

**A 类维修间隔；Category A Maintenance Interval**

经批准的限定维修期限。

The approved qualified repair deadline.

**安全；Safety**

风险可以接受的状态。

The state in which risk is acceptable.

**安全保障合作计划；Partnership for Safety Plan；PSP**

局方和申请人之间的书面"顶层"协议，它规定用以规划产品合格审定、建立一般期望或操作规范，并确定可交付成果的通用程序。PSP 还规定用于规划和管理合格审定项目的通用纪律和方法。PSP 的签署不是强制要求，由局方和申请人根据需要协商决定。

A written "top level" agreement between the Authority and the applicant that defines a common procedure for planning product certification，establishing general expectations or operating specifications, and determining deliverables. The PSP also defines common disciplines and methods for planning and managing certification programs. The signing of the PSP is not a mandatory requirement and shall be decided by the Authority and the applicant through consultation as necessary.

## 安全风险管理；Safety Risk Management；SRM

安全管理体系内的一个正式过程，由系统和工作分析、危险源识别、风险分析、风险评估和风险控制组成。风险管理过程处于提供产品或服务的过程中，不是一个独立的或特殊的过程。

A formal process within the SMS that describes the system, identifies the hazards, analyzes the risk, assesses the risk, and controls the risk. The SRM process is embedded in the processes used to provide the product/service; it is not a separate/distinct process.

## 安全管理；Safety Management

应用工程和管理理论、准则和技术来优化安全，是一项整体和综合的工程成果。

The application of engineering and management principles, criteria and techniques to optimize safety. It is an integrated and comprehensive engineering effort.

## 安全管理体系；Safety Management System；SMS

管理安全的系统做法，包括必要的组织机构、政策、问责制和程序。

A system method for safety management, include necessary organization, policy, accountability system and procedure.

## 安全目标等级；Target Level of Safety；TLS

在特定情况下被视为可以接受的风险水平的统称。

A generic term representing the level of risk which is considered acceptable in particular circumstances.

## 安全事件;Safety Issue

导致或可导致不安全结果的原因、诱因或问题。安全决策着眼于问题和原因,而不是事件。例如,调查一个非指令的飞行操作面移动(一个事件),可能发现原因是自动飞行计算机的电路故障。对安全影响来说,电路故障是需要评估的安全问题和原因,并应采取相应的对策。

Cause(s), contributing factor(s), or finding(s) that led to, or could lead to, an unsafe outcome. Safety decisions are rendered on issues/causes, not events. For example, investigation of an uncommanded flight control surface movement (an event) might reveal that the cause was a circuit failure in the autopilot's computer. Circuit failure is the safety issue/cause to evaluate for safety implications, and corrective action should be taken.

## 安全寿命;Safe-life

飞机结构能够承受预期的变幅重复载荷作用而没有可觉察的裂纹的时间。

The time that the structure has been evaluated to be able to withstand the repeated loads of variable magnitude expected during its service life without detectable cracks.

## 安全系数;Factor of Safety

① 考虑到载荷可能大于假定值及设计和制造中的不确定性而采用的设计系数;② 最小破坏载荷与额定工作载荷的比值。

① A design factor used to provide for the possibility of loads greater than those assumed, and for uncertainties in design and fabrication;② the ratio of minimum breaking load and rated load.

## 安全运行;Safe Operation

飞机以强制的或推荐的方式运行,以符合适航要求。

The operation of the airplane in a manner that is mandatory, or is recommended, for compliance with the airworthiness requirements.

## 安装批准;Installation Approval; IA

安装批准需要满足以下三个特定准则:① 维修记录已完成;② 重要修理和

改装的表格已正确地执行;③ 如果改装导致在已批准的飞机飞行手册(AFM)中飞机的操作限制或飞行数据有任何改变,那么将需要适当修订。

Installation approval requires the following three specific criteria to be met: ① a maintenance record is accomplished; ② a Major Repair and Alteration form has been properly executed; and; ③ if the alteration results in any change in the aircraft operating limitations or flight data contained in the approved Aircraft Flight Manual (AFM), then appropriate revisions will be required.

## 安装评估;Installation Appraisal

对安装的完整性和安全性进行定性评估。

The qualitative appraisal of the integrity and safety of the installation.

## 安装资质;Installation Eligibility

基于产品的适航数据和构型,在通过型号认证的产品上安装一个部件的可接受性。

Acceptability of an article for installation on type-certificated product(s) based on airworthiness data and the configuration of the product.

# B

**B 类维修间隔;Category B Maintenance Interval**

3 个连续的日历日(72 小时),不包括在航空器维修记录/飞行记录本上进行故障记载的那一天。

3 consecutive calendar days (72 hours), excluding the day on which the failure was recorded in the aircraft maintenance records/flight log book.

**半自动飞行检查系统;Semiautomatic Flight Inspection System**

用配备专门设备的航空器进行定期的检查飞行,来评价导航辅助设施。

Evaluation of navaids by periodic check flights flown by aircraft specially equipped for the purpose.

**包机;Chartered Aircraft**

执行机构根据规定性能和一次性专用合同协议商业租用的飞机。商业机构运营和维护一架包机。包租是全面服务合同的一种形式。

An aircraft that an executive agency hires commercially under a contractual agreement specifying performance and one-time exclusive use. The commercial source operates and maintains a charter aircraft. A charter is one form of a full service contract.

**保护功能;Protection Functions**

用于保护航空器不超过飞行机动操纵极限而设计的系统功能。

Systems functions designed to protect an airplane from exceeding its flight maneuver limitations.

## 保留故障；Deferred Defect

航空器在飞行后和/或维修检查中发现的故障、缺陷，因工具设备、器材短缺或停场时间不足等，不能在起飞前排除的故障项目。

Faults and defects found after flight and/or maintenance checks of the aircraft, which can not be eliminated before takeoff due to shortage of tools, equipment and materials or sufficient parking times.

## 保湿系统；Running Wet System

能够提供足够的热量，防止撞击水滴在加热表面冻结的防冰系统。湿态运转的防冰系统不提供足以使撞击水完全汽化的热量。

Any anti-icing system that supplies enough heat to prevent impinging water drops from freezing on the heated surface. A running wet system does not supply enough heat for complete evaporation.

## 报废航空器；Destroyed Aircraft

严重损坏的航空器，无法恢复原有的适航状态的航空器。

An aircraft damaged to the extent that it would be impracticable to return the aircraft to an airworthy condition.

## 报废件；Scrap

被拥有者因修理不经济、残值低或其他航空原因而不可用等理由处理的航空产品和部件。报废件分成 4 类：① 航空产品和部件除基本材料外没有价值；② 航空产品和部件如果误用的话引起的安全风险是微不足道的；③ 航空产品和部件在航空应用中是低风险的，但可能将来用在非航空应用中；④ 航空产品和部件在航空应用中是与安全攸关的，但可能将来用在非航空应用中。

Products and articles an owner has disposed of because they are beyond economical repair, considered to be of little value, or unusable for any other aviation reason. Scrap products and articles are placed into four categories: ① products and articles that have no value except for the base material; ② products and articles whose misuse in aviation poses an insignificant safety risk; ③ products and articles that were used in low-risk safety aviation applications and may have future use in non-aviation applications; ④ products and articles that were used in safety-critical aviation applications and may have future use in non-aviation applications.

### 暴露时间；Exposure Time

某产品从某一项目最后一次正常运行到再次正常运行之间的间隔。

The period of time between when an item was last known to be operating properly and when it will be known to be operating properly again.

### 备用显示器；Standby Display

当主显示器故障时使用的备份显示器。

A backup display that is used if a primary display malfunctions.

### 被监管机构；Regulated Entity

民用航空安全和环境试验及批准的活动在法律和法规管辖下的自然人或法人。

Any natural or legal person whose civil aviation safety and environmental testing and approval activities are subject to the statutory and regulatory jurisdiction.

### 本场飞行；Local Flight

保持在离出发点 250 海里以内的飞行，或在出发点终止的飞行，或不包括停机持续时间超过 15 分钟的飞行。

A flight which remains no more than 250 nautical miles from the departure point，or which terminates at the point of departure，or which does not include a stop of a greater duration than 15 minutes.

### 便携式自动应急定位发射器；Automatic Portable ELT；ELT(AP)

紧固在航空器上，但易于从航空器上取下的自动启动的应急定位发射器。

An automatically activated ELT which is rigidly attached to an aircraft but readily removable from the aircraft.

## 变更或改装；Alteration or Modification

影响民用航空产品的结构、配置、性能、环境特点，或操作限制的变化。

A change to the construction, configuration, performance, environmental characteristics, or operating limitations of the affected civil aeronautical product.

## 标准程序；Standard Procedures

获得合格证或批准所使用的通用审定流程。

Standard procedures refers to obtaining certificates or approvals through usual certification processes.

## 标准大气；Standard Atmosphere

此种大气的条件如下：① 空气为干燥的理想气体；② 海平面温度为 59 华氏度[①]；③ 海平面压力为 29.92 英寸汞柱[②]；④ 从海平面至温度为 −69.7 华氏度的高度的温度梯度为 −0.003 566 华氏度/英尺[③]，在该高度之上为 0；⑤ 在①到④的条件下，海平面密度压力为 0.002 377×32（重力）= 0.076 064（单位为磅[④]/英尺）。

Atmosphere in which：① the air is a dry perfect gas；② the temperature at sea level is 59℉；③ the pressure at sea level is 29.92 in Hg；④ the temperature gradient from sea level to the altitude at which the temperature is −69.7℉ is −0.003 566℉/ft and zero above that altitude；⑤ the density pressure at sea level under the conditions described in subparagraphs ① through ④ is 0.002 377×32(gravity) =0.076 064(density in lb/ft³).

## 标准定位服务；Standard Positioning Service；SPS

任何 GPS 系统的用户在全球范围内连续使用的定位、速度和定时精度的标准规定水平，不受任何限制。

The standard specified level of positioning, velocity and timing accuracy

---

① 华氏度，温度单位，换算公式如下：1 华氏度＝1.8×摄氏度＋32。
② 英寸汞柱，压强单位，1 英寸汞柱＝3 386.39 帕斯卡。
③ 英尺，长度单位，1 英尺＝0.304 8 米。
④ 磅，质量单位，1 磅＝0.453 6 千克。

that is available, without qualifications or restrictions, to any user of the GPS system on a continuous worldwide basis.

### 标准化的评审准则;Standardized Evaluation Criteria

对每个评审子系统提出的评审问题,供飞机审定系统评估项目的评估组用来计划和记录评审情况。适用的规章要求、有关的咨询通告和指令、国际标准和规范及已建立的工业方法是提出这些评审问题的依据。

Questions developed for each system element that the ACSEP(Aircraft Certification Systems Evaluation Program) evaluation teams use to plan and document the evaluation. The applicable regulation requirements, appropriate advisory circulars (AC) and directives, international standards and specifications, and established industry practices are the basis for these questions.

### 标准进场路线;Standard Terminal Arrival Route; STAR

预先计划并编码的空中交通管制仪表飞行规则进场路线,以图表加文字的形式或仅以文字的形式预先打印好,供驾驶员使用。

A preplanned coded air traffic control IFR arrival routing, preprinted for pilot using graphic and textual or textual form only.

### 标准类别;Standard Category

正常、实用、特技、通勤或交通类别之一。

One of the normal, utility, acrobatic, commuter, or transport categories.

### 标准偏离;Standard Deviation

偏离量或分布变量的统计尺度,对算术平均值偏离的平方求算术平均,再对算术平均的结果取平方根,平方根的结果等于标准偏离。

A statistical measure of the dispersion or variation in a distribution, equal to the square root of the arithmetic mean of the squares of the deviations from the arithmetic means.

### 标准适航证;Standard Airworthiness Certificate

为按照 CCAR‑21 第 21.21 条取得型号合格证或者按照第 21.29 条取得型号认可证的正常类、实用类、特技类、通勤类、运输类航空器,载人自由气球,特殊类别航空器(如滑翔机、飞艇、甚轻型飞机和其他非常规航空器)颁发标准适航证。

The standard airworthiness certificates can be issued for an aircraft in the normal, utility, acrobatic, commuter or transport category, or for a manned free balloon, special class of aircraft (e. g. gliders, airship, very light airplane and other nonconventional aircraft) entitled in type certificate under §21.21 or validation of type certificate under §21.29.

### 标准数据包;Standard Data Package; SDP

包括现场批准检查单,改装或维修的文件复印件,以及重要修理和改装批准表。

SDP contains the field approval checklist, copies of any data describing the alteration or repair, and major repair and alteration approval form.

### 标准项目;Standard Items

该型号航空器所配备的设备和加载的液体,这些设备和液体不是某一特定构型航空器作为一个整体所必需的一部分,但对同型号的航空器是相同的,主要涉及但不局限于以下项目:① 不可用燃油和其他不可用液体;② 发动机滑油;③ 厕所用液体和化学用品;④ 灭火器、信号弹、紧急氧气设备;⑤ 厨房、餐吧和酒吧、小卖部内的结构;⑥ 附加的电子设备。

Equipment and fluids not considered an integral part of a particular aircraft and not a variation for the same type of aircraft. These items may include, but are not limited to, the following: ① unusable fuel and other unusable fluids; ② engine oil; ③ toilet fluid and chemical; ④ fire extinguishers, pyrotechnics, and emergency oxygen equipment; ⑤ structure in galley, buffet, and bar; ⑥ supplementary electronic equipment.

### 标准仪表离场;Standard Instrument Departure; SID

预先计划并编码的空中交通管制仪表飞行规则离场路线,仅以图表加文字的形式预先打印好,供驾驶员使用。

A preplanned coded air traffic control IFR departure routing, preprinted for pilot using graphic and textual form only.

### 冰脊;Ice Ridge

在防冰表面,特别是后部,形成冰突起的部分。

Formation of a ridge of ice typically aft of the ice protection surface.

### 冰脱落周期;Ice Shed Cycles

在给定的功率和结冰条件下,推进系统表面从聚积形成冰到冰脱落所需的时间。

The time period required to buildup and shed ice on a propulsion system surface for a given power and icing condition.

### 并行试飞;Combined Flight Test

中国民用航空局(CAAC)把同时作为申请人飞行试验和审定飞行试验的那些飞行试验称为"并行试飞"。在某些特定情况下,为了减轻申请人的负担,当审查组认为并行试飞是适当的且可行时,可以考虑进行并行试飞。并行试飞的典型例子包括一些低风险的项目,如电子设备安装等,因为此类飞行试验很可能成功地表明符合性。此外,并行试飞还可能包括但不限于最小离地速度($V_{mu}$)、地面最小操纵速度($V_{mcg}$)、空中最小操纵速度($V_{mca}$)、最大刹车能量和湿跑道飞行试验,这些试验的特征是不可重复的,所以对这些飞行试验,可能进行并行试飞,前提是要完成适当级别的风险管理评估。

The Civil Aviation Administration of China (CAAC) refers to those flight tests that are both applicant flight tests and certification flight tests as combined flight tests. In some specific cases, combined test flights may be considered when the review team considers them appropriate and feasible in order to reduce the burden on the applicant. Typical examples of combined flight tests include low risk items such as avionics installations, since such flight tests are likely to successfully demonstrate compliance. Additionally, combined flight test may include, but are not limited to, minimum ground clearance speed ($V_{mu}$), minimum control speed on ground ($V_{mcg}$), minimum control speed in the air ($V_{mca}$), maximum brake energy, and wet runway flight test. For these flight tests, may conduct combined flight test. The prerequisite is the completion of an appropriate level of risk management assessment.

### 补充型号合格证;Supplemental Type Certificate; STC

局方在确认申请人已经表明对经批准型号设计的更改符合相关适航要求后所颁发的证件。

A certificate issued by the authority after confirming that the applicant has shown that the changes to the approved type design meet the relevant airworthiness requirements.

### 补充型号合格证适用的规章;STC Applicable Regulation

适用于补充的型号审定合格证修订(也称为审定基础)的适航规章。

Portions of airworthiness regulations which apply to the STC modification (also called the certification basis).

### 补充型号检查报告;Supplemental Type Inspection Report; STIR

用以允许制造检查员记录申请 STC 改装部件的检查和试验结果的正式文件。

An official document that allows the manufacturing inspector to record the results of the inspections and tests conducted on modified products presented for STCs.

### 补充型号批准;Supplemental Type Approval

适航当局颁发的补充型号合格证或其他形式的等效批准文件。

A STC or equivalent issued by the airworthiness authority.

### 补充型号运行;Supplemental Type Operation

设计 31 座以上(按责任民航局颁发的航空器型号合格证确定)的航空器在任何国家的任何着陆机场实施的补充运行(全货运行除外)。

Any supplemental operation (except an all-cargo operation) conducted with an airplane designed for at least 31 passenger seats (as determined by the aircraft type certificate issued by a competent civil aviation authority) at any land airport in any state.

## 补充氧气；Supplemental Oxygen

额外的氧气供应，用于保障每位乘客应对过高的客舱高度造成的不利影响，并持续提供可接受的生理条件。

The additional oxygen required to protect each occupant against the adverse effects of excessive cabin altitude and to maintain acceptable physiological conditions.

## 不安全状况；Unsafe Condition

如果不经过纠正，可预期造成一起或多起严重损害的状况。

A condition which, if not corrected, is reasonably expected to result in one or more serious injuries.

## 不必要的风险；Unnecessary Risk

产生后不会对任务有任何实质贡献的任何风险。

Any risk that, if taken, will not contribute meaningfully to the task.

## 不符合项；Noncompliance

在生产批准书持有人或申请人处发现的与规章、局方批准质量系统文件或内部程序不一致的情况。在生产批准书持有人或申请人供应商处发现的与生产批准书持有人或申请人采购订单要求不一致的情况，也应作为生产批准书持有人或申请人的不符合项。

Discrepancies with regulations, Authority-approved quality system documents or internal procedures found in the production approval holder or applicant. Discrepancies with the requirements of the production approval holder or the applicant's purchase order found at the production approval holder or applicant's supplier shall also be treated as nonconformance by the production approval holders or the applicant.

## 不工作；Inoperative

某一系统或者其部件因发生故障而不能完成预定的任务，或者不能按照经批准的工作极限或容差范围持续正常工作。

Inoperative means that a system and/or component has malfunctioned to the extent that it does not accomplish its intended purpose and/or is not consistently functioning normally within its approved operating limits or tolerances.

### 不可能失效；Improbable Failures

在某个特定型号的某一架飞机的全寿命周期内不会出现的失效，但是可以在某个特定型号的所有飞机全寿命周期内出现。发生概率为 $1 \times 10^{-5}$ 或更小，但大于 $1 \times 10^{-9}$，失效或纠正措施不能影响飞机的安全飞行和着陆。

Improbable failures are not expected to occur during the total operational life of a random single airplane of a particular type，but may occur during the total operational life of all airplanes of a particular type. The probability of occurrence is on the order of $1 \times 10^{-5}$ or less，but greater than $1 \times 10^{-9}$. The consequence of the failure or the required corrective action must not prevent the continued safe flight and landing of the airplane

### 不可逆操纵系统；Irreversible Control System

操纵面的运动不能反向驱动驾驶舱内操纵装置的一种操纵系统。

A control system where movement of the control surface will not backdrive the pilot's control on the flight deck.

### 不可用件；Unserviceable Part

不合格和（或）不合法的航空器材。

Parts that may have exceeded approved repair limits，time，or cycle limits，etc.

### 不通风舱；Unventilated Compartment

空气交换时间未知或不通风超过 6 分钟的隔间。

A compartment where the air change time is not known or exceeds 6 minutes.

## 不完整 CTSO 件;Incomplete TSO Articles

仅能实现该项技术标准规定中所规定的"主要且独立的"性能或功能的 CTSO 件。

CTSO parts that can only achieve the major and independent performance or function specified in this technical standard.

## 布线;Wire Bundle Routing

下面是在设计电气线路互连系统(EWIS)安装时应该考虑的因素:① 电线束应安置在可触及的地方,以避免受人员、货物和维修活动的影响而损坏,在实际操作中,尽量不要将电线束安置在其可能被用作把手或个人设备的支撑的区域,或在设备移动期间其可能受到破坏的地方;② 应将电线固定在一起,以避免与设备及结构相接触,如果无法实现这一点,那么要采取锁环、摩擦带等额外手段固定电线束,如果电线不能被固定,则要使用保护性的锁环,以确保远离主体结构,电线不能对摩擦带或锁环产生预载;③ 在实际操作中,电线要尽量远离高温设备和管线,以防止绝缘层老化;④ 电线在经过可折面板时,应将线束进行固定,以确保其在面板上移动时,可以弯曲而不会弯折,如果不能实现这一点,那么弯曲半径一定不能小于其最小允许弯曲半径。

Following are some considerations that should go into the design of an EWIS installation: ① wire bundles should be routed in accessible areas that are protected from damage from personnel, cargo, and maintenance activity, as far as practicable they should not be routed in areas where they are likely to be used as handholds or as support for personal equipment or where they could become damaged during removal of aircraft equipment; ② wiring should be clamped so that contact with equipment and structure is avoided, where this cannot be accomplished, extra protection, in the form of grommets, chafe strips, etc., should be provided, wherever wires cannot be clamped, protective grommets should be used, in a way that ensures clearance from structure at penetrations, wire should not have a preload against the corners or edges of chafing strips or grommets; ③ as far as practicable, wiring should be routed away from high-temperature equipment and lines to prevent deterioration of insulation; ④ wiring routed across hinged panels should be routed and clamped so that the bundle will twist, rather than bend, when the panel is moved, when this is not possible, the bending radius must be in accordance with the acceptable minimum bundle radius.

## 布线标准实践;Wiring Standard Practice

详细说明电气、电缆束和同轴电缆的通用维护程序和实践的信息。

Information detailing procedures and practices for the universal repair and maintenance of electrical cable bundles, and coaxial cables.

## 部件;Component

系统运行所需的、执行特定功能的任何自包容性零件、零件组合、组件或装置。

Any self-contained part, combination of parts, subassemblies or units, that perform a distinctive function necessary to the operation of the system.

## 部件测试;Component Testing

针对具体零件、部件或组件的测试,以表明其功能按要求满足相应规章。

Testing of a detail part, component, or subassembly to demonstrate that it functions as required to meet the applicable regulations.

## 部件类别;Parts CATs

根据对安全的潜在影响,部件分为三类。它们被列在类别部件清单(CPL)中,用于建立和识别部件类别的标准。标准详细说明了局方参与对批准这类部件制造的必要程度,也规定了技术参数、质量控制体系、程序建立,以及证明每个类别的部件制造必要过程的水平。

Parts are classified into one of three CATs, depending on their potential effect on safety. They are listed on a category parts list (CPL). Criteria exists for establishing and identifying part CATs. The criteria details the level of Authority involvement necessary to approve the fabrication of such parts. It also specifies the level of technical data, quality control system, procedures development, and processes necessary to substantiate fabrication of such parts within each CAT.

# C

**CTSO 件;CTSO Article**

符合特定技术标准规定的零部件,包括依据技术标准规定项目批准书(CTSOA)或设计批准认可证(VDA)生产的零部件。

Parts that meet the requirements of specific technical standards, including parts produced under Chinese technical standard order authorization (CTSOA) and validation of design approval (VDA).

**CTSO 件制造人;CTSO Article Manufacturer**

对生产的 CTSO 件(或准备申请生产的 CTSO 件)的设计和质量,包括外购的零部件、工艺或服务实施控制的人。

The person who controls the design and quality of the CTSO article produced (or to be produced, in the case of an application), including any related parts, processes, or services procured from an outside source.

**C 类维修间隔;Category C Maintenance Interval**

10 个连续的日历日(240 小时),不包括在航空器维修记录/飞行记录本上进行故障记载的那一天。

10 consecutive calendar days (240 hours), excluding the day on which the failure was recorded in the aircraft maintenance records/flight log book.

**材料还原;Material Reversion**

固化的材料转回到它固化前状态的情形(如固化的灌注胶转回到黏性的液态样)。

The situation wherein a cured material reverts toward its original precured condition, (for example, a cured potting compound that reverts to a sticky, liquid-like consistency).

## 材料强度特性; Material Strength Properties

材料的属性,它定义了任何给定材料的强度相关特性。典型的材料强度特性包括压缩、拉伸、支承、剪切等的极限和屈服值。

Material properties that define the strength related characteristics of any given material. Typical examples of material strength properties are ultimate and yield values for compression, tension, bearing, shear, etc.

## 材料设计值; Material Design Values

基于 § 25. 613(b) 的要求,或通过 AC 25. 613 - 1 定义的其他方法建立的材料强度特性。这些值通常在设计时基于足够的数据来统计确定,且材料变异使结构失效的概率最小化。使用模量标准值。

Material strength properties that have been established based on the requirements of § 25. 613 (b), or by other means as defined in AC 25. 613 - 1. These values are generally statistically determined based on enough data that, when used for design, the probability of structural failure due to material variability will be minimized. Typical values for module are used.

## 采取纠正措施后的风险; Corrected Risk

采取纠正措施后仍存在的剩余风险。当采取高效的纠正措施后,剩余风险被认为是零。

Residual risk that remains after corrective action is taken. When highly effective corrective action is taken, residual risk is considered to be zero.

## 采取纠正措施后的风险系数; Corrected Risk Factor

整个世界机队(或者适用的受影响子机队)采取最终纠正措施后,预期发生事件的预测值。

The forecasted number of future events expected to occur after the entire worldwide fleet (or, if applicable, the relevant affected subfleet), incorporates the final corrective actions.

### 参比面；Reference Surface

用于表征所需监测表面的结冰情况。参比面可以直接或者间接地观察到，并且结冰情况必须先发生在参比面上或同时发生在参比面和监测面上。

The observed surface used as a reference for the presence of ice on the monitored surface. The reference surface may be observed directly or indirectly. Ice must occur on the reference surface before or at the same time as it appears on the monitored surface.

### 参比设备；Reference Facility

一种已知其性能的测试设备，可追溯至制造商的基线性能设备。

A designated test facility of known performance that is traceable to the OEM's baseline facility.

### 参考着陆速度；Reference Landing Speed

人工着陆确定着陆距离时，飞机在规定着陆形态下，下降穿过 50 英尺高度时的速度。

The speed of the aeroplane, in a specified landing configuration, at the point where it descends through the landing screen height 50 ft in the determination of the landing distance for manual landings.

### 残冰；Residual Ice

除冰系统启动后，紧贴着受保护表面的冰。

Ice that remains on a protected surface immediately after deicing system actuation.

### 操作检查；Operational Check；OPC

为了确定一个系统或部件是否在所有方面均功能正常，符合最低可接受的制造商设计规范操作测试。

An operational test to determine whether a system or component is functioning properly in all aspects in conformance with minimum acceptable manufacture design specifications.

**侧滑;Sideslip**

航空器航向与航空器在水平面运动方向之间的角度差。

The angle between the relative wind vector and the airplane plane of symmetry.

**测距机;Distance Measuring Equipment;DME**

用于测量航空器与导航辅助设施的倾斜距离的电子设备,距离以海里为单位。

Electronic equipment used to measure, in nautical miles, the slant range of the aircraft from a navigation aid.

**测试设备;Test Equipment**

制造厂家技术文件中推荐的专门用于某航空器或航空器部件维修的,以及用于确定航空器或航空器部件最终放行的设备。

Tools/equipments recommended in manufacturer's technical data, used for maintaining aircraft or its component, also used for determining the final return to service of aircraft or its component.

**差分 GPS;Differential GPS**

一项提高 GPS 系统精度的技术,判断 GPS 卫星对一个已知固定地点的定位误差,并将判断的错误或纠正因素传送给位于同一区域的 GPS 用户。

A technique used to improve GPS system accuracy by determining positioning error from the GPS satellites at a known fixed location and subsequently transmitting the determined error, or corrective factors, to GPS users operating in the same area.

**差异等级;Difference Level**

指定的有关训练方法或设备、检查方法和近期经历要求的级别,以满足 CCAR-121 有关差异的要求或型别等级条款要求。差异等级规定详细表明了 CAAC 要求,是与对衍生机型组间增加差异大小相对应的要求。按要求从低到高的顺序,CAAC 确定了五个差异等级 A、B、C、D 和 E 级,用于分别规定训练、检查和近期经历要求。

Specified training methods or equipment, inspection methods and recent experience level required to meet the requirements of CCAR – 121 differences or the terms of the type rating requirements. The level of variation specified details the CAAC requirements, which corresponds to the increasing magnitude of variation between groups of derived machines. Requirements are determined from the low to the high order, CAAC establishes five differences in grade A, B, C, D and E-Class for provisions to training, checking and recent experience requirements.

### 拆换率;Removal Rate

单位时间内航空器材的拆换次数。

The removal times of civil aircraft during a specified interval.

### 产品设计标准;Product Design Standards

在适航标准和运行规章范围内规定的对航空器、发动机和部件等物理特性和功能特性的要求。

Requirements delineated within the airworthiness standards and operating rules which regulate the physical and functional characteristics of airplanes, engines, and components thereof.

### 产品使用文件;Product Operating Documents

与由适航当局批准的且符合民航局批准的型号(或补充型号)设计定义有关的产品使用和使用限制的文件,如航空器的飞行手册、主最低设备清单和构型偏离清单等,以及发动机或螺旋桨的安装手册和使用手册等。

Each operating and operating limitation document approved by airworthiness authority and compliant with the type (or supplemental type) design definition approved by CAAC, such as aircraft flight manuals, master minimum equipment list (MMEL), weight and balance manuals, configuration deviation list (CDL) for aircraft, or installation manuals and operating manuals for engines or propellers.

### 场长极限重量;Field Length Limit Weigh

由跑道的长度、坡度、风和温度决定的最大起飞总重。

Maximum takeoff gross weight based on runway length，slope，wind and temperature.

### 场强；Field Strength

在自由空间传播的电磁波能量的幅度。用伏/米(V/m)表示。

Magnitude of the electromagnetic energy propagating in free space expressed in volts per meter (V/m).

### 超差；Out of Tolerance Condition

工作状态超出规定的容许限制范围，是一种令人不满意的状态。

The condition of operation outside of specified acceptable limits，an unsatisfactory condition.

### 超出许可生产范围的产品；Products Beyond the Approved Scope of Production

超出生产许可证持有人被授权范围而生产的零件。

Production parts which is manufactured beyond the scope of PC authorization.

### 超高频；Super High Frequency；SHF

3～30吉赫兹的频带。

Frequency band from 3 to 30 gigahertz.

### 超高速集成电路硬件描述语言；VHSIC Hardware Description (Design) Language；VHDL

用来描述集成电路的结构和性能的标准语言。

Standard language describes structure and performance of integrated circuit.

### 超规范修理；Beyond Specification Repairs

超出局方已批准或认可的规范、手册、标准或图纸等文件而进行的修理工作。

Repair works beyond the authority approval or endorsement specification， manuals，standards or drawings documents.

### 超障高度或超障高；Obstacle Clearance Altitude or Obstacle Clearance Height；OCA or OCH

用于确定符合相应超障准则的最低高度，或在相应跑道的入口标高或机场标高（视适用情况而定）之上的最低高度。包括：① 超障高度以平均海平面为基准，超障高以跑道入口标高为基准，在非精密进近中一般以机场标高为基准，但当跑道入口标高低于机场标高 2 米(6.6 英尺)以上时，以跑道入口标高为基准，盘旋进近中的超障高以机场标高为基准；② 在同时使用超障高度和超障高时，为方便起见，可用"超障高度/高"表示，缩写为"OCA/H"。

The lowest altitude or the lowest height above the elevation of the relevant runway threshold or the aerodrome elevation as applicable，used in establishing compliance with appropriate obstacle clearance criteria. Note：① obstacle clearance altitude is referenced to the mean sea level and obstacle clearance height is referenced to the threshold elevation or in the case of non-precision approaches to the aerodrome elevation or the threshold elevation if that is more than 2 m (6.6 ft) below the aerodrome elevation，an obstacle clearance height for a circling approach is referenced to the aerodrome elevation；② for convenience when both expressions are used they may be written in the form："obstacle clearance altitude/height" and abbreviated "OCA/H".

### 撤轮挡和放置轮挡时间；Block to Block Time

以小时和十分之一小时表示的总飞行时间，从飞机在自身动力装置的驱动下开始移动计算，直到完成飞行后动力停止和关闭，用于记录机组飞行时间。

Total elapsed flight time expressed in hours and tenths of an hour，calculated from the time when the aircraft begins to move under its own power until it comes to a rest and shuts down after completion of the flight，used for recording crewmember flight time.

### 撤销；Declassify

把不运行的航空器从航空器目录中删除。

To remove a non-operational aircraft from the aircraft inventory.

### 成员；Member

可能带有会引起失效或行为异常性错误的飞机/系统功能或项目（该定义仅对应"功能失效集"概念）。

An aircraft/system function or item that may contain an error causing its loss or anomalous behavior(this definition is limited to the functional failure set application herein).

### 承运人；Carrier

根据租约或其他协定直接从事航空运输的法定代表人。

A person who undertakes directly by lease，or other arrangement，to engage in air transportation.

### 乘务员；Flight Attendant；F/A

飞行机组之外,执行国内运行、载旗运行或补充运行合格证持有人按其运行规范规定的最低机组补充要求委派的,或者飞行中在飞机上值勤的人员,职责包括(但不限定)客舱安全的相关责任。

An individual，other than a flight crewmember，who is assigned by a certificate holder conducting domestic，flag，or supplemental operations，in accordance with the required minimum crew complement under the certificate holder's operations specifications or in addition to that minimum complement，to duty in an aircraft during flight time and whose duties include but not necessarily limited to cabin-safety-related responsibilities.

### 持续安全飞行和着陆；Continued Safe Flight and Landing

持续控制安全飞行和着陆并降落在一个合适的位置的能力,有可能使用紧急应变程序,但不需要特殊的驾驶员技能或力量。一些飞机损害可能与飞行中或着陆时的失效条件有关。

Capability for continued controlled flight and landing at a suitable location，possibly using emergency procedures，but without requiring exceptional pilot skill or strength. Some aircraft damage may be associated with a failure condition during flight or upon landing.

## 持续分析和监控系统；Continuing Analysis and Surveillance System；CASS

一种航空安全运营人质量保证系统。美国联邦航空管理局(FAA)已引入规章要求，要求航空运营人建立一个系统并使用该系统持续分析、监控检查和修理方案的性能和效果。

An air carrier quality assurance system. The FAA has introduced regulatory requirements that require air carriers to establish and use a system for the continued analysis and surveillance of the performance and effectiveness of inspection and maintenance programs.

## 持续适航；Continuous Airworthiness

确保航空器在其使用寿命内的任何时间都符合其型号审定基础的适航要求或登记国的要求，并始终处于安全运行状态的所有过程。

All processes to ensure that the aircraft meets the airworthiness requirements of its type certification basis or the requirements of the country of registration at any time during its service life and is always in safe operation.

## 持续适航管理；Continuous Airworthiness Management

在航空器满足初始适航标准和规范、型号设计要求，符合型号合格审定基础，获得适航证并投入运行后，为保持它在设计制造时的基本安全标准或适航水平，为保证航空器能始终处于安全运行状态而进行的管理。

The management to the aircraft that has already obtained the certificate of airworthiness in accordance with initial airworthiness standards and regulations，type design request，type certification basis. After operation in line，it make products always keep the TC or TDA safety standard or airworthiness level and is always in a safe state.

## 持续适航文件；Instructions for Continued Airworthiness；ICA

为保证航空器、发动机或螺旋桨持续适航，对相关重要维修项目进行说明和规定的文件。

A document describing and specifying important maintenance items to ensure the continued airworthiness of an aircraft，engine or propeller.

## 持证航线航空承运人；Certificated Route Air Carrier

航空承运人的一类，他们持有民用航空委员会颁发的"公众便利需要证书"。这些承运人获准从事规定航线上的定期航空运输和限量的非定期运营。

One of a class of air carriers holding Certificates of Public Convenience and Necessity issued by the Civil Aeronautics Board. These carriers are authorized to perform scheduled air transposition over specified routes and a limited amount of nonscheduled operations.

## 冲击极限；Impingement Limit

水滴撞击后机身最远的位置。这个距离通常通过测量表面前缘的表面长度来确定。

The location farthest aft on a body where water drops impinge. This distance is usually measured as the surface length from the surface's leading edge.

## (腐蚀)重复间隔；Repeat Interval

完成基准线方案中要求的、逐次计数的腐蚀任务的时间间隔，以年计算的日历时间。

The calendar time period in years between successive numbered corrosion task accomplishments as stated in the baseline program.

## 重复接地；Iterative Grounding

保护中性导体的一处或多处通过接地装置与大地再次连接的接地。

The grounding protecting one or more point of neutral conductor from grounding again through the grounding device.

## 出口单元；Exit Unit

同型出口撤离率相对Ⅲ型出口撤离率的无量纲比值，用出口撤离率除以35，四舍五入至 0.25 的倍数后得出。

A dimensionless number that is related to the exit rating of a pair of exits

of the same type and uses the rating of the Type Ⅲ exit (35) as the baseline.
The exit unit value of a pair of exits is determined by dividing the rating of the
exit by 35 and rounding the value down to the next quarter of a unit.

### 出口额定人数;Exit Rating (or Rating)

安装规定类型的出口所允许的乘客座位数的增加量。

The increase in passenger seating configuration allowed by the installation
of a pair of that type of exit.

### 出口国适航当局;Exporting Authority

认可批准证书申请人所在国的适航当局。

The airworthiness authority of an applicant within the exporting state.

### 出口偏移量;Exit Offset

拟定的出口中心线与计算的出口中心线之间的距离。

The distance between the proposed exit centerline and the calculated exit
centerline.

### 初步风险评估;Preliminary Risk Assessment

由安全问题引出的初步风险评估,通常用有限的数据或定性的信息来完成。
此评估是为了快速确定问题的潜在风险和紧迫性,当数据和条件允许的时候,评
估采用全面和定量分析,除非问题被视为涉及的风险很小。

An initial assessment of the risk posed by a safety issue, often performed
with limited data or qualitative information. This assessment is meant to
quickly determine an issue's potential risk and urgency, and is followed by
comprehensive and quantitative analysis as data and circumstances permit,
unless the issue is deemed to entail very little risk.

### 初步系统安全性评估;Preliminary System Safety Assessment

基于功能性危险评估和失效状态分类,对建议的系统架构及其实施进行系
统性评估,确定系统及软硬件的安全性要求。

A systematic evaluation of a proposed system architecture and its implementation, based on the functional hazard assessment and failure condition classification, to determine safety requirements for systems and items.

### 初级类航空器；Primary Category Aircraft

满足以下条件并且按照初级类航空器进行型号合格审定的航空器：① 无动力，或者由单台自然吸气式发动机提供动力、失速速度 $V_{SO}$ 等于或小于 113 千米/小时(61 节)的飞机，或者主旋翼桨盘载荷限制为 29.3 千克/米$^2$(6 磅/英尺$^2$)的旋翼航空器；② 最大重量不大于 1 224.7 千克(2 700 磅)或者对水上飞机，不大于 1 530.9 千克(3 375 磅)；③ 包括驾驶员在内，最大座位数不超过 4 人；④ 客舱不增压。

An aircraft that ① is unpowered, or is an airplane powered by a single, naturally aspirated engine with a 113 km/h(61 kn) or less stall speed, or is a rotorcraft with a 29.3 kg/square meter (6 lb/ft$^2$) main rotor disc loading limitations; ② weighs no more than 1 224.7 kg(2 700 lb), and for seaplane, weighs no more than 1 530.9 kg (3 375 lb); ③ has a maximum seating capacity of no more than four persons, including the pilot; ④ has an unpressurized cabin.

### 初始适航管理；Initial Airworthiness Management

在航空器交付使用前，适航部门依据各类适航标准和规范，对民用航空器的设计和制造所进行的型号合格审定和生产许可审定，以确保航空器和航空器部件的设计、制造是按照适航部门的规定进行的。

The management taken before aircraft delivery, according to airworthiness standards and regulations, type certification and production certification is done by airworthiness department, ensure the design and manufacture of aircraft and its part compliance regulation.

### 除冰；Ice Shedding

将飞机部件上积累的冰除去。

The act of separating or breaking away accreted ice from an aircraft part.

**除冰/防冰液保持时间;Deicing/Anti-Icing Fluid Hold Over Time; Deicing/Anti-Icing Fluid HOT**

除冰/防冰液可持续用于防止在飞机受保护表面结冰或者结霜和积雪的预计时间。保持时间开始于最后一次应用除冰/防冰液的时刻,结束于喷洒在飞机上的除冰/防冰液失效的时刻。

The estimated time that deicing/anti-icing fluid will prevent the formation of frost or ice and the accumulation of snow on the critical surfaces of an aircraft. HOT begins when the final application of deicing/anti-icing fluid commences and expires when the deicing/anti-icing fluid loses its effectiveness.

**厨房;Galley**

飞机上储存、冷藏、加热及配制食品和饮料的区域。

The area of the aircraft for storing, refrigerating, heating and dispensing food and beverages.

**处置日期;Disposal Date**

处置执行机构放弃航空器责任的日期,例如: ① 因该机构的出售或交换导致所有权变更;② 航空器退回至出租人或担保人;③ 航空器解密(航空器在展览会上的解密也被认为是"处置"行为,尽管该机构保留其所有权);④ 让渡监护权至另一机构。

The date that the disposing executive agency relinquishes responsibility for an aircraft, for example: ① when the agency transfers title in the case of a sale or exchange; ② returns the aircraft to the lessor or bailer; ③ declassifies it (for fairs, declassification is considered as a "disposal" action, even though the agency retains the property); ④ relinquishes custody to another agency.

**垂直间隔;Vertical Separation**

在空中交通管制中,通过指定不同高度或飞行高度层而确定的间隔。

In air traffic control, separation established by assignment of different altitudes or flight levels.

## 垂直偏移；Vertical Deviation

在显示器上显示的航空器高于或低于垂直剖面的偏移，在航空器低于垂直剖面时可以向上偏转。

The deviation of the aircraft above or below the vertical profile as displayed on an indicator such that deflections up when the aircraft is below the vertical profile.

## 垂直剖面；Vertical Profile

垂直平面上的一条直线或曲线，或由一系列相连接的直线或曲线组成。它是上升或下降的路径，开始或终结于一个指定的航路点或海拔，连接两个或多个指定的航路点或海拔。

A line or curve, or series of connected lines and/or curves in the vertical plane, defining an ascending or descending flight path either emanating from or terminating at a specified waypoint and altitude.

## 垂直起落航空器；Vertical Takeoff and Landing Aircraft；VTOL Aircraft

具有垂直起降能力的航空器，不仅限于直升机。

Aircraft that have the capability of vertical takeoff and landing, not limited to helicopters.

## 错误；Error

机组或维修人员犯下的忽略或不正确行为，或在（研制）要求、设计或实施过程中的差错。

An omitted or incorrect action by a crewmember or maintenance person, or a mistake in requirements, design, or implementation.

# D

**D 类维修间隔；Category D Maintenance Interval**

120 个连续的日历日（2 880 小时），不包括在航空器维修记录/飞行记录本上进行故障记载的那一天。

120 consecutive calendar days (2 880 hours)，excluding the day on which the failure was recorded in the aircraft maintenance records/flight log book.

**大气静温；Static Air Temperature; SAT**

使用相对于大气固定的温度传感器测量的温度。在其他文件中，温度同时也指"室外温度""真实室外温度"或"环境温度"。

The air temperature measured by a temperature sensor which is not in motion with respect to that air. This temperature is also referred to "outside air temperature" "true outside temperature" or "ambient temperature" in other documents.

**大气数据惯性基准系统；Air Data and Inertial Reference System; ADIRS**

向机组和飞机系统提供下列类型数据的系统：高度、空速、温度、航向和当前姿态。

ADIRS provides the following types of data to the flight crew and the aircraft system：height, airspeed, temperature, course, altitude and the current attitude.

**大气总温；Total Air Temperature; TAT**

空气流过航空器受到阻滞时流速降到零，动能转换为热能使局部温度升高，该温度加上空气静温称为大气总温。

The static air temperature plus the rise in temperature due to the air being brought to rest relative to the airplane.

### 大型航空器；Large Aircraft/Aeroplane

最大审定起飞重量超过 5 700 千克(12 566 磅)的航空器。

An aircraft/aeroplane that a maximum certificated takeoff weight of 5 700 kg (12 566 lb) or more.

### 大修；Overhaul/Major Repair

这种修理：① 若操作不当,可能显著影响重量、平衡、结构强度、性能、动力装置工作、飞行特性或影响适航性的其他品质；② 它不是按常规做法进行的,或不能用基本操作进行。

Major repair means a repair：① that if improperly done，might appreciably affect weight，balance，structural strength，performance，powerplant operation，flight characteristics，or other qualities affecting airworthiness；② that is not done according to accepted practices or cannot be done by elementary operations.

### 单传力路径；Single Load Path

所施加的载荷最终通过单个构件分布的路径,且其失效会导致承受载荷的结构性能丧失。

Single load path is where the applied loads are eventually distributed through a single member，the failure of which would result in the loss of the structural capability to carry the applied loads.

### 单独区域乘客容量；Individual Zone Passenger Capacity

单个区域内可乘坐的最大乘客数量。

The maximum number of passengers which may be seated in an individual zone.

### 单型号补充型号合格证；One-only STC

只适用于一个序列号的航空器、发动机、螺旋桨或设备的补充型号合格证。

STC applicable to only one serial number, for aircraft, engine, propeller, or appliance.

## 当局；Authority

负责在国家（地区）层面关注相关要求的组织或个人。

Organization or person responsible within the state (country) concerned with applicable requirements.

## 当量空速；Equivalent Air Speed；EAS

按具体高度进行绝热压缩气流修正的航空器校正空速，当量空速等于海平面标准大气条件下的校正空速。

The calibrated airspeed of an aircraft corrected for adiabatic compressible flow for the particular altitude. Equivalent airspeed is equal to calibrated airspeed in standard atmosphere at sea level.

## 导航雷达；Navar

雷达导航和交通管制装置的协调组合，利用波长为 10 厘米和 60 厘米的波的发射，在航空器上提供距一给定点的距离和方位，显示附近的其他航空器和发自地面的指令；还在地面显示器上提供附近所有航空器的地面显示及其高度和标志，以及发射某些指令的方式。

Coordinated series of radar air navigation and traffic control aid utilizing transmissions at wavelengths of 10 centimeters and 60 centimeters to provide in the aircraft distance and bearing from a given point, display of other aircraft in the vicinity, and command from the ground; also provide on the ground display of all aircraft in the vicinity, as well as their altitudes, identities, and means of transmitting certain commands.

## 登记国；State of Registry

航空器登记注册的国家。如一架属于国际运行机构的航空器不是以单个国家登记的，则组成这一机构的国家应共同并分别承担《芝加哥公约》要求登记国承担的义务。

The state where the aircraft is registered. In the case of the registration of aircraft of an international operating agency on other than a national basis, the states constituting the agency are jointly and severally bound to assume the obligations which, under the *Chicago Convention*, attach to a State of Registry.

## 等效安全水平；Equivalent Level of Safety，ELOS

虽不能表明符合条款的字面要求，但存在补偿措施并可达到等效的安全水平。

Equivalent level of safety means that although it cannot show that it meets the literal requirements of the clause, there are compensatory measures to achieve equivalent safety level.

## 等效显示；Equivalent Display

至少具有以下特征的显示：① 从低头到抬头不需要视觉注意力转变的头部演示；② 从驾驶员的外部视角显示来自传感器的图像；③ 允许同时显示来自EFVS传感器的图像、所需的飞机飞行图像以及外部视图；④ 显示特征及其动态适用于飞机的手动控制。

A display which has at least the following characteristics：① a head-up presentation not requiring transition of visual attention from head down to head up； ② displays sensor-derived imagery conformal with the pilots external view； ③ permits simultaneous view of the EFVS sensor imagery, required aircraft flight symbology, and the external view； ④ display characteristics and dynamics are suitable for manual control of the aircraft.

## 地标导航；Pilotage

目视参考地标的导航。

Navigation by visual reference to landmarks.

## 地面服务；Ground Handling

航空器在到达和离开机场时除空中交通服务以外的必要服务。

Services necessary for an aircraft's arrival at, and departure from, an airport, other than air traffic services.

## 地面结冰条件;Ground Icing Conditions

一般情况下指外界大气温度在 5 摄氏度以下,存在可见的潮气(如雨、雪、雨夹雪、冰晶、有雾且能见度低于 1.5 千米等);或者在跑道上出现积水、雪水、冰或雪的气象条件;或者外界大气温度在 10 摄氏度以下,外界温度达到或低于露点的气象条件。

Generally, when the ground freezes outside and air temperature is below 5℃, there is visible moisture (rain, snow, sleet, ice, fog and the visibility is less than 1.5 km); or any accumulation of water on the runway, weather conditions of snow, ice; or the outside air temperature is below 10℃, the outside temperature reaches or below the dew point of the meteorological conditions.

## 地面拦截点;Ground Point of Intercept; GPI

在地面拦截点处,下滑道拦截进近表面基线。地面拦截点表示为离降落临近点(LTP)的距离(单位为英尺)。

The glidepath intercepts the ASBL at the GPI. The GPI is expressed as a distance in feet from the LTP.

## 地面慢车状态;Ground Idling Conditions

采用最小发动机转速,此时飞机前进速度为零且处于最大排气温度。

The conditions of minimum rotational speed associated with zero forward speed and the maximum exhaust gas temperature at this speed.

## 地面能见度;Ground Visibility

接近地面的有效水平能见度。

Prevailing horizontal visibility near the earth's surface.

## 地面事故;Surface Incident

在活动区内发生未经授权或未经批准便移动的事件,或在活动区内发生的与可能影响飞行安全的航空器运行相关的事件。

Any event during which unauthorized or unapproved movement occurs within the movement area or an occurrence in the movement area associated with the operation of an aircraft that could affect the safety of flight.

### 地面试验；Ground Testing

在最后改装或安装中完成的结构、环境、燃油流量或类似的测试（飞行试验除外），以证明符合适用规章。

Structural, environmental, fuel flow, or similar tests (other than tests in flight) performed on the final modification or installation to demonstrate that it complies with the applicable regulations.

### 地面维修设备；Ground Maintenance Equipment；GME

在地面上直接用于航空器维修与测试、校验的设备。

The equipments are directly used to maintain, test and calibrate aircraft on the ground.

### 地面维修通信设备；Ground Maintenance Communication Equipment

在对航空器进行维修或保养时,使航空器维修现场多个区域维修人员之间以及维修人员与航空器驾驶舱和航空器下方区域维修人员之间进行通信联络与协调的设备。

The equipment used during aircraft maintenance or services to communicate and coordinate with each other between maintenance personnel at difference maintenance areas such as in aircraft cockpit or under the aircraft fuselage.

### 地面效应；Ground Effect

航空器贴近地面飞行,导致流经航空器的气流发生改变,空气动力特性发生的变化。

The change in aerodynamic characteristics due to the change in the airflow pasting the aircraft caused by the proximity of the earth's surface to the airplane.

### 地球参考系统；Earth Referenced System

一种惯性基准系统,它提供飞行轨迹在空间的显示。

An inertial-based system which provides a display of flight path through space.

## 地区维修单位;Regional Maintenance Organization

管理和维修设施在香港、澳门及台湾地区的维修单位。

A maintenance organization whose management and maintenance facility is located in Hong Kong，Macao and the Taiwan region.

## 地区维修培训机构;Regional Maintenance Personnel Training Organization

培训设施在香港、澳门及台湾地区的民用航空器维修培训机构。

The civil aircraft maintenance organization whose training facilities are located in Hong Kong，Macao and the Taiwan region.

## 地速;Ground Speed

航空器相对于地面的速度。

The speed of an aircraft relative to the surface of the earth.

## 地形感知和告警系统;Terrain Awareness and Warning System; TAWS

利用数字高度数据和仪表指示值预测航空器相对于地面的位置的系统。

A system works by using digital elevation data and airplane instrumental values to predict the likely future position of the aircraft intersecting with the ground.

## 地形感知显示;Terrain Awareness Display

显示与飞机有关的周边地形和障碍物。

A display of the surrounding terrain or obstacle（s）relative to the airplane.

## 地形数据库;Terrain Database

在 TAWS 中存储的地形或障碍物信息。

Terrain or obstacle information stored within a TAWS.

### 第二机长;Second in Command

被指定在飞行期间担任航空器上第二机长的驾驶员。

A pilot who is designated to be second in command of an aircraft during flight time.

### 第二阶段飞机;Stage 2 Airplane

已按规定表明符合第二阶段噪声级,而不符合第三阶段噪声级要求的飞机。

An airplane that has been shown under this part to comply with stage 2 noise level and that does not comply with the requirement for a stage 3 airplane.

### 第二阶段噪声级;Stage 2 Noise Level

处于或低于如下规定:① 飞越。最大重量等于或大于 272 000 千克(约 600 000 磅)时为 108 EPNdB,最大重量从 272 000 千克(约 600 000 磅)每减一半,则减少 5 EPNdB,直到最大重量为 34 000 千克(约 75 000 磅)或更小时为 93 EPNdB。② 横侧和进场。最大重量等于或大于 272 000 千克(约 600 000 磅)时为 108 EPNdB;最大重量从 272 000 千克(约 600 000 磅)每减一半,则减少 2 EPNdB,直到最大重量等于或小于 34 000 千克(约 75 000 磅)时为 102 EPNdB。但高于本规定中规定的第三阶段噪声限制的噪声级。① 飞越。(a) 多于二台发动机的飞机:最大重量等于或大于 385 000 千克(约 850 000 磅)时为 106 EPNdB;最大重量从 385 000 千克(约 850 000 磅)每减一半,则减少 4 EPNdB,直到最大重量等于或小于 20 200 千克(约 44 673 磅)时为 89 EPNdB。(b) 三台发动机的飞机:最大重量等于或大于 385 000 千克(约 850 000 磅)时为 104 EPNdB;最大重量从 385 000 千克(约 850 000 磅)每减一半,则减少 4 EPNdB,直到最大重量等于或小于 28 600 千克(约 63 177 磅)时为 89 EPNdB。(c) 少于三台发动机的飞机:最大重量等于或大于 385 000 千克(约 850 000 磅)时为 101 EPNdB;最大重量从 385 000 千克(约 850 000 磅)每减一半,则减少 4 EPNdB,直到最大重量等于或少于 48 100 千克(约 106 250 磅)时为 89 EPNdB。② 横侧。不管发动机的数量,最大重量等于或大于 400 000 千克(约 882 000 磅)时,为 103 EPN;最大重量从 400 000 千克(约 882 000 磅)每减一半,则减少 2.56 EPNdB,直到最大重量等于或小于 35 000 千克(约 77 200 磅)时为 94 EPNdB。③ 进场。不管发动机的数量,最大重量等于或大于 280 000 千克

（约 617 300 磅）时，为 105 EPNdB，最大重量从 280 000 千克（约 617 300 磅）每减一半，则减少 2.33 EPNdB，直到最大重量等于或小于 35 000 千克（约 77 200 磅）时为 98 EPNdB。

A noise level at or below the stage 2 noise limits but higher than the stage 3 noise limits. The stage 2 noise level limits：① Flyover. 108 EPNdB for maximum weight of 600 000 lb or more；for each halving of maximum weight （from 600 000 lb），reduce the limit by 5 EPNdB；the limit is 93 EPNdB for a maximum weight of 75 000 lb or less. ② Lateral and approach. 108 EPNdB for maximum weight of 600 000 lb or more；for each halving of maximum weight （from 600 000 lb），reduce the limit by 2 EPNdB；the limit is 102 EPNdB for a maximum weight of 75 000 lb or less. The stage 3 noise level limits. ① Flyover. （a） For airplanes with more than 3 engines：106 EPNdB for maximum weight of 850 000 pounds or more；for each halving of maximum weight （from 850 000 pounds），reduce the limit by 4 EPNdB；the limit is 89 EPNdB for a maximum weight of 44 673 pounds or less. （b） For airplanes with 3 engines：104 EPNdB for maximum weight of 850 000 pounds or more；for each halving of maximum weight （from 850 000 pounds），reduce the limit by 4 EPNdB；the limit is 89 EPNdB for a maximum weight of 63 177 pounds or less. （c） For airplanes with fewer than 3 engines：101 EPNdB for maximum weight of 850 000 pounds or more；for each halving of maximum weight （from 850 000 pounds），reduce the limit by 4 EPNdB；the limit i 89 EPNdB for a maximum weight of 106，250 pounds or less. ② Lateral. regardless of the number of engines. 103 EPNdB for maximum weight of 882 000 pounds or more；for each halving of maximum weight （from 882 000 pounds），reduce the limit by 2.56 EPNdB；the limit's 94 EPNdB for a maximum weight of 77 200 pounds or less. ③ Approach. regardless of the number of engines. 105 EPNdB for maximum weight of 617 300 pounds or more；for each halving of maximum weight （from 617 300 pounds），reduce the limit by 2.33 EPNdB；the limit is 98 EPNdB for a maximum weight of 77 200 pounds or less.

**第二类大改；Level 2 Major Changes**

除第一类大改之外的型号设计更改。

Changes to type design not classified as level 1 major changes.

## 第三阶段飞机;Stage 3 Airplane

已表明符合规定的第三阶段噪声级的飞机。

An airplane that has been shown under this part to comply with stage 3 noise level.

## 第三阶段噪声级;Stage 3 Noise Level

处于或低于规定的第三阶段噪声限制的噪声级,① 飞越。(a) 多于三台发动机的飞机:最大重量等于或大于 385 000 千克(约 850 000 磅)时为 106 EPNdB;最大重量从 385 000 千克(约 850 000 磅)每减一半,则减少 4 EPNdB,直到最大重量等于或小于 20 200 千克(约 44 673 磅)时为 89 EPNdB。(b) 三台发动机的飞机:最大重量等于或大于 385 000 千克(约 850 000 磅)时为 104 EPNdB;最大重量从 385 000 千克(约 850 000 磅)每减一半,则减少 4 EPNdB,直到最大重量等于或小于 28 600 千克(约 63 177 磅)时为 89 EPNdB。(c) 少于三台发动机的飞机:最大重量等于或大于 385 000 千克(约 850 000 磅)时为 101 EPNdB;最大重量从 385 000 千克(约 850 000 磅)每减一半,则减少 4 EPNdB,直到最大重量等于或少于 48 100 千克(约 106 250 磅)时为 89 EPNdB。② 横侧。不管发动机的数量,最大重量等于或大于 400 000 千克(约 882 000 磅)时,为 103 EPN;最大重量从 400 000 千克(约 882 000 磅)每减一半,则减少 2.56 EPNdB,直到最大重量等于或小于 35 000 千克(约 77 200 磅)时为 94 EPNdB。③ 进场。不管发动机的数量,最大重量等于或大于 280 000 千克(约 617 300 磅)时,为 105 EPNdB;最大重量从 280 000 千克(约 617 300 磅)每减一半,则减少 2.33 EPNdB,直到最大重量等于或小于 35 000 千克(约 77 200 磅)时为 98 EPNdB。

A noise level at or below the stage 3 noise limits. the stage 3 noise level. ① Flyover. (a) For airplanes with more than 3 engines: 106 EPNdB for maximum weight of 850 000 pounds or more; for each halving of maximum weight (from 850 000 pounds), reduce the limit by 4 EPNdB; the limit is 89 EPNdB for a maximum weight of 44 673 pounds or less. (b) For airplanes with 3 engines: 104 EPNdB for maximum weight of 850 000 pounds or more; for each halving of maximum weight (from 850 000 pounds), reduce the limit by

4 EPNdB; the limit is 89 EPNdB for a maximum weight of 63 177 pounds or less. (c) For airplanes with fewer than 3 engines: 101 EPNdB for maximum weight of 850 000 pounds or more; for each halving of maximum weight (from 850 000 pounds), reduce the limit by 4 EPNdB; the limit is 89 EPNdB for a maximum weight of 106 250 pounds or less. ② Lateral. regardless of the number of engines. 103 EPNdB for maximum weight of 882 000 pounds or more; for each halving of maximum weight (from 882 000 pounds), reduce the limit by 2.56 EPNdB; the limit is 94 EPNdB for a maximum weight of 77 200 pounds or less. ③ Approach. regardless of the number of engines. 105 EPNdB for maximum weight of 617 300 pounds or more; for each halving of maximum weight (from 617 300 pounds), reduce the limit by 2.33 EPNdB; the limit is 98 EPNdB for a maximum weight of 77 200 pounds or less.

## 第四阶段飞机;The Fourth Phase Plane

已按规定表明不超过规定的第四阶段噪声级的飞机。

According to this specified that does not exceed this specified in the fourth phase noise level airplane.

## 第四阶段噪声级;Stage 4 Noise Level

处于或低于规定的第四阶段噪声限制的噪声级,对于任何第四阶段飞机,其飞越、横侧和进场最大噪声级为国际民用航空公约附件16,第1卷第三版,2002年3月21日颁发的第7修正案中的第4章第4.4段和第3章第3.4段中规定的最大噪声级。

At or below this specified in the fourth stage noise limiting the noise level, for any fourth aircraft, which fly over, lateral across and approach the maximum noise level for the Convention on international civil aviation annex 16, volume first, Third version, issued in March 21, 2002 Seventh Amendment in Chapter fourth and chapter third, paragraph 4.4, specified in paragraph 3.4 of the maximum noise level.

## 第一阶段飞机;Stage 1 Airplane

尚未按规定表明符合第二或第三阶段飞机所需达到的飞越、横侧和进场噪声级的飞机。

An airplane that has not been shown to comply with the flyover, lateral or approach noise levels required for stage 2 or stage 3 airplanes.

### 第一阶段噪声级；Stage 1 Noise Level

飞越、横侧和进场超过了第二阶段噪声限定的规定,具体规定如下：① 飞越：最大重量等于或大于 272 000 千克(约 600 000 磅)时为 108 EPNdB；最大重量从 272 000 千克(约 600 000 磅)每减一半,则减少 5 EPNdB,直到最大重量为 34 000 千克(约 75 000 磅)或更小时为 93 EPNdB。② 横侧和进场：最大重量等于或大于 272 000 千克(约 600 000 磅)时为 108 EPNdB,最大重量从 272 000 千克(约 600 000 磅)每减一半,则减少 2 EPNdB,直到最大重量等于或小于 34 000 千克(约 75 000 磅)时为 102 EPNdB。

A flyover, lateral or approach noise level greater than the stage 2 noise limits prescribed in following. ① Flyover：108 EPNdB for maximum weight of 600 000 lb or more；for each halving of maximum weight (from 600 000 lb), reduce the limit by 5 EPNdB, the limit is 93 EPNdB for a maximum weight of 75 000 lb or less. ② Lateral and approach：108 EPNdB for maximum weight of 600 000 lb or more；for each halving of maximum weight (from 600 000 lb), reduce the limit by 2 EPNdB, the limit is 102 EPNdB for a maximum weight of 75 000 lb or less.

### 第一类大改；Level 1 Major Changes

影响审定基础的设计更改或者需要对已批准的手册或者依据证后管理指导原则的其他特殊标准进行修改。

Design changes that affect the certification basis, or require certain revisions to approved manuals, or meet other specific criteria according to the guidelines of the post-type validation principles.

### 点火源；Ignition Source

一个能点燃油/气混合气体的充足能量源。超过可燃蒸气的自动点火温度的表面也被认为是点火源。如果能释放足够的能量来点燃,那么电弧、电火花和摩擦火花也被认为是点火源。

A source of sufficient energy to ignite combustion of a fuel/air mixture. Surfaces that can exceed the auto-ignition temperature of the flammable vapor under consideration are considered to be ignition sources. Electrical arcs, electrical sparks and friction sparks are also considered ignition sources if sufficient energy is released to initiate combustion.

### 电磁干扰;Electromagnetic Interference; EMI

由于另外一个电源的工作而在电路中出现的不希望的电压或电流。

The presence of undesirable voltages or currents which appear in a circuit as a result of the operation of another electrical source.

### 电弧径迹;Arc Tracking

绝缘表面形成一个导电碳质槽的现象。这个碳质槽为电流提供了短接通路。该现象通常由电弧诱发。

A phenomenon in which a conductive carbon path is formed across an insulating surface. This carbon path provides a shortcircuit path through which current can flow. Normally a result of electrical arcing.

### 电路保护装置;Circuit Protective Device; CPD

通过自动中断电流的方式保护电气/电子电路元件,避免其电压或电流过载的装置。飞机上最常用的电路保护装置是断路器和保险丝。

A device used to protect electrical/electronic circuit components from an over-voltage or over-current condition, by automatically interrupting the current flow. The most common types of CPD used in aircraft are the circuit breaker and the fuse.

### 电气和电子工程师协会;Institute of Electrical and Electronics Engineers; IEEE

由美国无线电工程师协会(IRE,创立于 1912 年)和美国电气工程师协会(AIEE,创建于 1884 年)合并而成,是一个国际性的电子技术与信息科学工程师协会。

IEEE merged by the American Radio Engineers Association (IRE, founded in 1912) and the American Institute of Electrical Engineers (AIEE, established in 1884), is an international electronic technology and information science engineer's association.

### 电气线路互连系统;Electrical Wiring Interconnection System; EWIS

安装在飞机上的任意区域,用于在两个或多个端接点之间传输电能的导线、导线装置或两者的组件,包括端接装置。

Any wire, wiring device, or combination of these, including termination devices, installed in any area of the airplane for the purpose of transmitting electrical energy between two or more intended termination points.

### 电气线路互连系统预期服役寿命; EWIS Expected Service Life

EWIS 预期服役寿命,一般不比航空器结构件的预期服役寿命短。如果预期服役寿命要求在某个间隔更换所有或部分的 EWIS 部件,那么这个间隔必须在 ICA 中列出。

The expected service lifetime of the EWIS. This is not normally less than the expected service life of the aircraft structure. If the expected service life requires that all or some of the EWIS components be replaced at certain intervals, then these intervals must be specified in the ICA.

### 电子飞行显示器;Electronic Flight Display; EFD

导航数据的补充,但不能替代航路图。

A supplementary source of navigation data and can not replace enroute charts.

### 电子飞行仪表系统;Electronic Flight Instrument System; EFIS

安装在飞机驾驶舱中显示飞行信息的电子显示系统,一般由显像管(CRT)或液晶显示器(LCD)组成。

An electronic display system which is mounted in aircraft cockpit, displaying flight information and generally include tube (CRT) or liquid crystal display (LCD).

## 调查成因;Probable Cause

通过对所提供的材料、数据进行调查,事故调查组确认的导致事故/事件最有可能的原因。

Event(s) determined by the investigation team which is (are) the most likely cause of the accident/incident based on the available facts/data.

## 定量;Quantitative

使用数学方法来评价系统和飞机的安全性的分析过程。

Analytical processes that apply mathematical methods to assess system and airplane safety.

## 定期检测;Periodic Testing

为确保符合工艺规范而对每批原料或产品进行的检测。当检测结果具有持续一致性时,可根据质量保证抽样计划减少检测次数。

A test for each batch of raw materials or products to meet the technical specifications. The number of testing can be reduced according to quality assurance sampling plan when the test results have continued consistency.

## 定期检修;Scheduled Maintenance

根据适航性资料,在航空器或者航空器部件使用达到一定时限时进行的检查和修理。定期检修适用于机体和发动机/APU,不包括大修。

Inspections and repairs of civil aircraft or aircraft component, which have been in service to a defined time limit, according to the airworthiness data. Scheduled maintenance applies to airframe and engine/APU, but doesn't include "overhaul" defined herein for the purpose of this regulation.

## 定期载客运行;Scheduled Passenger-carrying Operation

航空承运人或者航空运营人以取酬或者出租为目的,通过本人或者其代理人以广告或者其他形式提前向公众公布的,包括起飞地点、起飞时间、到达地点和到达时间在内的任何载客运行。

Any passenger-carrying operation performed by an air carrier or air operator in order to reward or hire, published to the general public in advance in the form of advertisement or otherwise by the principal or its agent，including takeoff location，takeoff time，destination，and landing time.

## 定位信标式方向指引；Localizer Type Directional Aid；LDA

用途和精度与某种着陆航向信标台相当的一种设施，该种着陆航向信标台的扩散角超过 3 度，但不超过 30 度，不能用于对准跑道。

A facility of comparable utility and accuracy to a localizer that is not aligned with the runway with an angle of divergence exceeding 3 degrees but not exceeding 30 degrees.

## 定向机（DF，UDF，VDF，UVDF）；Direction Finder（DF，UDF，VDF，UVDF）

配备有方向分辨天线的无线电接收机，用以接收无线电发射机的方位。专用的无线电定向机在航空器上作为空中助航装置，另一些在地面上，主要用于确定坠毁的航空器的位置，或获得请求协助重新定位的驾驶员的坐标位置。确定坐标位置的方法如下：利用两个位置分开的 DF，在航图上画两条或更多条方位线，获得丢失的航空器发射机的坐标位置，或者由驾驶员画出其 DF 对两个位置分开的地面发射机（这两个发射机是在其航图上能够识别的）的方位指示。UDF 在超高频无线电广播频带内接收信号；VDF 在甚高频频带内接收信号；而 UVDF 在两种频带内接收信号。

A radio receiver equipped with a directional sensing antenna used to take bearings on a radio transmitter. Specialized radio direction finders are used in aircraft as air navigation aids，others are ground based primarily to locate crashed aircraft or to obtain a "fix" on a pilot requesting reorientation assistance. A location "fix" is established by the intersection of two or more bearing lines plotted on a navigational chart using either two separately located DFs to obtain a fix on a lost aircraft transmitter；or by a pilot plotting the bearing indications of his DF on two separately located ground based transmitters，both of which can be identified on his chart. UDF receives signals in the ultra high frequency radio broadcast band，VDF receives signals in the very high frequency band，and UVDF receives signals in both bands.

## 定向机航向(操纵);DF Course (Steer)

航空器相对于定向机的指示磁方向以及航空器在无风条件下为到达该定向机必须操纵的方向。

The indicated magnetic direction of an aircraft to the DF station and the direction of an aircraft must steer to reach the station in a no wind condition.

## 定性;Qualitative

以客观的、非数值方式评价系统和飞机安全性的分析过程。

Analytical processes that assess system and airplane safety in an objective, non-numerical manner.

## 动力或推力严重损失;Serious Loss of Power or Thrust

发动机工作异常,如发生不可恢复或重复性喘振、失速、反流或熄火,这些会导致发动机明显失去动力或推力。

Engine operating anomalies such as non-recoverable or repeating surge, stall, rollback or flameout, which can result in noticeable engine power or thrust loss.

## 动力装置;Power-unit

由一台或多台发动机及其辅助部件组成的系统,为航空器提供动力,每台发动机可以单独连续工作,但不包括产生短时推力的装置。

A system of one or more engines and ancillary parts which are together necessary to provide thrust, independently of the continued operation of any other powerunit(s), but not including short period thrust-producing devices.

## 动态模拟;Dynamic Simulation

根据航空器预定运动的模拟轨迹的数据进行的动力系统试验。

Testing of a dynamic system by introduction of data derived from the simulated tracks of predicted aircraft movements.

## 陡梯度飞行航空器;Steep-gradient Aircraft

能在很短的跑道和小的区域进行很陡的爬升、下降、起飞和着陆的航空器。

Aircraft that is capable of very steep climbs, descents, takeoffs, and landings from very short runways and small areas.

### 独立的维修单位；Independent Maintenance Organization

独立于航空运营人和航空器或者航空器部件制造厂家，并提供航空器或者航空器部件维修服务的维修单位。

The maintenance organization which is independent from the operator and civil aircraft or aircraft component manufacturer, provides maintenance service for the civil aircraft or aircraft components.

### 短距起落(STOL)航空器；Short Takeoff and Landing (STOL) Aircraft

在其经批准的 STOL 使用重量范围内的某一重量下，能按照适用的 STOL 特性、适航性、运行、噪声和污染标准，在 STOL 跑道上起降的一种航空器。

An aircraft which, at some weight within its approved range of STOL operating weight, is capable of operating from a STOL runway in compliance with the applicable STOL characteristics, airworthiness, operations, noise, and pollution standards.

### 短途飞行；Hop

航空器从机场到机场的短距离飞行。

A short trip by aircraft from airport to airport.

### 断路器；Circuit Breaker; CB

当通过它的电流过载时自动断开电路的一种保护装置。

A protective device for cutting off a circuit automatically when excessive current is flowing through it.

### 多边适航协议；Multilateral Airworthiness Agreement; MAA

由多国政府代表签署并通过互换外交照会生效的技术性执行协定或协议。

Technical executive agreement signed by multinational government representatives and became effective through the exchange of diplomatic notes.

## 多重(多机)STC(FAA);Multiple STC (FAA)

适用于多个序列号的航空器、发动机或螺旋桨的 STC。

STC applicable to more than one serial number, for aircraft, engine, or propeller.

## 多传力路径;Multiple Load Path

传力路径是指载荷从一个部件传递到另一个部件的路径,多传力路径指两个部件之间的传力路径不唯一。

The load path is the path that the load is transmitted from one component to another, and the multiple load path means the one not unique between two components.

## 多功能 CTSO 件;Multiple Function CTSO Article

同时满足多项技术标准规定的 CTSO 件。

CTSO article meeting multiple technical standards.

## 多时区飞行;Multi-time Zone Flight

在北纬 60 度和南纬 60 度之间,连续向东或向西跨越不少于 5 个时区的飞行。

An easterly or westerly flight or multiple flights in one direction in the same duty period that results in a time zone of which difference is 5 or more hours and is conducted in a geographic area that is south of 60 degrees north latitude and north of 60 degrees south latitude.

# E

**EASA 批准的资料；Data Approved by EASA**

由欧盟技术代表或由其批准的机构批准的资料，包括按照国际民航组织附件 1 互认的美国设计资料。

Data approved by the EU TA or by an organization approved by the TA，including United States' design data reciprocally accepted under Annex 1.

**ETOPS 备降机场；ETOPS Enroute Alternate**

当发动机飞机在延伸航程运行的航路上遇到发动机熄火等非正常或紧急情况时，能进行着陆的可用和合适的备降机场。

A suitable and appropriate alternate aerodrome at which an aeroplane would be able to land after experiencing an engine shutdown or other abnormal or emergency condition while flying en route in an ETOPS operation.

**ETOPS 维修实体；ETOPS Maintenance Entity**

经授权可用于 ETOPS 维修和完整 ETOPS PDSC 的实体，且该实体应该满足以下条件：① 经审查可参与 FAR‐121 运行；② 按 FAR‐145 审查合格的维修单位；③ 按照 FAR‐43.17(c)(2)要求授权的单位实体。

An entity authorized to perform ETOPS maintenance and complete ETOPS PDSC and that entity is：① certificated to engage in FAR‐121 operations；② repair station certificated under part FAR‐145 of this chapter；③ entity authorized pursuant to §43.17(c)(2) of this chapter.

**额定起飞加力推力；Rated Takeoff Augmented Thrust**

在涡轮喷气发动机型号合格审定中，指在标准海平面条件下，在 FAR‐33

规定的发动机工作限制内,流体喷射或燃料在独立燃烧室中燃烧,静态产生的经批准的喷气推力,其起飞运转的使用时间不超过 5 分钟。

With respect to turbojet engine type certification, rated takeoff augmented thrust means the approved jet thrust that is developed statically under standard sea level conditions, with fluid injection or with the burning of fuel in a separate combustion chamber, within the engine operative limitations established under FAR - 33 of this chapter, and limited in use to periods of less than 5 minutes for takeoff operation.

### 额定起飞推力;Rated Takeoff Thrust

在涡轮喷气发动机型号合格审定中,指在标准海平面条件下,于 FAR - 33 规定的发动机工作限制内,在独立燃烧室中没有燃料燃烧的情况下,静态产生的经批准的喷气推力,其起飞运转的使用时间不超过 5 分钟。

With respect to turbojet engine type certification, means the approved jet thrust that is developed statically under standard sea level conditions, wipeout fluid injection and without the burning of fuel in separate combustion chamber, within the engine operating limitations established under FAR - 33 of this chapter, and limited in use to periods of not over 5 minutes for takeoff operations.

### 额定载荷;Load Rating

在一定载荷区间内的最大允许载荷。

The maximum permissible load at a specific inflation pressure.

### 额定中间连续功率/推力;Intermediate Contingency Power and/or Thrust Rating

在发动机型号合格证数据单里规定的,系列或新翻修发动机在特定条件和相应的可接受限制下运行时的最小试车台接受功率/推力。

The minimum test bed acceptance power and/or thrust, as stated in the engine type certificate data sheet, of series and newly overhauled engines when running at the specified conditions and within the appropriate acceptance limitations.

### 额定最大连续加力推力;Rated Maximum Continuous Augmented Thrust

在涡轮喷气发动机型号合格审定中,指在标准大气条件下,指定高度上,在FAR-33规定的发动机工作限制内,通过流体喷射或在独立燃烧室中燃料燃烧,静态或飞行中产生的经批准的喷气推力,其使用时间经批准不受限制。

With respect to turbojet engine type certification, rated maximum continuous augmented thrust means the approved jet thrust that is developed statically or in flight, in standard atmosphere at a specified altitude, with fluid injection or with the burning of fuel in a separate combustion chamber, within the engine operating limitations established under FAR - 33 of this chapter, and approved for unrestricted periods of use.

### 额定最大连续推力;Rated Maximum Continuous Thrust

在涡轮喷气发动机型号合格审定中,指在标准大气条件下、指定高度上,于FAR-33规定的发动机工作限制内,在没有流体喷射和没有燃料在单独燃烧室中燃烧的情况下,静态或飞行中产生的经批准的喷气推力,其使用时间经批准不受限制。

With respect to turbojet engine type certification, means the approved jet thrust that is developed statically or in flight, in standard atmosphere at a specified altitude, without fluid injection and without the burning of fuel in a separate combustion chamber, within the engine operating limitations established under FAR - 33 of this chapter, and approved for unrestricted periods of use.

### 二次雷达;Secondary Radar;SECRA

被探测物体上装有无线电接收/发射机(应答机)形式的协同设备的一种雷达系统。从搜索发射/接收机(询问机)站发射出的无线电脉冲被协同设备接收并触发应答机发射一个识别信号。然后,地面发射/接收机站将收到应答机发射的信号(非反射信号)。

A radar system in which the object to be detected is fitted with cooperative equipment in the form of a radio receiver/transmitter (transponder). Radio pulses transmitted from the searching transmitter/receiver (interrogator) site are received in the cooperative equipment and used to trigger a distinctive transmission from the transporter. This latter transmission rather than a reflected signal is then received back at the transmitter/receiver site.

## 二级腐蚀；Level 2 Corrosion

通过连续检查发现的腐蚀现象，需要采取返工等手段，通常情况下通过 CPCP 判断，其腐蚀严重程度超过一级腐蚀。

Damage occurring between successive inspections that requires rework or blend-the AD，in general，requires CPCP adjustments for corrosion exceeding Level 1.

## 二级航空器维修基地；Secondary Aircraft Maintenance Base；SAMB

从事航空器和航空电子设备定期与不定期维修工作的机构，其维修等级低于航空器维修基地。

Agency occupied in scheduled and unscheduled aircraft and avionics maintenance，its maintenance level is lower than that of aircraft maintenance base.

# F

**FAA 地区航空规章组;FAA Regional Airspace and Procedures Team**

建立在 FAA 地区管理局的团队,协调和处理新的或者改装飞行要求。

A team established at each FAA region for the purpose of coordinating and processing requests for new or modified flight procedures and related airspace matters.

**FAA 飞行计划;Flight Program**

为满足任务要求,由 FAA 按照 Order 4040.9E 制订的飞行计划。

An FAA flight program, established by the administrator, in accordance with Order 4040.9E to meet mission requirements (also referred to an FAA flight program).

**FAA 航空可靠性报告;FAA Aeronautical Reliability Report**

FAA 的一种内部系统,旨在连续报告 FAA 航空器与设备的失效、故障和缺陷。

Internal agency system for the continuous reporting of failures, malfunctions, and defects of FAA aircraft and equipment.

**FAA 航空器审定办公室;FAA Aircraft Certification Office; ACO**

在某一区域范围内,指导航空器审定的相关活动的 FAA 现场办公室。

FAA field office that serves certain geographic area for guidance on aircraft certification related activities.

**FAA 航空器审定司 ; FAA Aircraft Certification Service; AIR**

FAA 航空器审定司负责管理规范民用航空产品的设计、生产和适航性的安全标准；监督设计、生产和适航审定程序，以确保其符合规定的安全标准；提供一个安全性能管理系统，以确保航空器的持续运行安全；并且与局方、制造商和其他股东一起工作，帮助他们提高国际航空运输系统的安全性。

The office is responsible for governing the design, production, and airworthiness safety standards of civil aeronautical products; overseeing design, production, and airworthiness certification programs to ensure compliance with prescribed safety standards; providing a safety performance management system to ensure continued operational safety of aircraft; and, working with aviation authorities, manufacturers, and other stakeholders to help them successfully improve the safety of the international air transportation system.

**FAA 航空人员合格证书 ; Airman Certificate**

FAA 局方颁发的证件，证明局方已确认证件持有人能遵守有关管理规章，它规定了该合格证授权持有人作为航空人员从事与航空器相关活动应具备的能力。

A document issued by the Administrator of the Federal Aviation Administration, certifying that he has found the holder to comply with the regulations governing the capacity in which the certificate authorizes the holder to act as an airman in connection with aircraft.

**FAA 监督办公室 ; FAA Oversight Office**

航空器审定办公室或运输类航空器管理局办公室，负责监管 FAA 颁发的相关型号合格证、补充型号合格证及其制造商。

The aircraft certification office or office of the Transport Airplane Directorate with oversight responsibility for the relevant type certificate, supplemental type certificate, or manufacturer, as determined by the FAA.

**FAA 批准的资料 ; Data Approved by FAA**

根据双边协议(BASA)/适航操作程序(IPA)，由 FAA 或 FAA 委任代表批准的资料。

Data approved by the Administrator or the Administrator's designated representative in accordance with the Bilateral Aviation Safety Agreement (BASA)/Implementation Procedures for Airworthiness (IPA).

### FAA 委任代表;FAA Designees

具有工程批准信息授权的委任工程代表(DER),以及有权代表 FAA 批准技术资料或补充型号合格证的设计机构批准(DOA)和设计保证系统(DAS)持有人。

Designated engineering representatives .(DER), who have delegated authority to approve engineering information, delegation option authorization (DOA) and designated alteration station (DAS) authorization holders, who have delegated authority to approve either technical data or supplemental type certificates on behalf of the FAA.

### FIS 数据链接;FIS Data Link; FISDL

通过各种广播数据链提供 FIS 数据,包括显示在飞机驾驶舱内的天气和操作信息。

The provision of FIS data over various broadcast data links, including weather and operational information for display in the cockpits of appropriately equipped aircraft.

### 发动机;Engine

用于或准备用于飞机推进的装置。发动机至少应包括运行和控制所需的部件和设备,但不包括螺旋桨(如适用)。

A unit used or intended to be used for aircraft propulsion. It consists of at least those components and equipment necessary for functioning and control, but excludes the propeller.

### 发动机/APU 失效模型;Engine and APU Failure Model

设计飞机时用于分析的一种描述发动机/APU 转子碎片的尺寸、质量、飞散角、能量等级和数目的模型。

A model describing the size, mass, spread angle, energy level and number of engine or APU rotor fragments that is considered when analyzing the airplane design.

### 发动机背对背测试；Engine Back-to-back Test

相同发动机在相同设施上进行的比较性能参数的测试，通常在对测试设施和/或发动机测试硬件进行修理或改装之后进行。

A test comparing performance parameters measured on the same engine in the same facility, typically done before and after a repair or modification to a test facility and/or engine test hardware.

### 发动机喘振；Engine Surge

发动机在空气压缩过程中所产生的气流严重堵塞或反流现象。

The response of the entire engine which is characterized by a significant flow stoppage or reversal in the compression system.

### 发动机的制造日期；Date of Engine Manufacture

验收检查记录表明发动机已完工并满足民航局批准的型号设计的日期。

The date on which inspection acceptance records show that the engine has been completed and meet CAAC approved type design date.

### 发动机电子控制系统；Electronic Engine Control System

发动机电子控制系统是发动机电子/电气控制系统的一般组成部分，发动机电子/电气控制系统包括数字发动机全权控制，监督控制及其他辅助控制。

The generic family of electrical/electronic engine control systems, including full authority digital engine controls, supervisory controls, and derivatives of these.

### 发动机额定输出；Rated Output

由民航局批准的发动机在标准大气条件下起飞时的最大功率或推力，包括加力作用（若适用），但不包括喷水作用和应急功率或推力额定值。

The maximum power/thrust available for takeoff at standard day conditions as approved for the engine by the Civil Aviation Administration of China, including reheat contribution where applicable, but excluding any contribution due to water injection and any emergency power/thrust rating.

### 发动机额定温度;Rated Temperature

发动机在额定功率和转速条件下运行时的最大涡轮进气口或排气口温度。

The maximum turbine inlet or exhaust gas temperature when the engine can operate at rated output and speed.

### 发动机额定压力比;Rated Pressure Ratio

发动机在额定输出下运行时,燃烧室入口压力与发动机入口压力之比。

The ratio between the combustor inlet pressure and the engine inlet pressure achieved by an engine operation at rated output.

### 发动机关键参数;Critical Engine Parameter

测量发动机推力最精确的参数。

The parameter that is the most accurate measure of propulsive force.

### 发动机机匣烧穿;Engine Case Burnthrough

烧穿发动机内机匣,使高压、高温燃气流从发动机逸出的着火状态。

A fire condition within the engine which burns through the engine case, allowing gas stream of high pressure and high temperature to escape from the engine.

### 发动机基准设施;Engine Baseline Facility

原始设备制造商指定的作为发动机型号审定标准的设施。

A facility designated by the OEM as the standard for certification of an engine model.

### 发动机加速基准状况;Engine Acceleration Datum Conditions

在型号耐力测试中,将发动机从指定功率加速到起飞功率的95%时发动机的状况,如发动机的转速、扭矩和排气温度等。

The engine conditions, such as rotational speed, torque, exhaust gas temperature, as appropriate, from which, during the type endurance test, the specified accelerations to 95% of take-off power and/or thrust is timed.

### 发动机排气温度;Exhaust Gas Temperature; EGT
从活塞发动机的汽缸或涡轮发动机的涡轮机部分排出的废气的温度。

The temperature of the exhaust gases as they leave the cylinders of a reciprocating engine or the turbine section of a turbine engine.

### 发动机启动失败;Engine False Start
发动机试图启动,但点火失败。

An attempt to start in which the engine fails to light up.

### 发动机启动中止;Engine Abortive Start
发动机点火成功但未加速。

An attempt to start, in which the engine lights up, but fails to accelerate.

### 发动机失速;Engine Stall
气流在一个或多个压气机叶片型面上分离。

Flow breakdown at one or more compressor airfoils.

### 发动机手册;Engine Manual; EM
提供维护发动机潜在的隶属于发动机的最大数量组件信息(不考虑设计责任)的技术数据要求。

Manual specification which is to provide technical data requirements for information needed to maintain the engine and the maximum potential number of parts that could, regardless of design responsibility, remain with the engine.

### 发动机图解零件目录;Engine Illustrated Parts Catalog; EIPC
用于识别和更换发动机部件和组件的手册,是发动机手册的配套手册。

EIPC is intended for using in the identification and requisition of replaceable engine parts and units. It is a companion document to the engine manual.

### 发动机熄火；Flameout

涡轮发动机非预期的燃烧终止。

Unintended loss of combustion in turbine engines.

### 发动机型别；Engine Model

具有相同的总序号、排气量和设计特性，并由同一型号合格证批准的所有航空涡轮发动机类别。

Engine type refers to all aviation turbine engines which have the same main serial number, exhaust capacity and design characteristics, and approved by the same type certificate.

### 发动机压力比；Engine Pressure Ratio；EPR

进气压力与排气压力的比值，某些发动机以此作为设定推力的主要依据。

The ratio of inlet pressure to exhaust pressure measurements. The ratio is used as the primary parameter to set thrust on some engines.

### 发动机影响件；Engine Influencing Parts

一种发动机零部件，这类零部件与发动机限寿件（CCAR‐33‐R2 第 33.70 条款）安装在同一个复杂的系统中，并会对限寿件的环境和工作条件造成影响。

Engine parts that are installed in the same complex system as the engine life limiting parts (CCAR‐33‐R2 §33.70) and have an impact on the environmental and operating conditions of the life limiting parts.

### 发动机裕度；Engine Margin

发动机压缩部件工作点距离喘振边界的量度。

The margin between working point of engine compression unit and surge limit.

### 发动机指示和机组告警系统；Engine Indicating and Crew Alerting System；EICAS

能提供差异化优先级系统信息的系统（如警告、警戒、咨询状态和维修）。

A system which is capable of providing different priority levels of system information messages ( such as warning, caution, advisory status and maintenance).

### 发动机自动恢复系统；Engine Auto-recovery Systems

包括自动重新点火系统、失速恢复系统,或任何其他在发动机熄火、喘振、失速或上述三者同时发生的情况下,旨在恢复发动机可操作性的发动机系统。

The engine system which typically includes auto-relight systems, stall recovery systems, or any other engine system intended to recover the operability of an engine following a flameout, surge, stall, or a combination of these.

### 发生事故；Launch Incident

在飞行中或其他时间发生的意外事件和事故,包括飞行安全系统、安全关键系统的故障,或许可证持有人、被许可的安全机构、设计或运营所发生的故障。

An unplanned event occurring during the flight of a launch vehicle, other than a launch accident, involving a malfunction of a flight safety system or safety-critical system or failure of the licensee or permitee's safety organization, design or operations.

### 发现问题；Finding

经过证据证明的结论存在一个不符合所制定标准的过程或产品。

A conclusion, supported by evidence, that has been or is a process or product which is not in compliance with an established standard.

### 发烟指数；Smoke Number；SN

表示烟排放量的无量纲参数。

The dimensionless term quantifying smoke emissions.

### 翻新（轮胎）；Retreading（Tire）

通过重新修整轮胎帘线区或帘线区及其侧壁,来修复磨损轮胎的方法。修理包含在轮胎翻新工艺中。

The methods of restoring a worn tire by renewing the tread area or by renewing the tread area plus one or both sidewalls. Repairs are included in the tire retreading process.

### 翻修；Overhaul；O/H

根据适航性资料,通过对航空器或者航空器部件进行分解、清洗、检查、必要的修理或者换件、重新组装和测试来恢复航空器或者航空器部件的使用寿命或适航性状态。

Restoring civil aircraft or aircraft component，by disassembling，cleaning，inspecting，necessarily repairing or replacing，reassembling and testing in accordance with the airworthiness data，to its serviceable life or airworthy conditions.

### 翻修产品；Rebuilt Product

产业制造商进行如下处理,使产品满足新部件的公差与限制或经批准的超差：① 分解、清洁、检查、修理、重新装配；② 经试验使其具有与新产品相同的限制和公差。

A product for the sake of conforming to new part's tolerances and limits or approved oversized or undersized dimensions，has undergone the following treatments by the original manufacturer：① has been disassembled，cleaned，inspected，repaired as necessary，and reassembled to the extent possible；② has conformed to the same tolerances and limits as a new product after test.

### 翻修发动机；Rebuilt Engine

对使用过的发动机完全分解、检查、修理、重新装配和试验,与新发动机一样进行批准,并且与其具有相同的限制与公差。

A used engine that has been completely disassembled，inspected，repaired as necessary，reassembled，tested，and approved in the same manner and conformed to the same tolerances and limits as a new engine with either new or used parts.

**翻修间隔时间；Time Between Overhaul；TBO**

由发动机制造商计算，是发动机在确定的发动机参数条件下可以可靠地运行的小时数，不得超过主要零部件（如曲轴、凸轮轴、气缸、连杆和活塞）大修时的使用磨损极限。

TBO is computed by the engine manufacturer and is a reliable estimate of the number of hours the engine could perform reliably within the established engine parameters and still not exceed the service wear limits for major component parts such as the crankshaft，cam shaft，cylinders，connecting rods，and pistons.

**返航；In Flight Turn Back；IFTB**

航空器起飞后未到达目的地机场并返回出发机场。

The airplane does not arriving the destination airport and return to the departure airport after take-off.

**防爆；Explosion Proof**

在正常运行及任何失效情况下，设计和制造的部件不会点燃其周围的可燃蒸气或液体。

Components designed and constructed will not ignite flammable vapors or liquids surrounding under any normal operating condition and any failure condition.

**防爆电气设备；Explosion-protected Electrical Apparatus**

按规定条件设计、制造而不会引起周围爆炸性气体环境爆炸的电气设备。

Electrical apparatus designed and constructed according to specified requirement so they will not lead to explosion of surrounding explosive gas environment.

**防冰；Anti-icing；AI**

在限定时间（保持时间）内防止飞机的表面形成霜、冰和积雪的预防保护措施。防冰液通常在不加热的情况下用于清洁的飞机表面，但也可在加热的情况下使用，包括：① SAE Ⅰ型液；② 浓缩或混合水与 SAE Ⅰ型液的混合物；③ 浓缩或混合水与 SAE Ⅱ型液的混合物；④ 浓缩 SAE Ⅲ型液；⑤ 浓缩或混合水与 SAE Ⅳ型液的混合物。

A procedure used to provide protection against the formation of frost or ice and accumulation of snow or slush on clean surfaces of the aircraft for a limited period of time (holdover time). Anti-icing fluids are normally applied unheated on clean aircraft surfaces, but may be applied heated, and include: ① SAE Type Ⅰ fluid; ② concentrates or mixtures of water and SAE Type Ⅰ fluid; ③ concentrates or mixtures of water and SAE Type Ⅱ fluid; ④ concentrates of SAE Type Ⅲ fluid; ⑤ concentrates or mixtures of water and SAE Type Ⅳ fluid.

### 防冰系统；Ice Protection System；IPS

能够防止飞机某些关键部件积冰的系统。

A system that protects certain critical aircraft parts from ice accretion.

### 防火；Fireproof

① 对用来将火限制在指定火区的材料和零件，指相应于其使用目的的尺寸而言，承受该区内持续燃烧产生的热的能力至少与钢相同；② 对其他的材料和零件，指相应于其使用目的的尺寸而言，承受火焰所产生的热的能力至少与钢相同。

① With respect to materials and parts used to confine fire in a designated fire zone, fireproof means the capacity to withstand at least as well as steel in dimensions appropriate for the purpose for which they are used, the heat produced when there is a severe fire of extended duration in that zone; ② with respect to other materials and parts, fireproof means the capacity to withstand the heat associated with fire at least as well as steel in dimensions appropriate for the purpose for which they are used.

### 防静电工作服；Antistatic Coverall

能避免产生静电危害的工作服。例如，编织碳纤维衣料的电晕能消除静电荷，全棉衣料在相对湿度 20% 以上时亦有防静电功能。

Coveralls which can avoid electrostatic hazards. For example, woven carbon fiber cloth corona can eliminate electrostatic charge, cotton cloth in more than 20% relative humidity also has anti-static function.

### 防滞系统;Anti-skid

一种在任何跑道状况下,都能提供最大的刹车能力而不使轮胎打滑的系统。防滑部件是刹车系统的一部分。

A system for providing maximum braking without skidding for any runway condition. Anti-skid components are part of the brake system.

### 防撞系统;Collision Avoidance System; CAS

一种机载系统,它执行所有必要的功能,以使其输出的是一种在适当时刻才指示或开始采取相应规避机动动作的信号。

An airborne system that performs all the necessary functions such that its output is a signal indicating or initiating an appropriate avoidance maneuver at a suitable time.

### 放射性;Radioactivity

来自不稳定同位素的原子核的自发辐射,一般是 α 粒子或 β 粒子,往往伴随着伽马射线。

The spontaneous emission of radiation, generally alpha or beta particles, often accompanied by gamma rays, from the nuclei of an unstable isotope.

### 放行人员;Certifying Staff

维修单位中确定航空器或者航空器部件满足经批准的标准,并签署批准放行的人员。

The person appointed within a maintenance organization, to ensure the compliance of civil aircraft or aircraft component with the approved standards, and to issue a certificate of release to service for the civil aircraft or aircraft component.

### 飞机;Aeroplane

一种由动力驱动的重于空气的航空器。其飞行升力主要由在给定飞行条件下翼面上的空气动力反作用获得。

A power-driven flight vehicle which is heavier than air, deriving its lift in flight chiefly from aerodynamic reactions on surfaces which remain fixed under given conditions of flight.

### 飞机安全评估；Aircraft Safety Assessment；ASA

为表明已满足相关安全要求，对飞机进行的系统而全面的评估。

A systematic, comprehensive evaluation of the aircraft to show that the relevant safety requirements are met.

### 飞机操作手册；Aircraft Operating Manual

运营人所在国可以接受的手册，它包含正常、非正常与应急程序、检查单、限制、性能资料、航空器系统的详细内容以及与航空器运行有关的其他材料。飞机操作手册是运行手册的一部分。

A manual, acceptable to the state of the operator, containing normal, abnormal and emergency procedures, checklists, limitations, performance information, details of the aircraft systems and other material relevant to the operation of the aircraft. The aircraft operating manual is part of the operations manual.

### 飞机地面勤务；Aircraft Ground Service

包括航空器的进出港指挥、停放、推、拖、挡轮挡、拿取和堵放各种堵盖，为航空器提供电源、气源、加（放）水、加（放）油料、充气、充氧，航空器清洁和除冰、雪、霜，航线腐蚀预防与维护等工作。

Aircraft ground service includes aircraft inbound and outbound command, park, push, drag block round file, pick up and blocking put various blocking cover, and provides the power for the aircraft, the air source, the work of plusing (puting) water, plusing (puting) oil, air and oxygen charging, aircraft cleaning and ice, snow, frost sweeping, routes corrosion prevention and maintenance.

### 飞机地面站；Aircraft Earth Station；AES

部署在飞机上的，属于航空移动卫星服务体系的移动地面站。

A mobile earth station in the aeronautical mobile-satellite service located on board.

### 飞机防冰手册；Aircraft Icing Handbook；AIHB

包括地面和机载防冰设施的参考材料、模拟程序和分析技术的文件。

A document containing reference material on ground and airborne icing facilities, simulation procedures, and analytical techniques.

### 飞机飞行包线；Airplane Operating Envelope

申请人按照 25 部 G 分部确定的运行限制。

The operating limitations defined by the applicant under subpart G of part 25.

### 飞机飞行手册；Aircraft Flight Manual；AFM

含有飞机安全运行所需限制、程序、信息与数据的一种文件。

A document containing the limitations, procedures, information, and data necessary for the safe operation of aircraft.

### 飞机功能；Aircraft Function

由飞机的硬件和软件系统提供给飞机的能力。

Capability of the aircraft provided by the hardware and software of the systems on the aircraft.

### 飞机恢复手册；Aircraft Recovery Manual；ARM

包含以最迅速的方式恢复飞机的详细信息，同时保证人员安全和飞机不受损伤的手册。

The manual which contains information in sufficient detail to recover in the most expeditious manner while maintaining personnel safety and preventing from additional damage to the aircraft.

### 飞机紧急救援现场；Aircraft Emergency Rescue Site

飞机因事故在机场飞行区丧失了滑行和牵引能力，影响航班正常起降的场所。

A place where aircraft loses towing and taxing abilities because of accidents in the airport flight assisting area, influencing the normal flight takeoff and landing.

### 飞机实验协会;Experimental Aircraft Association; EAA

总部位于美国威斯康星州的奥什科什的航空爱好者国际组织。

An international organization of aviation enthusiasts based in Oshkosh, Wisconsin.

### 飞机事故;Aircraft Accident

供局方内部使用的定义：① 作为航空器运行的结果，任何人（乘员或非乘员）死亡或重伤，或任何航空器遭受 CAB 条例第 320 部 320.2(d)所述的严重损伤；② 航空器在飞行中相撞；③ 航空器晚点并被认为已处于事故之中。

As defined for within Authority use：① as a result of the operation of an aircraft, any person (occupant or nonoccupant) receives fatal or serious injury, or any aircraft receives substantial damage; ② aircrafts collide in flight; ③ an aircraft is overdue and is believed to have been involved in an accident.

### 飞机停场;Aircraft on Ground; AOG

因飞机不满足放飞条件而不能完成预定飞行计划的现象。

The situation that the aircraft doesn't accomplish scheduled flight plan because of non-compliance with the flying conditions.

### 飞机图解零件目录;Aircraft Illustrated Parts Catalog; AIPC

在识别和领用可替换的飞机零部件和组件时使用，是飞机维修手册的随同文件，应包含维修工作涉及的所有零件信息。

AIPC is intended for use in the identification and requisition of replaceable aircraft parts and units, as well as companion document to the aircraft maintenance, manual, and shall contain all parts' information for which maintenance practices coverage has been provided.

**飞机维护任务保障系统;Maintenance Task Oriented Support System; MTOSS**

允许使用电子化、信息化的维修数据的系统。

The system permitting the use of electronic data processing (EDP) of maintenance data.

**飞机维修基地;Aircraft Maintenance Base; AMB**

从事飞机和航空电子设备各种等级的定期与不定期维护,但不包括重要检查和翻修的维修企业。

Maintenance enterprises occupied in maintenance of all levels of scheduled and unscheduled aircraft and avionics maintenance, excluding major inspection and overhaul.

**飞机维修区;Aircraft Maintenance Area**

供飞机维修用的全部场地和设施,包括停机坪、机库、车间、航材库、危险品库、办公楼、停车场以及有关的道路。

All venues and facilities for aircraft maintenance, including airport apron, hangar, workshop, spare store, dangerous goods store, office buildings, parking lot and the relevant roads.

**飞机组类;Aircraft Set of Classes**

为方便机组成员和飞行签派员的训练管理,根据飞机动力装置的区别对飞机划分的种类。在本规则中,将飞机分为两个组类:① 组类Ⅰ为以螺旋桨驱动的飞机,包括以活塞式发动机为动力的飞机和以涡轮螺旋桨发动机为动力的飞机;② 组类Ⅱ为以涡轮喷气发动机为动力的飞机。为满足飞行机组成员训练需要,根据飞机最大起飞总重,再将组类Ⅱ飞机分为5 700 千克(含)至136 000 千克(含)和136 000 千克(不含)以上两个种类。

To facilitate the training management for crewmembers or dispatchers, classify the planes according to power plant, the plane is divided into two classes: ① classes Ⅰ is propeller-driven plane, including piston engine-powered planes and turboprop powered planes; ② classes Ⅱ is turbojet engine powered plane. To meet the needs of training, class Ⅱ plane is divided into more than 5 700 kg (including) to 136 000 kg (including) and 136 000 kg (not including) above, according to maximum takeoff gross weight(MTOGW).

### 飞行安全文件系统;Flight Safety Documents System

由合格证持有人制订,用于规定或指导合格证持有人和地面运行人员的日常安全运行所需的相关资料,其中应当包括 CCAR - 121 G 章规定的手册及内容,手册的保存、分发、获取、修订及有效性控制的程序和方法。

A set of necessary related documents are established by operation certificate holder to regulate or guide daily safe operation for the flight and ground operation personnel of operation certificate holder, which shall include the manuals and contents, the procedures and methods of manual save, distribution, acquisition, revised and effectiveness of control stated in CCAR - 121 chapter G of this regulation.

### 飞行安全系统;Flight Safety System

机载/地面系统,涉及但不局限于保护公共安全所需的商用运输航空器的跟踪、评估或命令和控制。例如包括在商业活动过程中使用的所有跟踪、显示和飞行终止系统。

A generic term referring to airborne/ground systems involved in, but not necessarily limited to, tracking, evaluating, or command and control of a commercial launch vehicle that is necessary to protect public safety. Examples include all tracking, display, and flight termination systems used during a commercial activity.

### 飞行安全重要设备;Essential to Safety in Flight

为符合 FAR - 25.1357(d)要求,若电路保护装置(CPD)断开会导致重大的、危险的或灾难性故障,则应当作保障飞行安全的必要设备。

For the purpose of compliance with § 25.1357(d), a CPD is considered to be essential to safety in flight if its disconnection would result in a major, hazardous, or catastrophic failure.

### 飞行标准区域办公室(美国);Flight Standards District Office; FSDO

FAA 区域办公室。

The local FAA office.

### 飞行标准委员会;Flight Standard Board; FSB

由飞行标准司为型号合格审定项目组建的机构,主要负责为提出型号定级申请的新的或改型航空器确定定级要求、制定用于相关机组考核鉴定的最低限度培训要求。在合格审定期间,飞行标准委员会要检查、确认航空器及其系统运行满足适航要求,飞行机组培训设施要求,驾驶员的型号定级要求,任何特定的或专门的培训要求和折叠式座椅、机组休息室和睡眠隔间的适当性。飞行标准委员会还要检查确认应急撤离的能力、飞行标准问题纪要的关闭情况以及完成其他相关的工作。

FSB determines the aircraft type rating requirement for both new and modified models of aircraft that require type ratings. It also develops the minimum training requirements used for flight crewmember qualification. The FSB determines the operational suitability of the aircraft and its systems; requirements for flight crew training aids; type rating requirements for pilots; any unique or special training requirements; and the suitability of jump seat, flight crew rest, and sleeping quarters. The FSB also determines emergency evacuation capability, the closure of flight standards issue papers, and other tasks as appropriate.

### 飞行标准职能部门;Functional Department of Flight Standard

中国民用航空局、地区民用航空管理机构及其派遣机构的飞行标准主管部门。

Flight standards department in charge of the Civil Aviation Administration of China, the regional civil aviation authorities and its dispatched institutions.

### 飞行操纵系统;Flight Control System

飞行操纵系统由控制面、驾驶舱操控装置、铰链以及必要的机械机构组成,用以操纵航空器飞行。

Flight control system is composed of control panel, cockpit manipulate device, hinge and necessary mechanical mechanism, and used to control aircraft flight.

### 飞行程序;Flight Procedures

为离港、在途和/或到达而设计的仪表和目视飞行程序,不管是否基于地面导航设备、卫星或可视导航。例如航空公司、路线、标准和特殊仪表进近程序,离港程序,区域导航程序,航图目视飞行程序和标准终端到达。

Instruments and visual flight procedures designed for departure, en route, and/or arrival purposes whether based on ground navigational aids, satellite, or visual navigation such as airways, routes, standard and special instrument approach procedures, departure procedures, RNAV procedures, charted visual flight procedures, and standard terminal arrivals.

### 飞行高度;Flight Altitude

航空器在海平面以上的运行高度。对无增压座舱的航空器,座舱气压高度和飞行高度是相同的。

The altitude above sea level at which the airplane is operated. For airplanes without pressurized cabins, "cabin pressure altitude" and "flight altitude" mean the same thing.

### 飞行管理系统;Flight Management System; FMS

机载计算机系统,可集成各个子系统的输入,辅助驾驶员进行飞机的横向和纵向控制。

An onboard computer system which integrates inputs from various subsystems to aid the pilot in controlling the airplane's lateral and vertical paths.

### 飞行活动和机组跟踪系统;Flight Activity and Crew Tracking System; FACTS

飞行程序使用的工具,用于全面管理和跟踪机组和有资质的非机组人员的系统。

A system that may be used by flight programs as a tool for overall management and tracking of crewmembers and qualified non-crewmembers, including the ability to manage and track flightcrew activity and currency.

### 飞行机组;Flight Crew

飞行机组负责飞机飞行控制、发动机和系统的操作和管理,包括但不限于机长、副驾驶、第二观察员（飞行工程师）。

Flight crew are responsible for the operation and management of the aircraft flight controls, engines, and systems, including but not limited to pilot in command (captain), first officer (copilot), second officer (flight engineer).

### 飞行机组操作手册；Flightcrew Operating Manual；FCOM

由制造商开发的详细描述飞机或其系统的特点和操作的文件。

A document developed by a manufacturer that describes, in detail, the characteristics and operation of the airplane or its system.

### 飞行机组成员；Flight Crew Member

飞行期间在航空器驾驶舱内执行任务的驾驶员、领航员、飞行通信员和飞行机械师。

Pilots, flight navigators, flight radiomen or flight mechanics who perform their duty in aircraft cockpit during flight time.

### 飞行机组响应；Flightcrew Response

由于发生告警，飞行机组完成的行为，如行动、决定、优先级或者寻找额外的信息。

The activity accomplished due to the presentation of an alert such as an action, decision, prioritization, or searching for additional information.

### 飞行记录器；Flight Recorder

记录航空器飞行性能或在飞行中遭遇的情况的所有仪器或装置的通用术语。飞行记录器可以记录给定飞行的空速、外界大气温度、垂直加速度、发动机转速、管道压力或任何其他有关的变数。

A general term applied to any instrument or device that records information about the performance of an aircraft in flight, or about conditions encountered in flight. Flight recorders may make records of airspeed, outside air temperature, vertical acceleration, engine speed, manifold pressure, or any other pertinent variable for a given flight.

### 飞行技术差错；Flight Technical Error; FTE

由驾驶员（或自动驾驶仪）的能力造成的导航误差，利用显示的指引信息来跟踪所需要的飞行路径。

Navigation error introduced by the pilot's (or autopilot's) capability to utilize displayed guidance information to track the desired flight path.

### 飞行检查；Flight Inspections

对空中导航设施进行飞行中调查和评估，以确认或验证其符合规定的容差。

An inflight investigation and evaluation of an air navigation aid in order to ascertain or verify that it meets established tolerances.

### 飞行教员（航空器）；Flight Instructor (Aircraft)

对特定型别、级别或者类别的航空器，有资格在其航空器、飞行模拟机或者飞行训练器上实施教学的人员。

A person who is qualified to instruct in an aircraft, in a flight simulator, or in a flight training device for a particular type, class, or category aircraft.

### 飞行结冰探测警告系统；Advisory Flight Ice Detector System; AFIDS

向飞行人员提供存在积冰或结冰条件的警告系统。

AFIDS provides an advisory alert to flight crews about the presence of ice accretion or icing conditions.

### 飞行结冰探测系统；Flight Ice Detection Systems; FIDS

探测飞机表面结冰状况的系统。

FIDS may either directly detect the presence of ice on an airplane reference surface or detect that the airplane is in icing conditions.

### 飞行经历时间；Pilot Time

机组必需成员在其值勤岗位上执行任务的飞行时间，即在座飞行时间。

The flight time when a necessary crewmember is on duty in his or her duty position, namely the seated flight time.

## 飞行模拟机指令;FSTD Directive

出于安全原因导致飞行模拟设备应当改装时,由局方颁发给飞行模拟机运营人的有关文件。

A document issued by the authority to an FSTD sponsor requiring a modification to the FSTD due to a flight safety issue and amending the qualification basis for the FSTD.

## 飞行模拟培训装置;Flight Simulation Training Device; FSTD

能够在地面模拟飞行条件的下列三种装置中的任何一种:① 飞行模拟机能精确复现某型航空器的驾驶舱,逼真地模拟出机械、电气、电子等航空器系统的操纵功能,飞行机组成员的正常环境及该型航空器的性能与飞行特性;② 飞行程序训练器能提供逼真的驾驶舱环境,模拟航空器的仪表反应和机械、电气、电子等航空器系统的简单操纵功能,以及特定级别航空器的性能与飞行特性;③ 基本仪表飞行训练器能够配备适合的仪表,在仪表飞行条件下,能够模拟空中飞机驾驶舱环境。

Any one of the following three types of apparatus in which flight conditions are simulated on the ground: ① a flight simulator which provides an accurate representation of the flight deck of a particular aircraft type to the extent that the mechanical, electrical, electronic aircraft systems control functions, the normal environment of flight crew members, and the performance and flight characteristics of that type of aircraft are realistically simulated; ② a flight procedures trainer which provides a realistic flight deck environment, and which simulates instrument responses, simple control functions of mechanical, electrical, electronic aircraft systems, and the performance and flight characteristics of aircraft of a particular class; ③ a basic instrument flight trainer which is equipped with appropriate instruments, and which simulates the flight deck environment of an aircraft in flight in instrument flight conditions.

## 飞行培训装置;Flight Training Device; FTD

敞开的驾驶舱区域或封闭的航空器驾驶舱复制件内的航空器仪表、设备、各类仪表板以及操纵器件的复制品,包括表示航空器(或一组航空器)在地面和飞行条件下运行所必需的设备和计算机程序,具有 FAR－60 所规定的该装置内的各类安装系统的满量程能力,并达到相对应具体 FTD 鉴定等级的鉴定性能标准(QPS)。

The aircraft instruments, equipments, all kinds of instrument panels and the duplicate of control devices in an open cockpit area or closed aircraft cockpit duplicate, including the required equipments and computer programs for aircraft (or a group of aircraft) operation on the ground and flight conditions, possessing the full scale ability of kinds of installation systems in this device stated in FAR－60, and achieving the identification performance standards corresponding to the specific FTD qualification level.

### 飞行日;Flight Day

24 小时的协调世界时间(UCT)或当地时间(从午夜到午夜),由运营人制订,且在此期间受影响的飞机至少飞行一次。

A 24－hour period (from midnight to midnight) of either universal coordinated time (UCT) or local time, as established by the operator, during which at least one flight is initiated for the affected aircraft.

### 飞行时间;Flight Time; FT

自航空器为准备起飞而借自身动力开始移动时起,直到飞行结束停止移动的时间。

The time that commences when an aircraft moves under its own power for the purpose of flight and ends when the aircraft comes to rest after landing.

### 飞行事故;Flight Accidents

民用航空器在运行过程中发生人员伤亡、航空器损坏的事件。

The occurrence of casualties and aircraft damage during the process of operation.

### 飞行手册;Flight Manual

与适航审定相关的手册,包含飞机适航的限制,以及机组人员安全操作飞机所需的说明和信息。

A manual, associated with the certificate of airworthiness, containing limitations within which the aircraft is to be considered airworthy, and instructions and information necessary to the flight crew members for the safe operation of the aircraft.

## 飞行数据分析;Flight Data Analysis

为提高飞行运行安全而对记录的飞行数据加以分析的过程。

A process of analyzing recorded flight data in order to improve the safety of flight operations.

## 飞行通报;Notice to Airmen; NOTAM

有关国家空域系统危险情况或其任何组成部分(设施、服务或程序)的建立、状况或变化的信息(尚未提前用其他手段公布的信息)的一种通知,与飞行运行有关的人员必须及时了解这些信息。① NOTAM(D):(除当地传播外)通过电传打字机将已知的 NOTAM 向飞行服务站责任区域以外做远距离传播,这些 NOTAM 应存储并每小时重复,直到撤销为止;② NOTAM(L):通过声音(适用时通过电传打字机)以及各种手段(如传真电报、电传打印机、传真复印、"热线"、电传复写机、电报电话),将已知的 NOTAM 在当地进行传播,以满足当地用户的要求;③ NOTAM 摘要:以简略明了的语言对现行 NOTAM(D)做的汇编。

A notice containing information (not known sufficiently in advance to publicize by other means) concerning the establishment, condition or change in any component (facility, service or procedure) of, or hazard in the National Airspace System, the timely knowledge of which is essential to personnel concerned with flight operations. ① NOTAM (D): a NOTAM given (in addition to local dissemination) distance dissemination via teletypewriter beyond the area of responsibility of the flight service station, these NOTAMs will be stored and repeated hourly unto cancelled; ② NOTAM (L): A NOTAM given local dissemination by voice, (teletypewriter where applicable), and a wide variety of means such as: telautograph, teleprinter, facsimile reproduction, "hot line", telecopier, telegraph telephone to satisfy local user requirement; ③ NOTAM summary: a compilation of current

NOTAMS(D) in abbreviated plain language.

### 飞行运行活动(FAA);Flight Operations Activity

FAA 组织架构中的要素,主要用于大量使用 FAA 飞机,其人员的主要职能是驾驶飞机。

An element of the FAA's organizational structure whose primary program responsibilities require significant use of FAA aircraft and which has personnel whose primary function is piloting aircraft.

### 飞行运行评审委员会;Flight Operations Evaluation Board; FOEB

由具体负责航空器型号审定的局方人员组成的委员会,由运行、电子、维修监察员和航空器型号审定专家组成,负责制定或修订 MMEL。

FOEB is composed of Authority personnel who are operations, avionics, airworthiness, and aircraft certification specialists. The FOEB develops an MMEL for a particular aircraft type under the direction of the AEG.

### 飞行运行质量保证;Flight Operational Quality Assurance; FOQA

日常收集和分析飞行运行数据的一个自愿性项目,目的是提供更多关于整个飞行运行环境的信息并更深入地理解这些信息。该项目将上述数据与其他来源的数据以及运行经验结合起来,建立增强安全性、培训效果、运行程序、维护和工程程序、空中交通管制程序的客观信息。

A voluntary program for the routine collection and analysis of flight operational data to provide more information about, and greater insight into, the total flight operations environment. A FOQA program combines these data with other sources and operational experience to develop objective information to enhance safety, training effectiveness, operational procedures, maintenance and engineering procedures, and air traffic control (ATC) procedures.

### 飞行值勤期;Flight Duty Period

从飞行机组成员开始值勤(该时刻紧接休息期的结束,并在执行下一次飞行或一系列飞行任务之前)起,到完成这一次或这一系列飞行,解除一切任务为止的总时间。

The total time from the moment when a flight crew member commences duty (immediately subsequent to a rest period and prior to making a flight or a series of flights), to the moment when the flight crew member is relieved of all duties and has completed such flight or series of flights.

## 飞行指引仪;Flight Director; FD

以目视方式指示驾驶员按照预选飞行计划对航空器进行正确操纵的一种机载设备;正常的仪表数据输入计算机进行变换、组合和计算,从而形成适合简单目视显示的输出信号。

An airborne device which indicates the pilot, by visual means, the correct control application for the operation of an aircraft in accordance with a preselected flight plan; normal instrument data is fed to a computer for modification, combination, and calculation leading to an output to a simple visual display.

## 飞行咨询服务;Flight Advisory Service

由设备提供的咨询和信息,协助驾驶员安全飞行和移动航空器。

Advice and information provided by a facility to assist pilots in the safe conduct of flight and aircraft movement.

## 非 CTSO 功能;Non-CTSO Function

未被技术标准规定批准的最低性能标准(MPS)所涵盖的、不支持也不影响"承载 CTSO 件"中 CTSO 功能的并且在 CTSO 件之外可以实现的功能。

Functions that are not covered by the approved minimum performance standards (MPS) specified in the technical standards, do not support or affect the CTSO functions in the CTSO parts, and can be technically realized outside the CTSO parts.

## 非关键件;Non-critical Part

除关键件之外的零部件。

Components other than critical part.

**非计划需求；Unanticipated Scheduling Requirements**

短期需要的、不可预知的需求，如可用航空器。

Requirements needed by short term, unforeseen circumstances, such as airplane availability.

**非精密进近程序；Non-precision Approach Procedure**

不提供电子下滑道的标准仪表进近程序。

A standard instrument approach procedure in which no electronic glide slope is provided.

**非精密仪表进近；Non-precision Instrument Approach**

使用全向信标（VOR）、导航台（NDB）或者航向台（LLZ）（ILS 系统下滑台不工作）等地面导航设施，只提供方位引导，不具备下滑引导的进近。

An approach in which only azimuth guidance is provided by means of ground navigation facilities such as VOR, NDB, ILZ (when ILS system glide slope is not working), and no glide guidance is provided.

**非精密仪表着陆用跑道；Non-precision Instrument Runway**

具有现有仪表进近程序的跑道，使用仅有水平制导的空中导航设施，其直接非精密仪表进近程序已获批准。

A runway having an existing instrument approach procedure utilizing air navigation facilities with only horizontal guidance for which a straight-in nonprecision instrument approach procedure has been approved.

**非例行工作；Nonroutine Task**

不出自运营人/制造商的维修大纲中计划任务的工作。

A task which is not a planned/scheduled task coming from the operator's/manufacturer's maintenance program.

**非例行维修；Unscheduled Maintenance**

对已知或可疑的故障提供改正方法，以使其恢复到满意状态的非计划性维修工作。

Maintenance performed to restore an item to a satisfactory condition by providing correction of a known or suspected malfunction and/or defect.

## 非商业运输运营人；Non-commercial Transport Operators

运营人按照规定取得局方颁发的商业非运输运营人运行合格证和运行规范，使用民用航空器实施公共航空运输之外的，以取酬或出租为目的的商业航空飞行的航空器运营人。

The operator which obtain the commercial non-transport certificate and operations specifications issued by the authority in accordance with the rules, using civil aviation aircraft to do the implementation of commercial activity except common transportation for the purpose of emuneration or hire.

## 非寿命件；Non-life-limited Part

除寿命件之外的零部件。

Components other than life-limited part.

## 非影响件 Non-influencing Parts

除影响件之外的零部件。

Components other than influencing parts.

## 非正常飞行条件；Abnormal Flight Conditions

正常运行包线以外的飞行条件。

Flight conditions outside the normal operating envelope.

## 分贝；Decibel (dB)

噪声强度的相对单位，如声压级、噪声级和功率级，用从 0（平均最低能感知级别）到 130（平均痛苦级别）的范围描述。

The unit in which the relative levels of intensity of acoustical quantities, such as sound pressure levels, noise levels and power levels, are expressed on a scale from zero (for the average least perceptible level) to about 130 (for the average pain level).

### 分层；Delamination

基体失效导致的层压材料的层间分离。

The separation of layers in a laminate through failure of the matrix.

### 分基地；Satellite Base

航空运营人在主基地之外驻扎飞机运行，并且承担驻扎飞机具体维修计划和控制职责（即执管飞机）的各分（子）公司所在地点。分基地执管的飞机不受驻扎和维修地点转移的影响，但维修计划和控制职责的转移将造成执管的转移。

The sub-companies location where the aircrafts are stationed outside of the main base of the air carriers, and takes charge of aircraft operating and stationed aircraft maintenance planning and control responsibilities (control aircraft). The control aircraft of the satellite base is not transfer when the stationed and maintenance location changed，but the maintenance plan and control responsibilities will result in the transfer of the control aircraft.

### 分时显示；Time-shared Display

显示器综合显示地形和其他系统的附加信息。

A display that shows terrain information，plus additional information from other systems.

### 分销商；Distributors

对安装于具有 TC 证的航空器、航空发动机、螺旋桨和电器上的部件进行销售的经纪人、经销商、代理商或其他个人或机构。

Brokers，dealers，resellers，or other persons or agencies engaged in the sale of parts for installation in TC aircraft, aircraft engines, propellers, and appliances.

### 风切变；Windshear

因短距离内风力或风向的改变而形成的撕拉效应和剪切效应。可发生在水平方向或垂直方向上，有时会在两个方向上同时出现。

The tearing and shearing effect caused by a change in wind speed and/or direction in a short distance. It can exist in a horizontal or vertical direction and occasionally in both.

### 风切变脱离机动；Windshear Escape Maneuver

偶然遭遇风切变时，驾驶员采用的一种改出技术。其方法是运用必要的推力，以俯仰操纵使飞机达到初始的目标姿态。这种改出技术的目的是使飞机能做尽可能长的飞行，以求脱离风切变。这种机动是用来使飞机脱离已遭遇的风切变的一种操作技术，它有效、简单、容易记忆，而且具有通用性。

A pilot recovery technique used when an inadvertent windshear encounter is experienced. It is achieved by pitching toward an initial target attitude while using necessary thrust. The objective of the recovery technique is to keep the airplane flying as long as possible in hipe of exiting the windshear. the maneuver is an operational technique to be used to escape from the encounter that was developed to be effective, simple, easily recalled, and to have general applicability.

### 风扇噪声；Fan Noise

涡轮风扇发动机的风扇内产生的噪声的通用术语，包括离散频率噪声和随机噪声。

General term for the noise generated within the fan stage of a turbofan engine, including both discrete frequencies and random noise.

### 风险；Risk

以事件的严重程度和概率表示其对不期望事件的影响。

Expression of the impact of an undesired event in terms of event severity and probability.

### 风险分析；Risk Analysis

分析危险源严重性和可能性的过程。这个过程可以是定性的，也可以是定量的。

Process whereby hazards are objectively characterized for their severity and probability. The process can be either qualitative or quantitative.

### 风险分析和管理；Risk Analysis and Management

识别航空运营中的风险及其替代方案的一种系统过程。选择将最大限度提高航空安全的替代方案。

A systematic process for identifying risks associated with alternative courses of action involved in an aviation operation. Choosing the alternative courses of action that will promote optimum aviation safety.

### 风险管理；Risk Management；RM

识别危害，系统地量化或限定其对暴露的个人、人群或资源构成的风险程度，并努力降低或消除风险的过程。

The process of identifying hazards, systematically quantifying or qualifying the degree of risk they pose for exposed individuals, populations, or resources, and working to reduce or eliminate risk.

### 风险管理计划；RM Plan

利用可接受的风险管理过程制订具体项目的计划。

A project-specific plan formulated using an accepted RM process.

### 风险评估；Risk Assessment

对暴露的个体、人群或资源造成危险的程度进行的系统的、定量和定性的确定危害的过程。

The process of identifying hazards and systematically quantifying or qualifying the degree of risk they pose for exposed individuals, populations, or resources.

### 风险系数；Risk Factor

一种定量评估输出，其等于在给定时间内预期发生的事故平均值。

A quantitative assessment output equal to the average number of future events expected to occur within a given time.

### 风险要素；Risk Elements

航空决策的风险要素应考虑下列 4 种基本情况：驾驶员、飞机、环境和包括任何给定飞行情况的运行种类。

Risk elements in ADM take the four fundamental risk elements into consideration: the pilot, the aircraft, the environment, and the type of operation that comprise any given aviation situation.

## 风险准则;Risk Guideline

可接受的风险上限,其可以帮助航空安全工程师决定是否需要颁发适航指令或采取其他强制性的改正措施和基于暴露的风险而可能采取的备选改正措施。

The upper limit of acceptable risk which assists the ASE in determining the need for AD or other mandatory corrective actions and the adequacy, in terms of risk exposure, of a proposed candidate corrective action.

## 服务通告;Service Bulletin; SB

航空产品的设计和生产厂家根据自身和用户信息,改进所生产的航空产品的可靠性或使用的安全性,是对用户的一种技术服务措施和对自身生产技术改进的要求,以及对航空产品实施检查、重复检查、改装或使用寿命更改等的技术要求。

A document issued by designer or manufacturer of aeronautical product in order to improve the reliability or using safety of product base on the information collected from his customer or himself. It is a technical service provided to his customer and improve his manufacturing technical itself. It is also a technical requirement for product to perform check, repeat check, modify or alteration of its service life.

## 服务文件;Service Documents

型号合格证持有人或附件或部件制造商的出版物,用于沟通与安全性、产品改进、经济和操作和/或维修实践相关的有用信息。典型的出版物形式包括服务通告、所有操作人员文件、服务简讯和服务文摘或杂志。不包括如飞行手册和某些维修手册等 FAA 型号合格审定或批准要求的出版物。

Publications published by a type certificate holder or appliance or component manufacturer that communicate useful information relative to safety, product improvement, economics, and operational and/or maintenance practices. Typical forms of publications include service bulletins, all-operators' letters, service newsletters and service digests or magazines. Publications, such as flight manuals and certain maintenance manuals, that are required for FAA type certification or approval are excluded.

### 服务信函;Service Letter; SL

用于通知航空公司未在服务通告中列出的信息的一类文件。

The document used to notify the Airlines of the types of information that are not outlined under service bulletins.

### 符合性;Compliance

民用航空产品和零部件的设计符合规定的适航规章和要求。

The design of the civil aviation products and articles are found to satisfy the prescribed airworthiness regulations and requirements.

### 符合性方法;Means of Compliance; MOC

在型号合格审查过程中,为了获得所需的证据资料以表明适航条款的符合性,申请人通常需要采用的不同方法。

In order to obtain the required evidence to show the compliance of the airworthiness regulations in the process of type certification, the different means that the applicant usually needs to adopt.

### 符合性检查;Conformity Inspection

针对硬件的规范性审查或物理检查,以验证特定部件或改装符合相应规章要求。

Specification review and physical inspection of hardware to verify that a particular component or modification complies with the requirements of the applicable regulations.

### 符合性检查清单;Compliance Check List; CCL

按审定基础确定的规章条款逐条列出表明条款符合性的符合性方法、相关型号资料及批准情况的汇总性文件,用于记录和检查型号合格审定项目的完成情况。

A list of documentation in which itemized the compliance means, related type data and approval status to show compliance with the regulation items specified by the applicable certification basis, used to record and check the progress of a type certification project.

## 符合性进度表;Compliance Schedule

执行指定纠正措施(如维护、检查、零件更换、改装等)的时间表,以减轻识别出的不安全状态的时间表。

A timetable for performing specified corrective actions ( e. g. , maintenance, inspections, part replacements, alterations, etc. ) to alleviate an identified unsafe condition.

## 符合性决定;Compliance Determination

审定或认可适航当局对申请人已经表明对指定的单个适航标准的符合性做出决定。

Determination, by either the CA or the VA, that the applicant has demonstrated compliance with identified, individual airworthiness standards.

## 符合性判定;Compliance Finding

局方对申请人已表明对所有确定适用的适航标准的符合性做出法定判定。

Official act, which the responsible authority makes a legal finding that the applicant has demonstrated compliance with all identified applicable airworthiness standards.

## 符合性判定委派;Assignment of Finding of Compliance

认可局方委托审定当局按认可当局的合格审定基础进行符合性判定。

The validating authority entrusts the certificating authority to make findings of compliance to the validating authority (VA) certification basis.

## 符合性试验;Compliance Testing

也称为合格审定试验,参见"部件试验""地面试验"和"飞行试验"。

Also called Certification Testing, see " Component Testing", " Ground

Testing" and "Flight Testing".

### 符合性数据；Compliance Data

为表明改装设计符合审定基础要求的数据。这些数据通常包括符合性检查单、程序、过程、各种分析（如载荷分析、结构分析、材料选用分析、许用值分析、重量平衡分析、电负载分析、安全性分析等）报告，从零部件到改装包的各个验证性试验的大纲及其报告等。

A type of design data which shows the data of modify design compliance certification basis, including the compliance checklist, procedures, processes, analysis (load、structure、material、values、weight and balance、electrical load、safety analysis and so on) report, the compliance test program and report of component and modify package.

### 符合性替代方法；Alternative Means of Compliance；AMOC

符合性替代方法的目标是允许运营人或制造商提出替代 AD 中规定的纠正措施，这种措施并不一定要与 AD 中发布的相一致，但能满足 AD 中所要求的安全性水平。

The objective of AMOC is to allow an operator or manufacturer to propose an alternative corrective action to that prescribed in an AD. The intent of the AMOC is to enable approval of options that are not necessarily conceived of at the time of AD issuance, but which provide an equivalent level of safety to that afforded by the AD.

### 符合性验证资料；Substantiation Data

用于证明或表明型号设计符合审定基础的资料，包含试验大纲、计算或分析报告、试验报告等。

Substantiation data is used to prove or indicate that the type design conforms to the certification basis, including test plan, calculation or analysis report, test report, etc.

### 辅助动力装置；Auxiliary Power Unit；APU

飞机内的小涡轮发动机，在地面或空中给各系统提供电源和气源。

A small turbine engine in the airplane for supplying electrical and pneumatic power for systems operation on the ground or in flight.

### 辅助动力装置循环；APU Cycle

APU 启动至关停的一个完整过程。

A complete operating process of an APU from start to stop.

### 辅助燃油系统；Auxiliary Fuel System

辅助燃油系统是装在飞机内，使飞机具有额外可用燃油从而增加航程的系统。"辅助"的意思是该系统相对于飞机主燃油系统是次要的，主燃油系统为其提供支持，也就是说，当辅助燃油系统发生失效或燃油意外耗尽时，主燃油系统的功能应立即可用并启动，无须机组成员立即采取控制措施。实质上，对装有辅助燃油系统的飞机，即使该系统不使用也能安全飞行，即该系统中所存燃油对短航程飞行不是必需的。

An auxiliary fuel system is a system installed within the airplane which makes additional fuel available for increasing the flight range of the airplane. The term "auxiliary" means that this system is secondary to and backed by the airplane's essential fuel system，that is to say，the functions of the essential fuel system are immediately available and operative without immediate supervision by the flightcrew in the event of failure or inadvertent depletion of fuel in the auxiliary fuel system. In essence，an airplane equipped with an auxiliary fuel system is capable of safe flight even when the auxiliary fuel system is not used，which means its fuel storage capacity is not required for short range flight.

### 辅助燃油箱；Auxiliary Tanks

安装在飞机上，用于提供额外燃油以增加飞机航程的燃油箱。"辅助"的意思是该燃油箱相对于飞机主燃油箱是次要的，也就是说，在辅助燃油箱发生失效或燃油意外耗尽时主燃油箱的功能应立即运转且无须机组立即采取措施。辅助燃油箱安装在飞机的不同位置，包括中央翼结构、水平安定门、机翼和货舱，通常在飞行中按计划耗尽可用燃油。

Fuel tanks installed on the aircraft，make additional fuel available for increasing the flight range. The term "auxiliary" means that the tank is secondary to the airplane's main fuel tanks, that is to say, the functions of the main tanks are immediately available and operate without immediate supervision by the flightcrew in the event of failure or inadvertent depletion of fuel in an auxiliary tank. Auxiliary tanks are usually intended to be emptied of usable fuel during flight and have been installed in various locations including center wing structure, horizontal stabilizers, wings, and cargo compartments.

### 腐蚀；Corrosion

物质(通常为金属)由于与环境作用而产生退化。

The deterioration of a substance (usually metal) caused by a reaction with its environment.

### 腐蚀等级；Corrosion Level

确定针对特定腐蚀问题的腐蚀预防与控制大纲的有效性的一种方法，用腐蚀严重程度及其对运营人机队持续适航的潜在后果来表示。

A method of determining the effectiveness of a CPCP relative to a given corrosion finding in terms of the severity of corrosion and the potential consequences to continuing airworthiness in the operator's fleet.

### 腐蚀率；Corrosion Rate

腐蚀侵袭的速度或速率，以单位时间内的材料质量损失表示。

The speed or rate of a corrosion attack expressed in material weight loss per unit of time.

### 腐蚀疲劳；Corrosion Fatigue

金属因暴露于腐蚀环境中，导致其承受循环应力的能力降低。

A reduction in the ability of a metal to withstand cyclic stress by its exposure to a corrosive environment.

### 腐蚀预防与控制大纲；Corrosion Prevention and Control Programs; CPCP

由航空运营人制订的，使航空器结构的承载能力不会降低到低于航空器适航性所需水平的一种全面而系统的腐蚀控制方法,包括执行、记录和报告要求。

A comprehensive system corrosion control methods made by Aviation operators, in order to make the carrying capacity of the aircraft structure not be reduced to a level lower than that required for aircraft airworthiness, including the implementation of the recording and reporting requirements.

### 腐蚀预防与控制基本大纲;Basic Program of Corrosion Prevention and Control

型号合格证或补充型号合格证持有人为某具体机型制订的腐蚀预防与控制大纲。它包括对航空器每一部位和区域的基本腐蚀检查任务、腐蚀级别定义、执行检查门槛值和重复间隔,并给出当任何部位或区域的腐蚀损伤超过 1 级时采取的特殊步骤。在没有型号合格证或补充型号合格证持有人制订的基本大纲的情况下,也可以由航空运营人自己或联合其他航空运营人制订。

The basic program of corrosion prevention and control is a program made by the type certificate or supplemental type certificate holders for specific models. It includes basic corrosion inspection tasks, corrosion-level definition, and implementation inspection threshold and repetitive intervals in each part of the aircraft and region, and gives the special steps for the corrosion damage which exceed first level at every part and region. In the case of there is no basic program made by the type certificate or supplemental type certificate holders, it can be made by the air carriers themselves or in combination with other air carriers to develop.

### 复飞;Go-around; GA

驾驶员在最后进近阶段,决定不落地而做的机动动作。复飞时需加油门,收起落架,以使飞机爬升至安全高度。

Airplane maneuver taken when the pilot is in the final approach phase of flight and decides not to land. Thrust is applied and the landing gear retracted so as to let the airplane climb to a safe altitude.

### 复飞功率或推力设定;Go-around Power or Thrust Setting

在性能资料中规定的飞行中最大允许功率或推力设定。

The maximum permissible power or thrust setting in flight are specified in the performance documents.

### 复训；Recurrent Technical Training

为保证技术能力或者满足特定技能开展的技术能力再培训。对维修领域的培训而言，复训主要是指为保证维修基地的雇员能够维持完成指定工作能力所开展的培训。

Recurrent technical training for specific tasks or functions to ensure currency in existing or added capabilities. Recurrent maintenance training commonly includes training known as refresher training，to ensure that a repair station employee remains capable of properly performing the assigned job.

### 复杂电子硬件；Complex Electronic Hardware

复杂硬件项目的一个子集；不被认为简单的自定义微编码组件。

A subset of a complex hardware item；a custom micro-coded component that is not considered to be simple.

### 副驾驶；Co-pilot/First Officer；F/O

从事除机长以外任何飞行职位的驾驶员，不包括在飞机上仅为接受飞行培训，获取执照或评级为目的的驾驶员。

A pilot serving in any piloting capacity other than as pilot-in-command or commander，but excluding a pilot who is on board for the only purpose of receiving flight instruction for a license or rating.

### 富燃油蒸气/空气混合；Rich Fuel Vapor/Air Mixture

燃油蒸气浓度高于支持燃烧浓度的混合气体。

A fuel vapor/air mixture that contains a concentration of fuel molecules that support combustion.

# G

**改进的发动机；Modified Engine**

之前已获批准的发动机，包含目前未获批准的修改。

An engine, previously approved，in which hitherto unapproved modifications have been embodied.

**改正措施；Corrective Action**

任何缓解安全问题的措施，包括纠正不安全状况的强制措施，如适航指令和规章更改；非强制措施或建议，如特别适航信息通告和航空警示；既直接纠正安全问题，也/或缓解运行限制的措施，如从即将起飞的航班上卸载某一产品。

Any action to mitigate a safety issue, including mandatory actions like Ads and rule changes，to correct an unsafe condition；and non-mandatory actions and recommendations like SAIB and Aviation Alerts；and actions that either directly corrects the safety problem and/or mitigates risk with operational limitations or restrictions，like grounding a product from further flight.

**改装；Modification**

在航空器及其部件交付后进行的超出其原设计状态的任何改变，包括任何材料和零部件的替代。

Any change beyond its original design state after the delivery of the aircraft and its components，including any substitute of materials and components.

**改装包；Modification Kit**

改装所用全套软、硬件及其必备的安装、使用和维护资料等的总称。

A general term for a full set of software and hardware with installation, use and maintenance documents required by the modification.

### 改装件；Modification Article

不包含在民用航空产品的原始型号设计中，而是通过对民用航空产品型号的设计大改或小改获得批准的零部件。改装件的常见来源是补充型号合格证（STC）或改装设计批准书（MDA）。

Parts not included in the original type design of civil aviation products，but approved through major or minor design changes to the civil aviation products. A common source for modification article is the supplemental type certificate (STC) or the modified design approval (MDA).

### 改装设计批准；Approval of Changes in Type Design

民航局对重要改装的设计批准。

Design approval of major changes by CAAC.

### 改装设计委任单位代表；Designated Modification Design Organization Representative；DMDOR

根据 AP-183-AA-2018-11《适航委任单位代表管理程序》的规定获得民航行政机关的授权，代表民航行政机关执行部分或全部改装或修理设计审核、检查或批准工作的单位或机构。改装设计委任单位代表是委任单位代表的一种。

According to AP-183-AA-2018-11 Airworthiness Designated Organization Representative Management Procedure，the organization or institution authorized by the civil aviation administrative organ to carry out part or all of the modification or repair design review，inspection or approval on behalf of the civil aviation administrative organ. It is a kind of designated organization representative.

### 改装者；Modifier

经局方批准，可在 TSO 物件上实施已批准的设计更改的人。

The person to whom the Authority approves to implement an approved design change to a TSO article.

### 干租;Dry Lease

由出租人(可能是航空运营人、银行或租赁公司)通过协议向承租人(航空运营人)仅提供航空器而不提供飞行机组的租赁。干租运营控制通常由承租人承担。

Any agreement by the lessor (possibly air operators, banks, or aircraft rental company) providing only aircraft without flight crew leasing to the lessee (air operators). Dry lease operation control is usually undertaken by the lessee.

### 高度保持能力;Height-keeping Capability

通过正确的操作措施和维护,在正常的运行环境下,航空器预期的高度保持性能。

The aircraft height-keeping performance that can be expected under nominal environmental operating conditions with proper aircraft operating practices and maintenance.

### 高度保持性能;Height-keeping Performance

观测到的航空器对飞行高度层的符合性能。

The observed performance of an aircraft with respect to adherence to cleared flight level.

### 高度表的指示基准;Indicated Datum of the Altimeter

当一个理想的绝对压力施加到高度表的传感元件,并且没有校正仪表误差(刻度误差)和校正静态误差时,高度表所显示的高度。

The altitude displayed by the altimeter when an ideal absolute pressure is applied to the sensing member of the altimeter, neither corrected for instrument error (scale error), nor corrected for static source error.

### 高度表校准数据;Calibrated Datum of the Altimeter

通过适用于特定高度计的特定校准卡进行的校正,仅用于校正仪器误差(刻度误差)。

The correction applied via a specific calibration card applicable to a specific altimeter to correct for instrument error (scale error) only.

### 高度测量系统误差;Altimetry System Error; ASE

以国际标准大气压(ISA)的标准地面气压设定(29.92英寸汞柱/11 013.25百帕)为基准,显示给飞行机组的气压高度与自由大气的气压高度的差值。

Based on the International Standard Atmosphere (ISA) — standard ground pressure setting (29.92 in Hg/11 013.25 hPa), ASE is the pressure altitude difference between displayed to flight crew and the free atmosphere.

### 高度告警;Altitude Alert

当飞机接近或偏离选定高度时,所出现的驾驶舱告警。

A flight deck alert occurring when the airplane approaches or departs a selected altitude.

### 高度告警系统;Altitude Alerting System; AAS

每个高度告警系统必须能够:① 为驾驶员提供告警。(a)接近预选高度时上升或下降,通过一系列的听觉和视觉信号使驾驶员有足够的时间在预定的高度上建立水平飞行;(b)接近预选高度上升或下降,通过一系列的视觉信号使驾驶员有足够的时间在预定的高度上建立水平飞行,并通过一个听觉信号来显示向上还是向下偏离预定高度。② 提供飞机所需的从海平面到批准的最高飞行高度的信号。③ 预设定海拔增量以与飞机飞行高度相匹配。④ 在不使用特殊设备条件下进行检测,以确定告警信号是否正确运行。⑤ 进行必要的气压设置。

Altitude alerting system must be able to:① alert the pilot. (a) to rise or descend near a preselected altitude. A series of auditory and visual signals is used to give the pilot enough time to establish level flight at a predetermined altitude;(b) upon approaching a preselected altitude in either ascent or descent, by a sequence of visual signals in sufficient time to establish level flight at that reselected altitude, and when deviating above and below the preselected altitude, by an aural signal. ② Provide the required signals from sea level to the highest operating altitude approved for the airplane in which it is installed. ③ Preselect altitudes in increments that are commensurate with the altitudes at which the aircraft is operated. ④ Be tested without special equipment to determine proper operation of the alerting signals. ⑤ Accept necessary barometric pressure settings if the system or device operates on barometric pressure.

### 高高原机场;Very High Elevation Airport

海拔高度在 2 438 米(约 8 000 英尺)及以上的机场。

The airport of which altitude exceeds 2 438 m (8 000 ft).

### 高空发动机;Altitude Engine

自海平面至某一规定的高度都可以产生起飞额定功率的航空发动机。

The reciprocating air engine which can produce a take-off rated power from sea level to a certain provisions higher height.

### 高频;High Frequency; HF

3 至 30 兆赫的频带。

Frequency band from 3 to 30 MHz.

### 高频通信;HF Communications

利用高频(HF)的通信,曾广泛用于空地声音通信,但现在由于空中交通管制使用甚高频(VHF)和超高频(UHF),作用已相对减小,不过仍用于海上的点对点和空/地通信。

Communications using high radio frequencies (HF), once used extensively for air-to-ground voice communication but now reduced to a relatively minor role due to the use of VHF and UHF for air traffic control, still used for overseas point-to-point and air/ground communication.

### 高强度辐射场脆弱性;HIRF Vulnerability

一个系统的磁化率特征,在它受一个高强度辐射场环境影响执行其预定的功能时,会导致其遭受不利影响。

Susceptibility characteristics of a system that make it to suffer adverse effects when performing its intended function as a result of having been subjected to a HIRF environment.

### 高强度辐射场(HIRF)环境；High-intensity Radiated Fields (HIRF) Environment

高功率射频能量传输到自由空间中存在的电磁环境。

Electromagnetic environment that exists from the transmission of high power RF energy into free space.

### 高原机场；High Elevation Airport

包括一般高原机场和高高原机场两类。

High elevation airport includes general high elevation airport and very high elevation airport.

### 根本原因；Root Cause

系统性或重复性问题的深层次原因，通常通过结构化分析来识别。

The underlying cause of a systemic or recurring noncompliance, usually identified through structured analysis.

### 更改类适航指令；Correction Airworthiness Directive

对非实质性内容进行更改后的适航指令。

An AD change is a correction of nonsubstantive material.

### 更改指令(航空器)；Change Order (Aircraft)；CO(A)

为批准并指导对现有局方航空器或不可从机上卸下的系统进行改装或安装新的航空器系统而颁发的文件。

Document issued to authorize and direct modification of an existing agency aircraft or system not removable from the aircraft, or for the installation of new aircraft systems.

### 工厂；Facility

PAH 或重要零件供应商完成所有或部分与局方颁发的批准书权限有关的

评审子系统功能的实际场所。

A physical location where a PAH or associate facility performs all or part of the system element functions relevant to the approval authority granted by Authority.

### 工具设备手册；Tool and Equipment Manual；TEM

机体、发动机翻修使用的特殊工具和设备要求。

TEM includes special tools and equipment required for the airframe and engine overhaul.

### 工业事故；Industrial Accident

除了不以飞行为目的之外，满足飞机事故标准的事件。

An occurrence that meets the criteria for an aircraft accident，except that there is no intention of flight.

### 工业指导委员会；Industry Steering Committee；ISC

由航空器、发动机、螺旋桨、设备制造商及航空运营人的代表组成。其任务是研究并制定 MRBR 建议书的政策和程序手册（PPH），指导工作小组（WG）的工作，准备 MRBR 建议书。它受由航空器制造人和航空运营人推荐产生的 ISC 主席领导。

The ISC membership should comprise representatives from aircraft，engine，propeller，appliance manufacturers and intended air carriers. Its role is to develop and establish policy with regard to procedural matters for the development of the proposed MRBR，to direct the activities of the working groups(WG)，and to prepare the MRBR proposal. It is under the direction of the ISC chairperson who is recommended by the appliance manufacturers and intended air carriers.

### 工作单卡；Maintenance Task Card

设定并记录工作顺序和步骤的文件，它不必列出维修工作的实施方法和标准，在实际使用中可以称为工作单、工作卡、工作指令、数据记录单等或其组合的工作包。

Maintenance task card is used to define and record work sequences and steps, it may not involve accomplishment methods and standard concerning the maintenance tasks. It can be named as work sheet, work card, work order, data record sheet or one of any combinations.

### 工作模式;Operating Modes

与飞行各阶段相关的多传感器设备模式,即海洋、航路、终端和非精密进近。

Multi-sensor equipment modes associated with various phases of flight, that is to say, oceanic, en route, terminal, and non-precision approach.

### 工作任务清单;Job Task Listing

为完成运行工作所需的所有任务、子任务、知识和技能的清单。

The lists of all tasks, subtasks, knowledge, and skills required for accomplishing the operational job.

### 工作组;Working Group; WG

由 ISC 选定的制造人和航空运营人的代表组成,其任务是按照 PPH,用 MSG - 3 的逻辑方法确定 MRBR 建议书中的维修任务和维修间隔。

WG is composed of represents selected by ISC from manufacturers and operators. Its task is to determine the maintenance tasks and maintenance interval in proposal MRBR by method of MSG - 3 ways, according to PPH.

### 工作组顾问;Work Group Consultant

由 MRB 主席选派的 MRB 成员组成,负责向工作组提供建议。

Work group consultant is assigned by MRB chairman and made up of MRB member, offering suggestions for work group.

### 公认标准;Consensus Standard

就审定轻型-运动航空器而言,指一种由工业方制定的公认标准,用于航空器的设计、制造和适航。它包括(但不限于)航空器的设计和性能标准、必要的设备、制造厂商的质量保证系统、生产验收试验程序、使用说明、维修和检查程序、重大修理和重大改装的标志和记录,以及持续适航。

For the purpose of certificating light-sport aircraft (LSA), an industry-developed consensus standard applies to aircraft design, production, and airworthiness. It includes, but is not limited to, standards for aircraft design and performance, required equipment, manufacturer quality assurance systems, production acceptance test procedures, operating instructions, maintenance and inspection procedures, identification and recording of major repairs and major alterations, and continued airworthiness.

### 功率设定;Power Setting

涡喷和涡扇发动机以千牛为单位的功率或推力输出,或涡桨发动机以千瓦为单位的轴功率。

The power or thrust output of an engine in terms of kN thrust for turbojet and turbofan engines, or shaft power in terms of kilowatts for turboprop engines.

### 功率消耗不稳定性;Power Loss Instabilities

发动机工作异常,如不可恢复或重复喘振、失速、反流或熄火,这些可能导致发动机功率或推力循环。

Engine operating anomalies such as non-recoverable or repeating surge, stall, rollback, or flameout, which can result in engine power or thrust cycling.

### 功能;Function

一个连续的工作过程,它是更大的过程的主要组成部分,并能从此更大的过程中分离出来。对组织来说,指给予组织某一部门独立的、可单独明确的责任分配。

A continuing work process which is a major segment of a larger process and can be separated from the larger process. With reference to organization, it means a discrete, separately definable assignment of responsibility to an organizational element.

### 功能独立性;Functional Independence

一种属性,其中的功能是不同的,以便最大限度地减小发生共同需求错误的可能性。

An attribute where the functions are different in order to minimize the likelihood of a common requirement error.

### 功能飞行检查;Functional Flight Check

由持有被检查机型的当前Ⅱ类驾驶员授权的驾驶员进行的检查。

A check conducted by a pilot holding a current category Ⅱ pilot authorization for the type airplane being checked.

### 功能软件;Functional Software

将被批准作为功能 TSO 授权的一部分或型号合格审定的一部分的软件应用。有时也被称为操作软件、应用软件或飞行软件。

Software applications that will be approved as part of a functional TSO authorization or as part of a type certification effort, sometimes called operational software, application software, or flight software.

### 功能危险性评估;Functional Hazard Assessment; FHA

对飞机和系统功能进行的系统、全面的研究,用来确定那些由功能故障或失效导致的潜在的、小的、重大的、危害的和灾难性的失效状态。

A systematic, comprehensive examination of airplane and system function to identify potential minor, major, hazardous, and catastrophic failure conditions that may arise as a result of malfunctions or failures to function.

### 功能研制保证等级;Function Development Assurance Level; FDAL

为功能而实现的研制保证任务的严重程度等级。

The level of rigor of development assurance tasks performed to functions.

### 共模错误;Common Mode Error

能同时影响若干元件的错误,如果没有这种错误,则这些元件被认为是彼此独立的。

An error which affects a number of elements otherwise considered to be independent.

## 共模分析；Common Mode Analysis；CMA

为验证飞机安全性评估/系统安全性评估中确定的失效事件在实际执行中是独立的而进行的分析。

An analysis performed to verify that failure events identified in the ASA/SSA are independent in the actual implementation.

## 共识；Consensus

一个总协定，但不一定是一致同意，包括试图解决有关各方异议的程序。此流程是意见解决流程，要求所有意见都已得到公平考虑，并且每个反对者都已被告知其异议的处理方式及其原因。此外，共识机构成员可在审查意见后更改其投票。

A general agreement，but not necessarily unanimity，that includes a process for attempting to resolve objections by interested parties. This process is a comment resolution process，which requires that all comments have been considered fairly and each objector is advised of the disposition of his or her objection(s) and the reasons. Also，consensus body members may change their votes after reviewing the comments.

## 供应商；Supplier

向生产批准书持有人提供零件或服务的任何法人。

Any person contracted by a PAH to provide articles or services (at any tier).

## 构型基线；Configuration Baseline

已知飞机/系统/产品的一种构型，更改过程可以此构型为基础进行。

A known aircraft/system/item configuration against which a change process can be undertaken.

## 构型控制；Configuration Control

对基本构型的所有更改进行系统评估、协调和所有更改的批准或不批准。

The systematic evaluation，coordination，approval or disapproval of all changes to a baseline configuration.

**构型控制指令；Configuration Control Directive**

国家空域系统规划办公室批准基本构型及其随后的所有更改的决定记录。

The record of a decision of the National Airspace System Program Office approving a baseline configuration and all subsequent changes therewith.

**构型偏离清单；Configuration Deviation List; CDL**

由负责型号设计的机构编制的、经设计国局方批准的清单，它指明在飞行开始时某航空器型号上可以缺少哪些外部部件，并在必要时包含相关的使用限制和性能修订的任何资料。

A list established by the organization responsible for the type design with the approval of the state of design which identifies any external parts of an aircraft type which may be missing at the commencement of a flight, and which contains, where necessary, any information on associated operating limitations and performance correction.

**构型、维修和程序；Configuration, Maintenance and Procedure; CMP**

经局方批准的一种文件，包含某种飞机-发动机组合满足延程运行型号设计批准要求所必需的最低限度构型、使用和维修要求、硬件的寿命限制，以及型号最低限度设备清单等方面的限制规定。

Configuration, maintenance, and procedures (CMP) document means a document approved by the authority that contains minimum configuration, operating, and maintenance requirements, hardware life-limits, and master minimum equipment list (MMEL) constraints necessary for an airplane-engine combination to meet ETOPS type design approval requirements.

**构型项；Configuration Item**

构型控制下的飞机、系统、产品及其相关数据。

Aircraft, system, item and related data that is under configuration control.

**固有安全；Intrinsically Safe**

在正常运行条件、预期故障条件和可能导致燃油箱内出现点火源的环境条件下,不能释放足够电能或热能的任何仪器、设备或线路。

Any instrument, equipment, or wiring that is incapable of releasing sufficient electrical or thermal energy under normal operating conditions, anticipated failure conditions, and environmental conditions which could cause an ignition source within the fuel tank.

## 故障;Fault

组件或系统内可能导致失效的异常情况。

A manifestation of an error in an item or system that may lead to a failure.

## 故障标志(故障旗);Failure Flag

可视故障参数指示。

A local visual means of indicating the failure of a displayed parameter.

## 故障隔离手册;Fault Isolation Manual; FIM

内含隔离并排除航空器故障程序的手册,供航空公司维修机械员使用。

A manual contains procedures for isolating and correcting reported airplane faults, used by airline mechanics.

## 故障集;Functional Failure Set; FFS

单个部件或特定的一组部件,被认为独立于导致最高级失效条件的另一个部件(不局限于一个系统)。

A single member or a specific group of members that are considered to be independent from one another that lead(s) to a top level failure condition.

## 故障模式;Failure Mode; FM

系统或部件发生失效的方式。

The way that the failure of a system or item occurs.

## 故障模式与影响分析;Failure Mode and Effect Analysis; FMEA

一种结构化、引导化、自下而上的分析过程,用于评估装于发动机上的控制系统运行的每个可能失效所引起的后果。

A structured, inductive, bottom-up analysis used to evaluate the effects of each possible failure of the control system on engine operation.

### 关键动力装置(一个或多个);Critical Power-unit(s)

其失效对航空器的特性造成最不利影响的一个或多个动力装置。有些航空器可能装有多台同等重要的关键动力装置。在这种情况下,关键动力装置指其中一个关键动力装置。

The power-unit(s) failure of which gives the most adverse effect on the aircraft characteristics relative to the case under consideration. On some aircraft there may be more than one equally critical power-unit. In this case, the expression "the critical power-unit" means one of those critical power-units.

### 关键飞机冰形构型;Critical Aircraft Ice Shape Configuration

在规定的飞行安全要求下,导致最坏后果的冰污染飞机构型。

The ice-contaminated aircraft configuration that results in the most adverse effects for specific flight safety requirements.

### 关键件;Critical Part

失效会对继续安全飞行和着陆产生直接危害性影响的零部件。

A part of which failure results in a direct hazardous effect to continued safe flight and landing.

### 关键结构;Critical Structure

一种承载结构/元件,其完整性对保证飞机整体飞行安全至关重要。

A load bearing structure/element whose integrity is essential in maintaining the overall flight safety of the aircraft.

### 关键设计构型控制限制; Critical Design Configuration Control Limitations; CDCCL

保持阻燃系统或燃油系统关键设计特征的限制要求,该要求在整个飞机型号寿命期间对设计满足 25. 981 条的性能标准是必要的。关键设计构型控制限制的目的是在改装、修理或维护措施引起构型变化时，提供指南以保持其关键特征(局限性)。

A limitation requirement to preserve a critical design feature of a flammability reduction system or of the fuel system design that is necessary for the design to meet the performance standards of §25. 981 throughout the life of the airplane model. The purpose of the CDCCL is to provide instructions to retain the critical features during configuration changes that may be caused by alterations, repairs, or maintenance actions (only for here).

**管理程序;Aviation Procedure; AP**

民航局各职能部门下发的有关民用航空规章的实施办法或具体管理程序，是民航行政机关工作人员从事管理工作，以及法人、其他经济组织或者个人从事民用航空活动应当遵守的行为规则。

Implementation methods and detailed management procedures related to civil aviation regulations issued by the functional departments of CAAC, which are the action rules of the CAAC staff engaged in civil aviation management, and that of the legal persons, other economic organizations or individuals engaged in civil aviation activities.

**惯性基准;Inertial Reference**

利用惯性器件测量相对惯性空间的线运动和转动运动而建立起来的基准或参考坐标系。

Reference or reference coordinate system established by using inertial devices to measure linear motion and rotation relative to inertial space.

**广播式自动相关监视;Automatic Dependent Surveillance — Broadcast; ADS‐B**

一种先进的监视技术，通过该技术，配置 ADS‐B 输出设备的飞机与空中交通管制中心以及其他配置 ADS‐B 输出设备的相关飞机共享位置、高度、速度以及其他信息。

An advanced surveillance technology which let aircrafts equipped ADS‐B out share position, altitude, velocity, and other information with aircrafts equipped ATC and other appropriately.

### 广播数据链; Broadcast Data Link

无须接收站发出初始请求的数据链路传输。此外, RTCA/DO - 267A 定义的广播数据传送不包括信息接收者的地址信息。

Data link transmission with no requirement for an initiating request from the receiving station. Further, the broadcast data link transmission, as defined in RTCA/DO - 267A, does not contain message recipient address information.

### 广布疲劳损伤; Widespread Fatigue Damage; WFD

在同一个结构元件上或在类似的相邻结构元件中的多个结构细节处同时存在裂纹,它们的尺寸和密度足以使其结构不再满足损伤容限的要求。

WFD is characterized by the simultaneous presence of small cracks in multiple structural details; where the cracks are of sufficient size and density, the structure can no longer sustain the required residual strength load level in the event of a primary load-path failure or a large partial damage incident.

### 规范; Specification

管理当局批准的文件,内容包括特殊的维修信息,如轮胎翻新。

Documents approved by the Administrator containing information for performing specialized maintenance, such as retreading of tires.

### 规划认可项目; Planning Grant Program; PGP

由 1970 年颁布的机场和航路发展法核准的一种计划项目,旨在促进机场的有效配置和发展,并提供对机场的主计划和系统计划的认可。

The program authorized by the Airport and Airway Development Act of 1970 to promote the effective location and development of airports and to provide grants for airport master plans and system plans.

### 规章研究责任部门; Regulations Research Responsibilities Department

经适航司授权,负责对某项规章进行持续研究和日常管理的适航部门,主要工作包括收集对该部规章的意见和建议,研究该部规章涉及的标准、政策及符合性方法,并定期向适航司提出相关的建议文稿等。

Airworthiness division authorized by airworthiness department which is responsible for continuing research and daily management for a regulation. Its main duties include gather the views and recommendations of the regulation, the research covered by the regulations of the ministry of standards, policies and compliance methods, and regularly submit the recommendations related documents to airworthiness division.

### 规章制定;Rulemaking

局方用来制定及颁发规章和法规的内部流程。

The internal process used by agencies to enact and issue rules or regulations.

### 滚转角;Angle of Roll

见倾斜角。

See angle of bank.

### 滚转预计距离;Roll Anticipation Distance;RAD

达到飞机转弯所需坡度角时飞机飞过的距离。

The distance traveled by the aircraft while rolling to the bank angle required for a turn.

### 国产全新民用航空器;Domestic Brand New Civil Aircraft

由中华人民共和国设计制造,通过合法方式被中华人民共和国公民或机构拥有或使用,并申请(或已经)在中华人民共和国注册登记的全新民用航空器。

Aircrafts designed and manufactured by the People's Republic of China, and owned by citizens or institutions of the People's Republic of China through legitimate possession, applying for (or already have) a brand new civil aircraft registered in the People's Republic of China.

### 国籍标志;Nationality Mark

用以标志航空器国籍的国际字母。

The international letter (or letters) used to identify the nationality of the aircraft.

### 国籍登记证书；Nationality Registration Certificate

由国家民用航空主管部门为依法进行国籍登记的民用航空器颁发的国籍证明文件，该证书一经颁发长期有效，且必须放置在航空器上，以备核查。

Nationality registration certificate is nationality document issued by the civil aviation authority of the country for the civil aircraft registered nationality in accordance with the law. Once the certificate is issued, it has a long-term effectivity, and must be placed on an aircraft for verification.

### 国际标准大气；International Standard Atmosphere；ISA

国际标准大气压是 ICAO7488/2 文件中规定的大气压。以取证规范为目的，下列是可接受的：① 空气是理想的干燥空气；② 海平面处的温度为 15 摄氏度；③ 海平面处的压力为 $1.013\,250 \times 10^5$ 帕（29.92 英寸汞柱，1 013.2 毫巴）；④ 从海平面到温度为 $-56.5$ 摄氏度的海拔处的温度梯度为每 500 米 3.25 摄氏度 (1.98 摄氏度/1 000 英尺)；⑤ 在上述条件下的海平面处密度 $\rho_0 = 1.225\,0$ 千克/米$^3$（约 0.002 378 斯勒格/英尺$^3$）；相对密度 $\rho/\rho_0$ 用 $\sigma$ 表示。

The atmosphere defined in ICAO Document 7488/2. For the purposes of certification specifications, the following is acceptable：① the air is a perfect dry gas；② the temperature at sea-level is 15℃；③ the pressure at sea-level is $1.013\,250 \times 10^5$ Pa (29.92 inHg, 1 013.2 mbar)；④ the temperature gradient from sea-level to the altitude at which the temperature becomes $-56.5$℃ is 3.25℃ per 500 m (1.98℃/1 000 ft)；⑤ the density at sea level $\rho_0$, under the above conditions is $1.225\,0$ kg/m$^3$ (0.002 378 slugs/ft$^3$)；$\rho$ is the density appropriate to the altitude and the relative density $\rho/\rho_0$ is indicated by $\sigma$.

### 国际飞行服务发射站；International Flight Service Transmitter Station；IFST

飞行咨询系统中有人管理并配备有无线电发射机，用以发射点对点通信和空地通信的一种设施。

A facility in the flight advisory system manned and equipped with radio transmitters used for the transmission of point-to-point and air/ground communications.

### 国际飞行服务接收站；International Flight Service Receiving Station; IFSR

飞行咨询系统中有人管理并配备有无线电接收机，用以接收点对点通信和空地通信的一种设施。

A facility in the flight advisory system manned and equipped with radio receivers used for the reception of point-to-point and air/ground communications.

### 国际飞行服务站；International Flight Service Station; IFSS

飞行咨询系统中有人管理并配有设备的一种中央控制设施，用以控制同在国际领域或公海上空飞行的驾驶员的航空点对点与空地无线电通信，提供飞行计划跟踪、天气信息、搜索营救行动及其他飞行支援活动。

A central operation facility in the flight advisory system，manned and equipped to control aeronautical point-to-point telecommunications，and air/ground telecommunications with pilots operating over international territory or waters，providing flight plan following，weather information，search and rescue action and other flight assistance operations.

### 国际机场；International Airport

① 已被财政部部长或海关税务司长指定为海关事务国际机场的入境机场；② 必须在预期使用前获得海关当局着陆特许证的着陆机场；③ 根据国际民航公约，指定供国际商业航空运输和（或）国际通用航空使用的机场；④ 就国际民航组织促进来往而言，由缔约国指定的任何机场在该国的领土上系国际空中交通的入境和离境机场，在该处办理与海关、移民、公共卫生、动植物检疫及类似程序有关的正式手续。

① An airport of entry which has been designated by the Secretary of Treasury or Commissioner of Customs as an international airport for customs service；② a landing rights airport at which specific permission to land must be obtained from customs authorities in advance of contemplated use；③ airports designated under the Convention on International Civil Aviation as an airport for use by international commercial air transport and/or international general aviation；④ as pertaining to ICAO facilitation，any airport designated by the

Contracting State in whose territory is situated as an airport of entry and departure for international air traffic, where the formalities incident to customs, immigration, public health, animal and plant quarantine and similar procedures are carried out.

### 国际民航组织;International Civil Aviation Organization; ICAO

联合国的一个专门机构,其宗旨是制定国际空中航行的原则与技术,以及促进国际民航运输的规划与发展。

A specialized agency of the United Nations whose objective is to develop the principles and techniques of international air navigation and to foster planning and development of international civil air transport.

### 国际民用航空公约;Convention on International Civil Aviation

习称"芝加哥公约",关于国际民用航空在政治、经济、技术等方面的问题的国际公约。

Convention on International Civil Aviation (also called "Chicago Convention") is an international convention which is about the international civil aviation in the political, economic, technical problems.

### 国家飞行数据中心(美国);National Flight Data Center; NFDC

由 FAA 建立的一个设施,用于收集、验证和传送航空人员信息。

A facility established by FAA for the collection, validation, and dissemination of airman information.

### 国家航空咨询委员会(美国);National Advisory Committee for Aeronautics; NACA

美国于 1915 年 3 月 3 日成立的联邦机构,负责航空科学研究的执行、促进与制度化。

The federal agency established by the United States on March 3, 1915, responsible for the execution of the aviation scientific research, promotion and institutionalization.

### 国家空域系统(美国); National Airspace System; NAS

由美国空域,导航辅助设施,通信设施和设备,空中交通管制设备和设施,航图和信息,规则、条例和程序,技术信息,以及 FAA 的人力和器材组成的公共网络。与军方联合共享的系统部分也包括在内。除所有指定的空域外,该系统还包括空中导航设施和机场。

The common network of U. S. airspace; navigation aids; communications facilities and equipments; air traffic control equipments and facilities; aeronautical charts and information; rules, regulations, and procedures; technical information; and FAA manpower and material. It includes system components shared jointly with the military. In addition to all designated airspace, the system includes the following components: air navigation facilities and airports.

### 国家运输安全委员会(美国); National Transportation Safety Board; NTSB

美国政府的独立机构,负责调查美国航空、公路、水路、管道以及铁路事故。国家运输安全委员会在美国国会的主持下调查美国的每一起民用航空事故。

An independent United States Government organization responsible for investigations of accidents involving aviation, highways, waterways, pipelines, and railroads in the United States. NTSB is charged by congress to investigate every civil aviation accident in the United States.

### 国内维修单位; Domestic Maintenance Organization

管理和维修设施在除香港特别行政区、澳门特别行政区或者台湾地区以外的中国境内的维修单位。

A maintenance organization whose management and maintenance facility is located in the People's Republic of China other than Hong Kong, Macao and the Taiwan region.

### 国内运行种类(美国); Domestic Type Operation

在下述情况下进行的国内运行: ① 设计载客量至少为31人的航空器(根据合格民用航空管理当局颁发的航空器型号合格证判定),在美国境内所有州、哥伦比亚地区或美国境内所有领土或领地的可用机场运行;② 设计载客量大于

9 人但少于 31 人的航空器(根据合格民用航空管理当局颁发的航空器型号合格证判定),在美国境内所有州(除阿拉斯加州)、哥伦比亚地区或美国境内所有领土或领地的可用机场运行。

Means any domestic operation conducted with: ① an airplane designed for at least 31 passenger seats (as determined by the aircraft type certificate issued by a competent civil aviation authority) at any land airport in any State of the United States, the District of Columbia, or any territory or possession of the United States; ② an airplane designed for more than 9 passenger seats but less than 31 passenger seats (as determined by the aircraft type certificate issued by a competent civil aviation authority) at any land airport in any State of the United States (except Alaska), the District of Columbia, or any territory or possession of the United States.

### 国外制造商;Foreign Manufacturer

非生产批准书持有人,在本国境外制造一个产品或物品的组织。

An organization other than an PAH who causes a product or article(s) to be produced outside the country.

### 过冷大水滴;Supercooled Large Drops;SLD

在温度低于 0 摄氏度时,直径大于 0.05 毫米的液滴,例如冻雨或冻细雨。

Liquid drops with diameter greater than 0.05 mm at temperatures less than 0℃, such as freezing rain or freezing drizzle.

### 过早下降预警;Premature Descent Alert;PDA

用于探测飞机是否在着陆阶段低于进近航道(大约为 3 度)的预警系统。

A warning system whose ability is to detect whether the aircraft is hazardously below the normal (approximately 3°) approach path for the nearest runway, and to provide a timely alert.

# H

**HIRF 不利影响;HIRF Adverse Effect**

在一定程度上导致了系统失效、故障,或有误导性的 HIRF 效应,在 HIRF 相关规章中对特定的飞机功能或系统来说是不可接受的。在确定系统或功能是否受到 HIRF 不利影响时,应考虑飞机整体及其运行相关的 HIRF 效应。

HIRF effect that results in system failure, malfunction, or misleading information to a degree that is unacceptable for the specific aircraft function or system addressed in the HIRF regulations. A determination of whether a system or function is adversely affected should consider the HIRF effect in relation to the overall aircraft and its operation.

**海平面发动机;Sea Level Engine**

只有在海平面才能产生额定起飞功率的航空发动机。

A reciprocating aircraft engine having a rated takeoff power that is producible only at sea level.

**海豚摆动;Porpoise**

① 用于陆上飞机时,指追随一条绕中间飞行路线上下波动的飞行航迹,常常是短周期纵向振荡的结果;② 用于水上飞机时,指沿水面运动反复地跳跃、上升和下降。

① As pertaining to airplanes, to follow an adulatory flight path about a medium line of flight, porpoise is often the result of short period longitudinal oscillations; ② as pertaining to seaplanes, to skip, rise, and plunge repeatedly while moving across the water.

### 罕见的正常情况；Rare Normal Condition

由于恶劣的环境条件（如大风、湍流或结冰），飞机很少经历的无故障状态。

A fault-free condition that is experienced infrequently by the airplane due to severe environmental conditions (for example, significant wind, turbulence, or asymmetric icing).

### 行业协调员（美国）；Industry Coordinator

经美国国家运输安全委员会或者 FAA 批准的代表运营人、协会或制造商的人，该人拥有事故调查所必需的技术知识或专业技能。

The person approved by NTSB or FAA to represent the operator, association, or manufacturer who possesses technical knowledge or expertise necessary to contribute to the accident investigation.

### 航材供应商；Air Material Supplier

向航空运营人提供民航局批准或认可的航空器部件和原材料的任何单位和个人。航材供应商应包括经批准或认可的航空器部件制造厂家、维修单位或者航材分销商，任何仅提供信息、运输、财务服务的代表、代理人（包括机构）不视为航材供应商。

Any unit or individual that provides aircraft parts and raw materials approved or recognized by the Civil Aviation Administration to air operators. The air material supplier shall include an approved or recognized aircraft component manufacturer, maintenance unit or air material distributor. Any representative or agent (including organization) who only provides information, transportation and financial services shall not be regarded as aviation material supplier.

### 航段；Track to Fix (TF) Leg

两个航站点之间的路径。

A TF leg is a geodesic path between two fixes.

### 航空安全办公室；Office of Aviation Safety；AVS

负责为航空器、运营人、承运人制定审定标准的局方组织。

AVS is the Authority organization responsible for establishing certification standards for aircraft, operators, and air carriers.

## 航空安全监察员(美国);Aviation Safety Inspector; ASI

建立、管理和执行航空安全规章标准的局方雇员。航空安全监察员向航空工业的很多部门和驾驶员提供建议和指导,以促进航空安全。

An Authority employee who develops, administers, and enforces the regulations and standards relating to aviation safety. ASI provides advice and guidance to many segments of the aviation industry and airmen in the interest of aviation safety.

## 航空出租航空器;Air Taxi Aircraft

由航空出租运营证持有人运营的航空器,该运营证批准按照 FAR 第 135 和 121 部为盈利目的而运载乘客、邮件或货物。

Aircraft is operated by the holder of an air taxi operating certificate which authorized the carriage of passengers, mail or cargo for revenue in accordance with FAR Parts 135 and 121.

## 航空出租运营人;Air Taxi Operator

提供定期或不定期航空出租服务或邮件服务的运营人。

An operator providing either scheduled or unscheduled air taxi service or mail service.

## 航空电子系统误差;Avionic Error; AVE

在将测量到的气压转化为电子化输出,使用静压源误差修正(SSEC)和显示对应高度的过程中产生的误差。

The error transformed from the measured air pressure to the electronic output, and the display corresponding to the errors generated in the height of the process using static source error correction (SSEC).

## 航空电子学;Avionics

应用于航空航天的电子学。

The electronics applied to aviation and astronautics.

### 航空固定通信线路；Aeronautical Fixed Circuit

构成航空固定通信服务一部分的线路。

A circuit forming part of the aeronautical fixed service.

### 航空货运；Air Freight

为空运货物而建立的系统或服务。

A system or service set up for the carrying of freight by air.

### 航空器；Aircraft

任何能够凭借空气的反作用力获得在大气中的支承力并由所载人员驾驶的飞行器械。

A device that is used or intended to be used for flight in the air, and that is supported in flight by the dynamic reaction of the air against its wings, piloted by the airman who onboard.

### 航空器部件；Aircraft Component

除飞机机体以外的任何装于或者准备装于航空器的部件，包括整台动力装置、螺旋桨和任何正常、应急设备等。

Any part and appliance installed or to be installed on aircraft other than the aircraft airframe, including the complete engine/APU, propeller and any operational/emergency equipment.

### 航空器等级；Aircraft Class

① 用于对航空人员颁证、定级、授权和限制，指具有相似运行特性的一类航空器中的航空器等级。例如：单发、多发、陆上、水上、旋翼航空器、直升机、飞艇和自由气球；② 用于航空器合格审定，指具有相似的推进、飞行或着陆特性的航空器大类。例如：飞机、旋翼机、滑翔机、气球、陆上飞机和水上飞机。

① As used with respect to the certification, ratings, privileges, and limitations of airmen, means a classification of aircraft within a category having similar operating characteristics. Examples include: single engine, multiengine, land, water, rotorcraft, helicopter, airship, and free balloon; ② as used with respect to the certification of aircraft, means a broad grouping of aircraft having similar characteristics of propulsion, flight, or landing. Examples include: airplane, rotorcraft, glider, balloon, landplane, and seaplane.

### 航空器电子协会；Aircraft Electronics Association；AEA

航空器电子协会(AEA)成立于 1957 年,从事 1 300 多个航空业务,包括专门从事通用航空器上的电子设备和系统的维护、修理和安装的修理站。

The Aircraft Electronics Association (AEA) was founded in 1957, representing more than 1 300 aviation businesses, including repair stations that specialize in maintenance, repair and installation of avionics and electronic systems in general aviation aircraft.

### 航空器发动机；Aircraft Engine

用于或拟用于推进飞机的发动机,包括涡轮增压器、附属装置和发动机工作所需的附件,但不包括螺旋桨。

An engine that is used or intended to be used for propelling aircraft. It includes turbosuperchargers, appurtenances, and accessories necessary for its functioning, but does not include propellers.

### 航空器飞行检查员；Check Airman（Aircraft）

有资格在航空器、飞行模拟机或者特定型别的航空器的飞行训练器上实施飞行检查的人员。

A person who is qualified to conduct flight checks in an aircraft, a flight simulator, or a flight training device for a particular type aircraft.

### 航空器服务基地；Aircraft Services Base；ASB

对局方机队及有关设备提供定期与不定期的航空器和航空电子设备维护、改装与翻修服务的企业。

An agency providing scheduled and unscheduled aircraft and avionics maintenance, modification, and overhaul services to the agency's fleet of aircraft and associated equipment.

### 航空器服务站;Aircraft Services Facility; ASF

从事航空器和航空电子设备各等级的定期与不定期维护,但不包括旨在支持地区与 SRDS 飞行大纲的翻修的企业。

An agency performing all levels of scheduled and unscheduled aircraft and avionics maintenance, excluding overhaul in support of regional and SRDS flight programs.

### 航空器改型;Aircraft Variant

用于机组的批准与操作,指与基本审定型号相同、有改装的航空器,这些改装会使设备和/或程序有明显改变,但不会明显改变操纵和/或飞行性能。

As used with respect to the licensing and operation of flight crew, aircraft variant means an aircraft of the same basic certificated type which contains modifications not resulting in significant changes of handling and/or flight characteristic, or flight crew complement, but causing significant changes to equipment and/or procedures.

### 航空器构型;Aircraft Configuration

描述影响航空器气动特性的各种部件相对位置的术语。

A term referring to the relative position of the various elements affecting the aerodynamic characteristics of an aircraft.

### 航空器合格审定系统评审大纲;Aircraft Certification Systems Evaluation Program; ACSEP

一种标准化的评审方法,用以评审生产许可证书持有人及其重要零部件供应商对经批准的型号设计进行控制和制造的生产活动状况。

A kind of standardized evaluation method, used to review how production certificate holder and its important parts suppliers control the approved type design and manufacturing production activity.

### 航空器空重;Aircraft Empty Weight; AEW

航空器的结构、动力装置、设施、系统和一些作为某一特定构型飞机的一个整体所必需的设备的重量,以及航空运营人认为该航空器所配标准项目的重量。

Aircraft structure, propulsion system, facilities, systems and equipments must be as a whole of a particular configuration of the aircraft's weight, as well as the weight of the aircraft with standard configuration.

### 航空器类别;Aircraft Category

① 当用于驾驶员的合格审定、定级、授权和限制方面时,指航空器的大的分类,例如固定翼飞机、旋翼航空器、滑翔机和轻于空气的航空器。② 当用于航空器的合格审定时,指根据航空器的预定用途或使用限制进行的分类,例如运输类、正常类、实用类、特技类、限制类、限用类和临时类。③ 当用于跑道分类时,指根据跑道进近程序进行的分类。

① As used with respect to the certification, ratings, privileges, and limitations of airmen, means a broad classification of aircraft such as airplane, rotorcraft, glider, and lighter-than-air. ② As used with respect to the certification of aircraft, means a grouping of aircraft upon intended use or operating limitations such as transport, normal, utility, acrobatic, limited, restricted, and provisional. ③ As used with respect to the categorization of runways, means a grouping according to the type of approach procedure for the runway.

### 航空器评审报告;Aviation Evaluation Group Report; AEGR

根据《民用航空产品和零部件合格审定规定》(CCAR - 21)开展的运行符合性评审结论所形成的报告。

A report based on the results of an operational compliance review conducted in accordance with Certification Rules for Civil Aviation Products and Articles(CCAR - 21).

### 航空器评审组;Aircraft Evaluation Group; AEG

由民航局飞行标准部门建立的,局方专业人员组成的工作组,负责在航空器型号审定过程中开展运行符合性评审,目的是为航空器获得型号合格证(认可证)后的运行审定和监察建立标准和规范,搭建起设计、制造与使用、维修之间的桥梁。

The working group established by the Flight Standards Department of the

Civil Aviation Administration and composed of the professional personnel of the Civil Aviation Administration. AEG is responsible for conducting operational conformity assessment in the process of aircraft type certification, with the aim of establishing standards and specifications for operational certification and supervision after the aircraft has obtained the type certificate (accreditation certificate), and building a bridge between design, manufacture and operation, maintenance.

### 航空器设备改型证;Aircraft Equipment Modification; ACEM

为批准和指导可从局方航空器上拆下的航空器设备和附件的改型工作而颁发的文件。

A document issued to authorize and direct the modification of agency aircraft equipments and accessories which are removable from the aircraft.

### 航空器审定管理小组;Aircraft Certification Management Team; ACMT

由航空器审定司于 1986 年建立,使有政策制定和实施责任的高管和经理参与解决影响服务有效性和效率的复杂的国家和国际管理问题。

ACMT was established in 1986 by the Director, Aircraft Certification Service, to enable senior executives and managers with policy formulation and implementation responsibilities to participate in the resolution of complex national and international management issues affecting the effectiveness and efficiency of the Service.

### 航空器审定中心;Aircraft Certification Center

业务上受适航司领导,是地区性职能机构。在适航司授权下,负责所辖地区内民用航空产品的初始适航审定和持证人的监督管理工作。

The regional functional organization whose function is under the jurisdiction of the airworthiness department. Under the authorization of the airworthiness department, ACC is responsible for the initial certification of civil aviation products and the supervision and administration of the holder within their jurisdiction areas.

### 航空器适航审定司；Aircraft Airworthiness Department

根据民航局的授权，具体负责中国民用航空器的适航管理。

The airworthiness department which is responsible for the Chinese civil aircraft airworthiness management in accordance with authorization of the Civil Aviation Administration of China.

### 航空器维修人员；Aircraft Maintenance Personnel

从事航空器、航空器部件维修和管理的所有人员。

Personnel engaging in aircraft maintenance，aircraft component repairs，and the associated management.

### 航空器系留点；Aircraft Tiedown

地面及航空器表面可用于固定航空器的位置。

Positions on the ground and aircraft surface that are available for securing aircraft.

### 航空器项目审定办公室；Project Aircraft Certification Office；PACO

负责项目审定的航空器审定办公室。如果该项目将进行 STC 或 PMA 证书的申请，则可能需要与 CMACO 协调。

The certification office which is in charge of project certification. PACO needs to coordinate with the CMACO，if the project is a follow-up certification activity，such as an STC or PMA.

### 航空器型号合格审定体系；Type Certification System；TCS

航空器型号合格审定体系是中国民用航空局（CAAC）针对航空器的型号合格审定而建立的工作体系。该体系包括授权的责任审定单位、责任审查部门、型号合格审定委员会（TCB）、型号合格审定审查组、项目工程师（project engineer，PE）、委任代表。

Type certification system is the certification working system set up by the Aircraft Airworthiness Department（AAD）of the CAAC for application of type certificate of a product. The system consists of Type Certification Board and its subordinate type certification team，ACD，project engineer（PE）， designated engineering representatives (DER) and designated manufacturing inspection representatives (DMIR) (as applicable).

### 航空器运行；Aircraft Operations

航空器在管制或非管制机场航站区及给定航路定位点附近,或能计数的其他地点的空中活动。运行分为本场和航线两类。航空器的本场运行包括下述情况：① 在本场起落航线 DX 机场视界内的运行；② 已知将在机场 20 英里半径内本场飞行训练区做离场或到场的飞行；③ 在机场进行模拟仪表进近或低空通过。航线运行指本场运行之外的所有航空器运行。

The airborne movement of aircraft in controlled or noncontrolled airport terminal areas and about given en route fixes or at other points where counts can be made. There are two types of operations, local and itinerant. Local operations are performed by aircraft which： ① operate in the local traffic pattern or within sight of the airport； ② are known to be departing for, or arriving from, flight in local practice areas within a 20 mile radius of the airport； ③ execute simulated instrument approaches or low passes at the airport. Itinerant operations are all aircraft operations other than local operations.

### 航空器执行运营人；Executive Aircraft Operator

经营自有或租用航空器的企业、公司或个人,航空器由主要责任是驾驶航空器的驾驶员来飞行,作为公司经营业务中运输人员或货物的手段。

A corporation, company, or individual which operates owned or leased aircraft, flown by pilots whose primary duties involve pilotage of aircraft, as a means of transportation of personnel or cargo in the conduct of company business.

### 航空器组；Aircraft Type Groupings

由一家制造商设计和组装,并且在所有可能影响高度保持性能准确性的细节方面具有名义上相同的设计和制造的一组航空器。

The same group of aircraft that are designed and assembled by one manufacturer and are of nominally identical design, building with respect to all details which could influence the accuracy of height-keeping performance.

## 航空燃气涡轮发动机;Aircraft Gas Turbine Engine

涡桨、涡扇或涡喷航空发动机。

A turboprop, turbofan, or turbojet aircraft engine.

## 航空人员;Airman

从事民用航空活动的空勤人员和地面人员,应当接受专门训练,经考核合格,取得国务院民用航空主管部门颁发的执照,方可担任其执照载明的工作。

The flight and ground person that engaged in aviation activities. They should receive special training, pass the assessment and get the license issued by the civil aviation department of The State Council before doing the associated work as the license stated.

## 航空事故调查委员会;Air Accident Investigation Board; AAIB

调查飞机事故和事件的机构。

An agency investigates aircraft accidents and incidents.

## 航空通信;Aeronautical Communication; AERCOM

航空部门之间利用电信设备进行联系,以传递飞机飞行动态、空中交通管制指示、气象情报和航空运输业务信息等的一种飞行保障业务。

A communication which is built between the aviation departments through telecommunications equipment, in order to transfer the aircraft flight dynamic, air traffic control instructions, weather information and air transport business information.

## 航空无线电公司;Aeronautical Radio Incorporated; ARINC

主要股东为美国定期的航空公司、航空运输公司、飞机制造厂和外国领队航空公司的组织。其目的是制定并出版电子设备和系统的规范。

ARINC is a corporation whose principal stockholders are United States scheduled airlines, air transport companies, manufacturers, and foreign flag airlines. ARINC envelops and publishes standards for electronic equipment and systems.

## 航空无线电技术委员会；Radio Technical Commission for Aeronautics；RTCA

由美国政府与工业界组成的一个航空组织。其致力于推动航空技术的发展，寻求在航空运营过程中使用电子设备和通信设备解决所遇到问题和深入的技术方法。其目标是通过其成员和参与组织来协商解决上述问题。

RTCA is an association of aeronautical organizations of the United States from both government and industry. Dedicated to the advancement of aeronautics, RTCA seeks sound technical solutions to problems involving the application of electronics and telecommunications to aeronautical operations. Its objective is the resolution of such problems by mutual agreement of its member and participating organizations.

## 航空无线电通信寻址和报告系统；ARINC Communication Addressing and Reporting System；ACARS

通过提供可寻址的数据通信链路，使飞机与地面操作中心可通过无线电网络进行数据和信息交换。

An addressable digital data link system which permits the exchange of data and messages between an aircraft and a ground-based operations center over a radio network.

## 航空信息报告（美国）；Aerospace Information Report；AIR

被编写为标准零件的各种配件的结构和恰当设计的定位和识别的辅助文件。

AIR is written as an aid for locating and identifying various fitting designs and configurations that are designated as standard parts.

## 航空信息出版物；Aeronautical Information Publication；AIP

国际版的驾驶员信息手册，其中包含供国际使用的国际机场的并行信息和具体信息。

An international version of AIM is called the aeronautical information publication that contains parallel information, as well as specific information on the international airports for use by the international community.

### 航空信息手册;Aeronautical Information Manual; AIM

为航空组织提供基本的航班信息和航空交通管制程序的手册。

AIM is designed to provide the aviation community with basic flight information and ATC procedures.

### 航空运输;Air Transportation

洲际、海外或与国外之间的航空运输或用航空器运输邮件。

Interstate, overseas or foreign air transportation, or the transportation of mail by aircraft.

### 航空运输承运人;Air Carrier

通过租约或其他协议直接从事航空运输的法人。

A person who undertakes directly by lease, or other arrangement, to engage in air transportation.

### 航空运营人;Air Operator; AO

在中华人民共和国登记的公共航空运输企业、通用航空企业和从事民用航空飞行活动的其他组织。

A public air transport enterprise, general air enterprise, or any other organization undertaking civil air flights registered in the People's Republic of China.

### 航空运营人的维修单位;Operator's Maintenance Organization

航空运营人建立的、主要为本运营人的航空器或者航空器部件提供维修服务的维修机构。航空运营人的维修单位在为其他航空运营人提供维修服务时,应视为独立的维修单位。

The maintenance organization set up by an operator to provide maintenance service primarily for the operator itself. The operator's maintenance organization shall be treated as an independent maintenance organization while providing maintenance services for the third-party organizations.

## 航空作业;Aviation Operations

使用航空器进行专业服务的运行,如农业、建筑、摄影、测量、观察与巡逻、搜寻与援救、空中广告等。

The operation of aircraft used for professional services, such as agriculture, construction, photography, surveying, observation and patrol, search and rescue, aerial advertisement.

## 航路;Air Route

两个可识别的航路点之间的可航行空域。

Navigable airspace between two points which is identifiable.

## 航路燃油计划;Enroute Fuel Planning

在飞行途中,确定飞完航段所需的燃油量。

The amount of fuel required to complete the remainder of a flight segment while in flight.

## 航图;Aeronautical Chart

为空中导航而特制的代表地球某一部分的地图。

A map representing a portion of the earth, made especially for use in air navigation.

## 航线可更换单元;Line Replaceable Unit; LRU

在航线维护操作中可以拆换的零件、组件或部件。

A part, assembly or component which can be replaced in line maintenance.

## 航线临界点;Point of No Return

沿飞行路线的一点,航空器到此点后靠自身的供油不能返航到其出发地点。

The point along a line of flight at which an aircraft cannot return to its place of departure on its fuel supply.

**航线维修；Line Maintenance**

按照航空运营人提供的工作单对航空器进行的例行检查，以及按照相应的飞机、发动机维护手册等在航线进行的故障和缺陷处理，包括换件和按照航空运营人机型最低设备清单、外形缺损清单保留故障和缺陷。

Routine inspections of civil aircraft in accordance with the worksheets provided by the operator and the rectification of malfunctions and defects encountered during line operation according to the applicable aircraft or engine maintenance manual，including LRU replacement and deferral of malfunctions and defects according to the operator's minimum equipment list/configuration deviation list of the particular type of aircraft.

**航线运行模拟；Line Operational Simulation；LOS**

一个模拟的航线飞行环境，在这个情景中的内容都设计用于测试机组成员的综合技术能力和机组资源管理能力。

A simulated line environment，the scenario content of which is designed to test integrating technical and CRM skills.

**航线运行评估；Line Operational Evaluation；LOE**

在模拟的航线飞行环境中，使用合格的高级训练大纲中的预定用途且经批准的设备，实施的训练或评估课程段。

A training or evaluation session，as applicable，that is conducted in a simulated line environment，using equipment qualified and approved for its intended purpose in an AQP.

**航向引导；Vector**

通过雷达为航空器提供导航引导的航向。

A heading issued to an aircraft to provide navigational guidance by radar.

**航行灯；Navigation Light**

机上所装的一组灯，目的是使航空器的大小、位置和运动方向在夜间或能见度差的情况下可见。

Any one of a group of lights mounted on an aircraft to make its dimensions, position, and direction of motion visible at night or under other conditions of poor visibility.

### 航行情况显示器;Navigation Situation Display; NSD

一种机载图像显示器,它将导航计算机给出的数据自动转换成显示在驾驶舱阴极射线管上的航行情况;闭路电视连线以光学方式将区域地图投射到同一个阴极射线管上,以显示航空器地面轨迹。

An airborne pictorial display which automatically converts data from navigational computers to a situation display on a cockpit CRT; a closed-circuit TV link optically projects area maps onto the same CRT to present a display of the ground track of an aircraft.

### 航行(灯光)信标;Aeronautical (light) Beacon

显示白色和(或)彩色闪光灯的目视助航设备,用于指示机场或着陆区、通向机场或着陆区的航路或可确定相对于机场或着陆区的方位的位置点或明显的地标或对飞行具有危险的物体或区域。为此而使用的主要灯光是光强相对高的旋转灯标,它可以辅以若干光强较低的非旋转闪光灯标。

A visual navaid displaying flashes of white and/or colored light which is used to indicate an airport or landing area, a route leading to an airport or landing area, or a point on which bearings can be taken to an airport or landing area, or an outstanding landmark, or an object or area presenting a hazard to flying. The principal light used is a rotating beacon of relatively high intensity, which may be supplemented by nonrotating flashing lights of lesser intensity.

### 航站 VOR;Terminal VOR; TVOR

甚高频航站全向信标台(位于或邻近机场,用作进近辅助设施)。

Very high frequency terminal omnirange station (located on or near an airport and used as an approach aid).

### 耗材;Consumable Materials

通常指在维护和修理飞机、发动机、设备、组件中用到的润滑剂、接合剂、化合物、油漆、化学制品、染料和补片等。

Lubricants，cements，compounds，paints，chemicals，dyes and patches used in the maintenance and repair of aircraft，engines，equipment and component.

### 合成视景；Synthetic Vision

计算机生成的从驾驶舱观察到的外部地形情景图像，通过航空器姿态、高精度导航解算，以及地形、障碍物和相关文化特征的数据库推导得出。

Images of external topography from the perspective of the flight deck generated by computer，derived from the database of the aircraft attitude，high precision navigation solutions，and terrain，obstacles and related cultural characteristics.

### 合成视景系统；Synthetic Vision System；SVS

向飞行机组显示外部地形情景综合可视图像的一种电子装置。

An electronic device which displays comprehensive visual image of external terrain scene for the aircrew.

### 合格审定规范(EASA)；Certification Specification (EASA)；CS

欧洲航空安全局（EASA）适航标准，对应于美国联邦法规 14 卷第 23、25、27、29、33、35 部等。

EASA airworthiness standards corresponding to 14 CFR parts 23，25，27，29，33，35，and so forth.

### 合格审定过程改进；Certification Process Improvement；CPI

通过促进设计申请批准人和局方之间更好地沟通、管理项目、划分责任以改善安全的举措。

An initiative to improve safety by fostering better communications，project management，and accountability between design approval applicants and Authority.

### 合格审定基础；Certification Basis

满足适用的适航、航空器噪声、燃油排泄和排气要求，专用条件或等效安全情况。申请人必须表明等效安全情况的符合性,或者获得表明符合性的豁免。

The applicable airworthiness, aircraft noise, fuel venting and exhaust requirements as appropriate, special conditions, and equivalent level of safety. The applicant must show the compliance of equivalent level of safety, or an exemption.

### 合格审定检查要求;Certification Check Requirement; CCR

设计上要求的一种重复的空勤或地勤检查,通过检查发现是否存在重大的潜在失效,并由此限制其发生时间以帮助表明对 FAR‐25.1309(b)和(d)(2)的符合性。这种潜伏失效与在安全性分析中确定的一个或多个特定失效或事件组合在一起便会导致产生危险的失效状态。

A recurring flightcrew or groundcrew check that is required by design to help show compliance with 25.1309(b) and (d) (2) by detecting the presence of, and thereby limiting the exposure time to, a significant latent failure that would, in combination with one or more other specific failures or events identified in a safety analysis, result in a hazardous failure condition.

### 合格审定审查项目;Certification Review Items; CRI

EASA/JAA 用于记录审定或认可中需要澄清的或重大技术、管理的问题,用于重大的或有争议的规章、技术政策和符合性方法的问题。

EASA/JAA documents used to record a certification or validation issue needing clarification or representing of a major technical or administrative problem, used for significant or controversial regulatory, technical policy, and means of compliance issues.

### 合格审定适航当局;Certification Authority; CA

负责初始型号合格审定和补充型号合格审定的局方。对美国的申请人/合格证持有人,局方为 FAA;对欧共体和 JAA 成员国的申请人及合格证持有人基于 JAA 程序的产品,局方为 EASA。对中国的申请人/合格证持有人,审定当局为中国民用航空局(CAAC)。局方也可称为出口国适航当局。

Aviation authority responsible for the original type certificate or supplemental type certificate. Certificating authority means the FAA for applicants/certificate holders located in the United States, and EASA for applicants/certificate holders located in the European Community and in JAA member states, for products under JAA procedures and CAAC for applicants/ certificate holders located in China. The certificating authority may also be referred to as the exporting authority.

### 合格审定项目;Certification Project

局方用于签发新的或修订的型号合格证以批准型号设计的重大更改的流程、方法和协调行为。审定项目由航空器审定办公室(ACO)或其委任授权机构发起。

The processes, methods, and coordination the Authority uses to issue new and amended type certificates and to approve major changes to type design. Certification projects are initiated by the ACO or ODA holder.

### 合格审定项目计划;Certification Project Plan; CPP

局方内部的项目计划,用于协调局方内部的人力资源、人员责任和进度。

The internal project plan of the Authority, which is used to coordinate the human resources, personnel responsibilities and progress within the Authority.

### 合格审定项目通知;Certification Project Notification; CPN

项目负责人用于通知责任理事会、航空器评审组(AEG)和证书管理航空器审定办公室(CMACO)关于新的审定或型号认可项目的程序。责任理事会、航空器评估组和证书管理航空器审定办公室也用该程序向航空器审定办公室(ACO)提供意见,并说明其在该项目中的参与程度。

The process used by the project manager to notify the accountable directorate, AEG, and CMACO of a new certification or type validation project. The accountable directorate, AEG, and CMACO also use a CPN process to provide comments to the ACO and to specify their level of involvement in the project.

### 合格审定项目通知程序;CPN Process

项目航空器审定办公室(PACO)向责任理事会、航空器评审组(AEG)和证书管理航空器审定办公室(CMACO)发送合格审定项目通知(CPN)文件时的通知和响应流程。

The process of notification and response when CPN data is sent by the PACO to the accountable directorate，the AEG，and the CMACO.

### 合格审定项目通知文件；CPN Data

为启动合格审定项目而编写并传递的文件。

The data that is documented and transmitted for the initiation of each certification project.

### 合格审定信用；Certification Credit

局方对某一过程、产品或验证满足审定要求的认可。

Acceptance by the Authority that a process，product，or demonstration satisfies a certification requirement.

### 合格审定行动项目；Certification Action Items；CAI

有时也称为 EASA 行动项目。EASA/JAA 用行动项目提出非敏感性问题。行动项目记录需要特别关注的非争议性问题。行动项目不需要记入合格审定审查项目(CRI)。

CAI sometimes is also called "EASA action items". EASA/JAA use the action item system to advance non-sensitive issues. AI documents non-controversial items that require special attention，and does not require a CRI.

### 合格证持有人差异要求；Operator Difference Requirements；ODR

合格证持有人对其所运行的具备相似特点的不同型别飞机之间差异的详细描述和所采取的符合性方法。混合机队的合格证持有人评估基本型别飞机和差异型别飞机之间的差别，可为制订经局方批准的减少训练时间的训练大纲提供依据。合格证持有人差异要求包括差异描述以及相关的培训，检查和满足近期经历的方法和程序，满足航空器评审组和 CCAR - 121 的要求。

Operator difference requirements are those operator specific requirements necessary to address differences between a base aircraft and one or more variants，when operating in mixed fleet flying，or when seeking credit in transition programs. ODR includes both a description of differences and a corresponding list of training, checking, and currency compliance methods which address pertinent FSB and CCAR – 121 requirements.

### 合格证书;Certificate of Conformity; CoC

证明商品符合合同中的规格的证书。

A certificate proving that the goods meet the specifications in the contract.

### 合格证书持有人;Certificate Holder

按中国民用航空规章第 121 部、第 135 部、第 141 部、第 142 部和第 91 部规定颁发的合格证的持有人。

Holders of certificates issued in accordance with the provisions of Parts 121, 135, 141, 142, and 91 of the Civil Aviation Regulations of China.

### 合格证书管理地区办公室;Certificate Holding District Office; CHDO

飞行标准地区办公室,负责证书的管理和对合格证持有人运行的全面监管。

Flight Standards District Office that has responsibility for administering the certificate and that is charged with the overall inspection of the certificate holder's operations.

### 荷兰滚(飘摆);Dutch Roll

飞机围绕纵轴的运动,特点是向一个方向滚转,向反方向做偏航运动。

An airplane movement about the longitudinal axis, characterized by a roll in one direction and a yaw in the opposite direction.

### 后掠;Sweepback

翼面或翼面的前缘或其他基准线从根部到梢部的后斜。

The backward slant from root to tip of an airfoil or the leading edge or other reference line of an airfoil.

### 后续的补充型号合格证;Follow-on STC

在第一种型号补充型号合格证之后颁发的补充型号合格证,为型号审定飞机的相似类型和型号而颁发。

A supplemental type certificate（STC）subsequent to the first-of-type STC，issued for a similar model of type certificated aircraft.

### 互换型飞行模拟机；Convertible FSTD

一种可以改变硬件和软件，使其成为同一机型不同型号航空器的飞行模拟设备。这种飞行模拟设备可利用相同的模拟设备平台、驾驶舱壳体、运动系统、视景系统、计算机以及必要的外围设备做一种以上机型或型号的模拟。

An FSTD in which hardware and software can be changed so that the FSTD becomes a replica of a different model，usually of the same type aircraft. The same FSTD platform，flight deck shell，motion system，visual system，computers，and peripheral equipment can be used in more than one simulation.

### 互换性；Interchangeability

在一个系统内用一个产品替换另一个产品并使系统按规范运行的能力。

The ability to substitute one part for another within a system and have the system perform to its specification.

### 花纹深度；Skid Depth

用模具测量的，轮胎表面花纹到最深凹槽之间的距离。

The distance between the tread surface and the deepest groove as measured in the mold.

### 滑橇式起落架飞机；Ski-Plane

装有滑橇以便在雪地或冰上起降的飞机。

An airplane equipped with skis for taking off and landing on snow or ice.

### 滑水现象；Hydroplaning

运动中的航空器轮胎因水、液态橡胶膜或水雾导致机械刹车效率降低而脱离跑道路面的现象。

The condition in which moving aircraft tires are separated from a pavement surface by a water or liquid rubber film, or by steam resulting in a derogation of mechanical braking effectiveness.

## 滑行/慢车(起飞);Taxi/Idle (Out)

在为滑行启动推进发动机,直至航空器转入起飞跑道期间,航空器的滑行和慢车工作状态。

Aircraft operations involving taxi and idle between the time of initial starting of the propulsion engine(s) used for the taxi and the turn onto the duty runway.

## 滑行/慢车(着陆);Taxi/Idle (In)

在着陆滑行到所有推进发动机最后停车期间,航空器的滑行和慢车工作状态。

Aircraft operations involving taxi and idle between the time of landing roll-out and final shutdown of all propulsion engines.

## 环保批准文件;Environmental Approval

出口国适航当局颁发的符合其噪声标准或燃油排泄及排气排出物标准的证件或等效文件。

An environmental certificate or equivalent issued by the exporting authority for finding that a product complies with its own standards concerning noise and/or fuel venting and exhaust emission.

## 环境测试;Environmental Testing

评估民用航空产品的设计或设计变更是否符合有关噪声、燃料通风或废气排放的适用标准和程序的过程。

A process by which the design or change to a design of a civil aeronautical product is evaluated for compliance with applicable standards and procedures concerning noise, fuel venting, or exhaust emissions.

## 环境感知;Situational Awareness

对在飞行前、飞行中和飞行后影响安全性的 4 个基本风险元素中所有因素和情况的精确感知和理解。

The accurate perception and understanding of all the factors and conditions within the four fundamental risk elements that affect safety before, during, and after the flight.

## 环境控制系统；Environmental Control System；ECS

为机组和乘客提供空气、温度控制和客舱压力的系统。

The system that provides air supply, thermal control and cabin pressurization for the crew and passengers.

## 恢复；Return to Service

使一个零件或部件适航，或在适航状态的技能。

The skill developed to make a part or component airworthy or to be in airworthy condition.

## 回流冰；Runback Ice

由从保护面流回非保护面的水冻结或再冻结而形成的冰。

Ice that forms from the freezing or refreezing of water leaving protected surfaces and running back to unprotected surfaces.

## 混合机队飞行；Mixed Fleet Flying；MFF

在两个型别等级之间交替进行的训练以及获取的近期经历，可以用于满足运行规章中规定的检查和近期经历要求。合格证持有人在一个型别上的定期训练和熟练检查可以替代合格证持有人在另一个型别上要求的相应的定期训练和熟练检查。

Mixed fleet flying is training that alternates between two types and obtains the recent experience to meet the inspection and recent experience requirements specified in the operating regulation. The certificate holder's periodic training and proficiency checks on one type may be substituted for the corresponding periodic training and proficiency checks required by the certificate holder on another type.

## 混合模型;Hybrid model

截掉前缘后部并设计使得前缘空气动力与整个翼面一致的风洞模型。

Wind tunnel model truncated aft of the leading edge and designed in order that the leading edge aerodynamic loading is the same as the full chord airfoil.

## 火控区;Controlled Firing Area

建立火控区是为了包容一类活动,这类活动如果不是在一个管制环境中进行将会危及不参与活动的航空器。

A controlled firing area is established to contain activities, which if not conducted in a controlled environment, would be hazardous to nonparticipating aircraft.

## 货运航空器;Cargo Aircraft

用于运载货物的非客运航空器。

An aircraft which is not a passenger-carrying aircraft, but used for the carriage of cargo.

## 豁免;Exemption

豁免是根据《民用航空产品和零部件合格审定规定》(CCAR - 21 - R4)第21.3条的规定,民航局同意受适航规章和环境保护要求中有关条款约束的人暂时或永久不用表明对某些条款的符合性的批准。

The exemption is based on the provisions of Article 21. 3 of Certification Rules for Civil Aviation Products and Articles(CCAR - 21 - R4), and the Civil Aviation Administration agrees that the person bound by the relevant provisions of airworthiness regulations and environmental protection requirements does not need to indicate the approval of compliance with certain provisions temporarily or permanently.

# I

**ICAO 建议措施；ICAO Recommended Practice**

ICAO 附件中含有的规范，为了国际空中航行的安全、有序和效率，统一应用这种规范被公认是需要的。

Any specification contained within an ICAO Annex，the uniform application of which is recognized as desirable in the interest of safety，regularity，and efficiency of international air navigation.

# J

**机动包线;Maneuvering Envelope**

除受到最大(静)升力系数的限制外,假定飞机经受对称机动而产生的限制载荷。

The airplane is assumed to be subjected to symmetrical maneuvers resulting in limit loads, except limited by maximum (static) lift coefficients.

**机动飞行;Maneuver**

用于航空器时,指有意地改变飞行中航空器的运动或姿态,如转弯、俯冲、改出、倾斜等,盘旋、横滚、跃升转弯半滚倒转等的变化或性能。

As pertaining to aircraft, maneuver is an intended change in the movement or attitude of an aircraft in flight, such as a turn, a dive, a pullout, a bank, and an evolution or performance such as a loop, a roll, a wing-over.

**机队;Fleet**

受同一安全问题影响的同一型号的在役飞机、发动机或螺旋桨产品。

Aircraft, engine or propeller products of a type currently in service affected by a certain safety issue.

**机队空重;Fleet Empty Weight;FEW**

构成一个机队的相同构型的航空器的平均空重。

The average aircraft empty weight of a fleet with the same configuration.

**机队平均可燃性暴露;Fleet Average Flammability Exposure**

指在世界范围的环境条件和燃油特性下一个机型的机队在全世界范围内运

行的各个航段距离范围内,每个燃油箱的空余空间处于可燃状态的时间占可燃性暴露评估时间(FEET)的比例。

The percentage of the flammability exposure evaluation time (FEET) that each fuel tank ullage is flammable for a fleet of an airplane type operating over the range of flight lengths in a world-wide range of environmental conditions and fuel properties.

### 机构委任授权;Organization Designation Authorization; ODA

局方对一个组织机构的授权,包括使用批准的程序的机构委任授权(ODA)单位,代表局方进行批准工作。

An authorization by the Authority for an organization,comprised of an ODA unit(s) using approved procedures,to make approvals on behalf of the Authority.

### 机构委任授权管理团队;ODA Management Team; OMT

由航空安全工程师(ASE)、飞行测试驾驶员和航空安全检查人员(ASI)组成。他们可能来自任何局方。局方必须位于机构委任授权(ODA)坐落的地理区域,或者大多数的授权功能将被执行的地点。

The OMT consists of aviation safety engineers (ASE),flight test pilots, and aviation safety inspectors (ASI). They may be from any of several offices. The office must be in the geographical area where the ODA is located and has a primary place of business — or where the majority of the authorized functions will be performed.

### 机构委任授权管理者;ODA Administrators

负责管理单元的活动和机构委任授权管理团队的沟通、协调。机构委任授权持有人可以为每个机构委任授权类型设置分开的机构委任授权管理员。

The focal points for the ODA holder who is responsible for managing the ODA unit's activities and communicating with the OMT. An ODA holder may have a separate ODA administrator for each ODA type.

### 机构委任授权机构;ODA Unit

在组织内由两个或两个以上的可识别的个人组成的机构,代表局方执行委托功能。

An identifiable unit of two or more individuals within an organization that performs the delegated functions on behalf of the Authority.

### 机内设施;Domestic Appliance

飞机上用于乘客服务的设施,如炉灶、烤箱、微波炉、咖啡壶、烧水器、冰箱和坐便器冲洗系统。

Item placed on the airplane to provide service amenities to passengers. Examples of domestic appliances are cooktops, ovens, microwave ovens, coffee makers, water heaters, refrigerators, and toilet flush systems.

### 机身长度系数;Fuselage Length Factor

客舱的长度与飞机上单位区域出口的总数的商。

A length determined by dividing the length of the passenger cabin by the sum of the exit units in each zone in the airplane.

### 机身结冰;Airframe Icing

飞机上除发动机外的结冰现象。

Airframe icing is ice accretions on the airplane, except for the propulsion system.

### 机身破坏点;Fuselage Break Points

沿着机身,事故数据显示机身外壳结构已经失效的那些点。它可能部分失效,断裂处的结构仍连接着,或两个分开的结构导致完全失效。这些通常归为两种条件类型:① 开孔和应力集中;② 最大载荷点,起落架的集中载荷或最大弯矩,例如在机翼和机身衔接处。

Fuselage break points are points along the fuselage where accident data has shown the fuselage shell structure failure. It can be a partial failure in which the structure across the break remains attached or a complete failure resulting in two separate pieces. These are generally attributable to two types of conditions：① cutouts and stress concentrations；② points of maximum load，either point loads by the landing gear or maximum bending moments，such as，at the juncture of the wing and body.

### 机身水平基准面；Fuselage Reference Plane；FRP

作为一个基本的参考，它沿着由机身上下部的交叉部分形成的机身尖处扩展。

FRP is used as a basic reference，extending along the fuselage cusps which are formed by the intersection of the upper and lower sections of the fuselage.

### 机体；Airframe

航空器的机身、尾梁、短舱、整流罩、整流片、翼面（包括旋翼，但不包括螺旋桨和发动机的旋转叶片）和起落架，以及它们的附件和操纵系统。

The fuselage，booms，nacelles，cowlings，fairings，airfoil surfaces (including rotors，but excluding propellers and rotating airfoils of engines)，and landing gear of an aircraft and their accessories and controls.

### 机载电子设备改型证；Airborne Electronic Equipment Modification；AEEM

为批准和指导可从局方航空器上拆下的机载电子（航空电子）设备的改型工作而颁发的文件。

A document issued to authorize and direct the modification of airborne electronics (avionics) equipment which is removable from agency aircraft.

### 机载设备；Appliance

安装或附加在航空器上，但不属于机体、发动机或螺旋桨组成部分的仪表、机构、设备、部件、仪器、辅助装置或附件，包括用于或拟用于操纵或控制航空器飞行的通信设备。

Any instrument，mechanism，equipment，part，apparatus，appurtenance，or accessory，including communications equipment，that is used or intended to be used in operating or controlling an aircraft in flight，that is installed in or attached to the aircraft，and is not part of an airframe，engine，or propeller.

### 机载增强系统；Aircraft-based Augmentation System；ABAS

用于验证 GNSS 位置输出完整性的机上算法的通用术语。

A generic term for an algorithm on the aircraft that verifies the integrity of the GNSS position output.

### 机载重量和平衡系统；Onboard Weight and Balance System；OBWBS

测量飞机和有效载荷的重量，并从机载设备上计算飞机重心。该系统主要显示飞机签派使用的实际重量和平衡信息。

OBWBS weighs the aircraft and payload, and computes the aircraft center of gravity from equipment onboard the aircraft. The system displays the actual weight, and balance information for use in primary dispatch of the aircraft.

### 机长；Pilot in Command；PIC

飞行过程中对航空器的运行和安全负责的驾驶员。

The pilot responsible for the operation and safety of an aircraft during flight.

### 机组成员；Crewmember

飞行期间在航空器上执行任务的航空人员，包括飞行机组成员和客舱乘务员。

Person assigned to perform duty in an aircraft during flight time, including flight crewmembers and cabin attendants.

### 机组成员的飞行时间；Crewmember Flight Time

机组成员在航空器飞行期间的值勤时间，包括在座飞行时间（飞行经历时间）和不在座飞行时间。

The duty period of a crewmember during the aircraft flight, including the seated flight time (pilot time) and unseated flight time.

### 机组程序委任代表；Aircrew Program Designee；APD

经授权，对运营人按照其经局方批准的培训大纲培训过的驾驶员实施某型航空器机组人员审定的代表。

APD is authorized to perform airman certification in one type of aircraft for an operator pilots who have been trained under the Authority is approved training program.

## 机组定期复训;Crewmember Recurrent Training

已取得资格的机组成员为了保持其资格和技术熟练水平,在规定的期限内按照规定的内容所进行的训练。

The training required for qualified crewmembers to remain adequately trained and currently proficient for their qualifications and techniques conducted in specified period on specified items.

## 机组资源管理;Crew Resource Management;CRM

机组对所有可用资源(包括机组成员之间)的有效利用,以实现安全高效的飞行。

The effective use of all the resources available to crewmembers, including the crewmembers each other, to achieve a safe and efficient flight.

## 积冰界限;Ice Accretion Limit

机身上附着冰的最远位置。

The location farthest aft on a body at which ice accretes.

## 基本 RVSM 飞行包线;Basic RVSM Envelope

在高度 FL290 到 FL410(或最大可用的高度层)之间,航空器预期运行最频繁的马赫数和总重量的范围。

The range of Mach numbers and gross weights within the altitude ranges FL290 to FL410 (or max available altitude) where an aircraft is expected to operate most frequently.

## 基本结构;Basic Structure

型号合格证持有人的原始制造结构。

Structure of the original manufacturer of the type certificate holders.

## 基本区域检查；Baseline Zonal Inspection；BZI

基本区域检查是一个工业术语，用于定义修理评估程序里的常规维护。对修理评估，常规维护是相对于补充检查进行定义的。

An industry term used to define "normal maintenance" for the repair assessment program. It is used to define normal versus supplemental inspections for repair assessment.

## 基本型飞机；Basic Aircraft

运营人指定的一架或一组飞机，作为与其机队的其他飞机进行差异比较的参照。

An operator designated aircraft or group of aircraft used as a reference to compare differences with other aircraft within an operator's fleet.

## 基本运行重量；Basic Operating Weight；BOW

飞机空重加上以下重量：正常的油量、厕所冲洗液、饮用水、所需机组成员和他们的行李、应急设备。

The empty weight of the aircraft plus the weight of the following: normal oil quantity; lavatory servicing fluid; potable water; required crewmembers and their baggage; and emergency equipment.

## 基于产品的系统审查；Product Based System Audit

通过检查某特定产品，以判定生产批准书持有人或申请人的质量系统是否符合适用规章要求以及经批准的质量系统文件的审查活动。这是一种有计划且生成记录的活动，最大可行地审查所选择的产品和零部件是否符合经批准的设计，PAH 是否符合质量系统要求（包括为满足这些要求所建立的程序和特种工艺等）。

Product based system audit is used to inspect a specific product to determine whether the production approval holder's or applicant's quality system complies with applicable regulatory requirements and the approved quality system document. This is a planned and documented activity to review, to the maximum extent practicable, selected products and components for conformance to approved design and PAH for conformance to quality system requirements (including procedures and special processes established to meet these requirements).

### 基于垂直引导的进近和着陆运行；Approach and Landing Operations with Vertical Guidance

使用横向和垂直引导但未达到精密进近和着陆运行要求的仪表进近和着陆。

An instrument approach and landing which utilizes lateral and vertical guidance but does not meet the requirements established for precision approach and landing operations.

### 基于风险的资源决策；Risk-based Resource Decision

一种结构化的过程，旨在支持航空器审定司确定风险，基于该风险分配资源以及确定多个项目的优先级。

A structured process designed to support Aircraft Certification Department in determining risk, assigning resources based on that risk, and prioritizing multiple projects.

### 基于风险的资源目标；Risk-based Resource Targeting；RBRT

为了支持 AIR 识别风险及基于风险分配资源，优化多项目管理而设计的一个机构化过程。

A structured process designed to support AIR management in determining risk, assigning resources based on that risk, and prioritizing multiple projects.

### 基于卫星的增强系统；Satellite-based Augmentation System；SBAS

一个覆盖广泛的增强系统，通过在地面对 GPS 信号进行完整性计算和校正，并利用地球同步卫星向 GPS/SBAS 用户发送数据。

A wide coverage augmentation system to GPS that calculates integrity and correction data on the ground and uses geostationary satellites to broadcast the data to GPS/SBAS users.

### 基准测试；Baseline Testing

针对基准座位执行的初始系列测试，作为初始审定的一部分，以验证该座位型号。

The initial series of tests performed on the baseline seat as part of the original certification to substantiate the seat family.

## 基准大气条件；Reference Day Conditions

修正气态排出物（碳氢化合物和烟雾）的基准气象条件。基准大气条件如下：温度＝15 摄氏度，绝对湿度＝0.006 29 千克水/千克干空气，压力＝101 325 帕。

The reference ambient conditions to which the gaseous emissions (HC and smoke) are to be corrected. The reference day conditions are as follows: temperature＝15℃, specific humidity＝0.006 29 kg $H_2O$/kg of dry air, and pressure＝101 325 Pa.

## 基准级差；Reference Level Difference

以分贝为单位，对于一个规定的频率，在某个级程上测得的相对于校准声压级电输入信号的级差，可根据级程做适当调整。

In decibels, for a specified frequency, measured on one level away for calibration sound pressure level of the electrical input signal level difference, appropriate adjustment can be made according to the level process.

## 基准级程；Reference Level Range

以分贝为单位，包含校准声压级，用于确定测量系统声学灵敏度的级程。

In decibels, contains calibration sound pressure level for the process of determining the level of sensitivity of the measurement system acoustics.

## 基准频率；Reference Frequency

以赫兹为单位，由声校准器产生的正弦声压信号的标称频率。

In hertz, the nominal frequency of a sinusoidal acoustic pressure signal generated by the acoustic calibration.

## 基准湿度；Reference Humidity

温度与基准湿度的关系规定如下：① 当温度等于或低于国际标准大气温度

时,相对湿度为 80%;② 当温度等于或高于国际标准大气＋28 摄氏度时,相对湿度为 34%;③ 当温度在国际标准大气与国际标准大气＋28 摄氏度之间时,相对湿度在各温度所规定的湿度之间线性变化。

The relationship between temperature and reference humidity is defined as follows: ① at temperatures at and below ISA, 80% relative humidity; ② at temperatures at and above ISA ＋ 28℃, 34% relative humidity; ③ at temperatures between ISA and ISA ＋ 28℃, the relative humidity varies linearly between the humidity specified for those temperatures.

### 基准座位;Baseline Seat

一类新型座位中设计并制造出来的第一个座位。

The first seat designed and manufactured within a new family of seats.

### 极不可能的失效;Extremely Improbable Failures

永远不会出现,除非工程判断要求考虑的失效,发生概率为 $1\times10^9$ 或更小。这类失效包括会妨碍飞机安全飞行和着陆的失效或失效组合。

The failures that never appear and don't need to be considered, unless engineering judgment would require their consideration, of which probability of occurrence is $1\times10^9$ or less. This category includes failures or combinations of failures that would prevent the continued safe flight and landing of the airplane.

### 极限载荷;Ultimate Load

限制载荷乘以适当的安全系数。

The limit load multiplied by the appropriate factor of safety.

### 急救氧气;First Aid Oxygen

为不能从过大的客舱高度恢复的乘客提供的额外氧气供应。

The additional oxygen provided for the use of passengers, who do not satisfactorily recover following subjection to excessive cabin altitudes.

### 集成;Integration

① 使系统/项目中的各元件协同运作的行为;② 使若干分散的功能集中到

单个实施过程之中的行为。

① The act of causing elements of a system/item to function together；② the act of gathering a number of separate functions within a single implementation.

### 几何垂直精度；Geometric Vertical Accuracy；GVA

在相关允许误差范围内，报告的垂直位置（几何高度）95％的精确度。

The 95％ accuracy of the reported vertical position (geometric altitude) within an associated allowance.

### 计算机仿真；Computer Modeling

使用基于计算机的有限元或多体瞬态分析方法，模拟相应适航标准的紧急着陆动态情形。

The use of computer based on finite element or multi-body transient analysis to simulate the emergency landing dynamic condition of the applicable airworthiness standard.

### 记录保存系统；Record-keeping System

该系统记录时寿件件号、序号和现行的寿命状况。时寿件每次从航空产品上拆下后，记录应与现行寿命状况一起更新。此系统可以包括电子、纸张或其他记录方式。

The system records time limited parts number, serial number, and the current life condition. When time limited parts are removed from the aviation products，records should be updated together with the current life situation. This system can include electronic，paper or other records ways.

### 记录管理；Records Management

一切与档案材料（包括记录性和非记录性材料）的编写、保管、维护和处置有关的活动。

All activities related to the creation，custody，maintenance，and disposition of file material，which consists of record and nonrecord materials.

### 技术标准规定;Technical Standard Order; TSO

局方颁布的在民用航空器上所用的某些材料、零部件或者机载设备(简称项目)的最低性能标准。

A minimum performance standard issued by the Administrator for specified materials, parts, processes, or appliances used on civil aircraft.

### 技术标准规定授权;Technical Standard Order Authorization; TSOA

由局方向已满足特殊技术标准规定的制造商的产品颁发的设计和生产许可。该制造商必须控制产品的设计和质量,确保产品符合技术标准规定的要求。该制造商同时要控制用于该产品设计和制造的部件或服务的供应商满足技术标准规定的条款要求。局方只向本国的制造商颁发技术标准规定授权。

TSOA is the design and production approval issued by FAA, to the manufacturer of an article that has been found to meet a specific TSO. A manufacturer is the person who controls the design and quality of the article produced to ensure that it meets the TSO requirements. The manufacturer also must control all the suppliers they use for parts or services in the design and production of the TSO article. Authority issues a TSOA only to the domestic manufacturer.

### 技术标准规定项目批准书;Technical Standard Order Approval; CTSOA

局方颁发给符合特定技术标准规定的零部件(以下简称 CTSO 件)的制造人的设计和生产批准书。除技术标准规定项目批准书的持有人外,任何人不得用 CTSOA 标记对 CTSO 件进行标识。按照技术标准规定项目批准书制造的零部件,只有得到相应的装机批准,CTSO 件才能安装到航空器上使用。

A Chinese technical standard order authorization (CTSOA) is an authority design and production approval issued to the manufacturer of an article that has been found to meet a specific technical standard order (hereinafter referred as CTSO article). No person, except the holder of a CTSOA, may identify its article with the applicable CTSO marking. Articles manufactured under a CTSOA may be used for aircraft installation only when the applicable approval for installation on the aircraft is granted.

## 加拿大民航局；Transport Canada Civil Aviation；TCCA

加拿大国家民航管理监督部门。

The authority of Canada in charge of civil aviation.

## 加速-停止距离；Accelerate-stop Distance；ASD

飞机加速到特定的速度，并假设发动机在达到 $V_1$ 速度的瞬间失效的情况下，将飞机停住所需的距离。

The distance required to accelerate an airplane to a specified speed，and assuming failure of the critical engine at the instant that speed（$V_1$）is attained，to bring the airplane to a stop.

## 加油安全区域；Fueling Safety Area

航空器以及航空器机头、翼尖、尾翼位置向外延伸 15 m 后在地面的投影形成的闭合多边形区域。

The closed polygon area formed by the projection on the ground which is the position of the aircraft，aircraft head，wing tip and tail extending to 15 m outward.

## 假冒零件；Counterfeit Part

未获授权的情况下仿照批准的零件生产或改装，并旨在以原件或真件的幌子进行误导和欺骗的零件。

A part made or altered to imitate or resemble an approved part without authority or right，and with the intent to mislead or defraud by passing as original or genuine.

## 驾驶舱；Flight Deck

飞机上的一个隔间，供飞行机组运行飞机使用。

The compartment of the aircraft arranged for the flight crew in operating the aircraft.

## 驾驶舱效应；Flight Deck Effects

因系统故障而在驾驶舱中出现的指示。

Indications on the flight deck caused by a system's malfunction.

### 驾驶舱资源管理；Cockpit Resource Management；CRM

多人机组配置的驾驶舱资源管理，是对机组所有可用的人员和实物资产的有效利用。驾驶舱资源管理强调良好的沟通和人际关系技能。

Cockpit Resource Management (CRM) in multiperson crew configurations is the effective use of all personnel and material assets available to a flight crew. CRM emphasizes good communication and other interpersonal relationship skills.

### 驾驶员报告；Pilot Report；PIREP

从驾驶员处收到的有关飞行细节（气象、航行）或航空器状况的消息。

A communication received from a pilot concerning details of his flight (meteorological, navigation) or status of his aircraft.

### 驾驶员警告仪表；Pilot Warning Instrument；PWI

一种机载仪表，其功能是警告驾驶员并提供适当的信息，协助他评估碰撞的危险。PWI 设备从功能上进一步分为以下两类：① 助视装置提醒驾驶员观察或告诉驾驶员观察何处，或两者均有，但此种装置要求驾驶员进行实际观察；② 代视装置将数据显示在某种形式的机内显示器上，驾驶员观察此显示器而不必观察实际的航空器。

An airborne instrument whose function is to warn a pilot and provide suitable information to assist him in evaluating a collision threat. PWI equipment is further functionally subdivided into two categories：① aids to vision devices which alert the pilot to look, or tell him where to look, or both, but require him to do the actual observation；② substitutes for vision devices which present data on some form of internal display, which the pilot observes in lieu of the actual aircraft.

### 监造；Construction Supervision

航空器在制造厂家制造组装的过程中，由航空器的运营人或其授权人对飞机生产过程的监督和对飞机实施的检查。

Aircraft operator or his authorized oversight of the aircraft production process and inspect the aircraft during the aircraft manufacturer and assemble process.

### 减距起落；Reduced Takeoff and Landing；RTOL

在比常规喷气运输机要求的短，但比短距起落（STOL）航空器要求的长的跑道上起降的未来喷气运输机。名义 RTOL 跑道长度为 4 000 英尺。

Future jet transport aircraft which will operate from runways that are shorter than those required by conventional jet transports，but longer than those required for STOL aircraft. Nominal RTOL runway length is 4 000 ft.

### 检查；Check

确定设备的物理完整性或功能性的检查或测试。

An examination (such as the inspection or test) determining the physical integrity or functional capability of an item.

### 检查/测试；Inspection/Test

不分解航空器部件，而根据适航性资料，通过离位的试验和功能测试来确定航空器部件的可用性。

Verifying serviceability of removed civil aircraft components without disassembling by examination and functional check in accordance with the standards specified in the airworthiness data.

### 检验和测试程序；Inspection and Test Program

工程检验和测试程序可确定对下列条件的符合性：① 在 FAA 批准的规章、技术标准规定或其他相关规范中所列出的适用的适航标准；② 军用标准或与有关联邦航空法规（FAR）一致的外国政府的符合验证。

An engineering inspection and test program determining compliance with：① applicable airworthiness standards set forth in regulations，technical standard orders，or other pertinent specifications approved by the Federal Aviation Administrator；② military standards or a foreign government's validation found to be compatible with the appropriate Federal Aviation

Regulations(FAR).

### 简单检查；Simple Inspection or Search

不需要特殊工具或设备的目视检查，并且被检查处没有子隔层。检查时，所有乘客带上飞机的物品都应从飞机上拿下。

A visual search in which no specialty tools or equipments are needed，and there are no sub-compartments. All items which are carried by passengers are assumed to have been removed from the airplane when this inspection is performed.

### 简单硬件项目；Simple Hardware Item

全面合并确定性试验和分析结果的项目，试验和分析结果适用于设计保证等级，在可预见的运行环境下，可保证其功能性能正常，没有异动。

The item with a comprehensive combination of deterministic tests and analyses appropriate to the design assurance level that ensures correct functional performance under all foreseeable operating conditions，with no anomalous behavior.

### 间歇最大结冰；Intermittent Maximum Icing

大气结冰的间歇最大强度（间歇最大结冰）由云中液态水含量、云中水滴平均有效直径、周围空气温度等变量，以及上述三个变量间的相互关系定义。

The intermittent maximum intensity of atmospheric icing conditions (intermittent maximum icing) is defined by the variables of the cloud liquid water content，the mean effective diameter of the cloud droplets，the ambient air temperature，and the inter-relationship of these three variables.

### 建议修订的通知(EASA)；Notice of Proposed Amendment；NPA

EASA 法规建议修订的通知。

A notice of a proposed amendment to a EASA Code.

### 交叉机组资格；Cross Crew Qualification；CCQ

在经过合格证持有人评估且局方批准，具备相似系统、操纵特点和程序的不同型别飞机之间,经合理缩减的改装训练。

The modified training in the certificate holders assessment, between different types of aircraft with similar systems, handling characteristics and procedures approved by the authority, after reasonable reduction.

### 交通咨询;Traffic Advisory; TA

TCAS Ⅱ 对机组发出的声音和显示信息,确定附近最低间隔标准的交通情况。

Aural voice and display information from TCAS Ⅱ to a flight crew, identifying the location of traffic nearby that meets the certain minimum separation criteria.

### 校准;Calibration

按照规定的文档程序实施的,比较被测仪器和标准计量设备测量示值的一组操作,目的在于检测、报告或通过调节消除被测仪器的误差。

A set of operations, performed in accordance with a definite document procedure, which compares the measurements performed by an instrument or standard, for the purpose of detecting and reporting, or eliminating by adjustment, errors in the instrument tested.

### 校准(标高);Alignment (Elevation)

起始于航线某一特定点的高于水平面的实际角度,用以指引高度。

The actual angle above a horizontal plane, originating at a specific point of a course used for altitude guidance.

### 校准(方位);Alignment (Azimuth)

航线的方位或实际磁方位。

The azimuth or actual magnetic bearing of a course.

### 校准标志;Calibration Markers

雷达显示器屏幕上的标记,它将距离刻度分成若干已知的间隔,用于确定距离或用于对机械式指示刻度盘、标尺或计数器进行校正。

Indications on the screen of a radar indicator which divide the range scale into known intervals for range determination or for checking against mechanical indicating dials, scales or counters.

### 校准高度；Calibrated Altitude

对压力高度表显示的指示高度进行静压误差、安装误差和仪表误差修正后确定的高度。

Altitude as determined by the application of corrections for static-pressure error, installation error, and instrument error to the indicated altitude shown by a pressure altimeter.

### 校准规范；Specification of Calibration

进行计量器具校准工作时应遵守的技术文件。

Technique documents which should be complied with during measurement instruments calibration.

### 校准空速；Calibrated Airspeed; CAS

进行过位置误差和仪表误差修正的航空器指示空速。在海平面标准大气条件下,校准空速与真实空速相等。

The indicated airspeed of an aircraft which is corrected for position and instrument error. Calibrated airspeed is equal to true airspeed in standard atmosphere at sea level.

### 校准误差(标高)；Alignment Error (Elevation)

实测航线角位置与正确航线角位置之差(以度数计)。

The difference in degrees between the position of the measured angle of the course and the correct angle for the course.

### 校准误差(方位)；Alignment Error (Azimuth)

选定航线位置和该航线正确陆方位之差(以度数计)。当该航线按顺时针方向偏离正确方位时,此误差为正。

The difference in degrees between the position of the selected course and the correct magnetic azimuth for that course. The error is positive when the course is clockwise from the correct azimuth.

### 阶段介入性评审；Stage of Involvement；SOI

局方为监控软件的生命周期过程，并确定其对 DO-178B 的符合性而定义的监控机载软件过程的评审方式，共分为 4 个阶段：SOI♯1 为软件计划阶段评审；SOI♯2 为软件开发阶段评审；SOI♯3 为软件验证阶段评审；SOI♯4 为软件最终审定评审。

SOI is defined by Authority to monitor the life cycle process of software and to determine its compliance with DO-178B. It is divided into four stages：SOI♯1 is software planning review；SOI♯2 is software development review；SOI♯3 is software verification review；SOI♯4 is final certification software review.

### 接收机自主完整性监视；Receiver Autonomous Integrity Monitoring；RAIM

确定 GPS 导航信号完整性的一种技术。

Technique used to assess integrity performance levels of GPS.

### 结冰条件；Icing Condition

能够引起在飞机或发动机上结冰的大气环境。

An atmospheric environment causing ice to form on the aircraft or in the engine(s).

### 结冰形式；Ice Formation

过冷水滴撞击在推进系统表面而形成，被分为以下几种。① 明冰，一种透明的、坚硬的冰，在温度接近（但不低于）冰点，液态水含量比较高，水滴直径比较大的空气中形成。水滴撞击到表面不会马上冻结，而是沿着表面溢流直到结冰。明冰通常没有符合空气动力学的形状，更容易受到空气动力学作用导致脱落。与霜冰相比，明冰同时具有低冻结系数和低附着性两个特点。相比较而言，明冰对静止部件有影响，而霜冰通常对转动部件有影响。② 霜冰，呈牛奶状，白色，在温度较低，空气中含有较低的液态水含量和较小的水滴直径时形成。典型的霜冰符合空气动力学原理的形状，既能在发动机旋转件上也能在静态部件上形成。霜冰的冻结系数比较高,数值上接近 1.0。典型的霜冰比起明冰有更高的附着性,但是密度更低。附着性随温度的降低而增加,直到温度降低不会产生额外收益。③ 混合冰。明冰与霜冰的混合,在明冰的尾部形成轻微的霜冰小片。混合冰形成的温度、液态水含量和水滴尺寸介于形成明冰和霜冰的条件之间。

Ice formations resulting from the impact of supercooled water droplets on propulsion system surfaces are classified as follows. ① Glaze ice, a clear, hard ice, which forms at temperatures close to (but below) freezing, in air with high liquid water content and large droplet sizes. Droplets impacting the surface do not freeze immediately, but run back along the surface until freezing occurs. Glaze ice typically has a non-aerodynamic shape and is more susceptible to aerodynamic forces that result in shedding. Glaze ice typically has both lower freezing fraction and lower adhesive properties than rime ice. Glaze ice is often a concern for static hardware while rime ice is often a concern for rotating hardware. ② Rime ice. A milky, white ice which forms at low temperatures, in air with low liquid water content and small droplet sizes. Rime ice typically forms in an aerodynamic shape, on both rotating and static engine hardware. The freezing fraction is high for rime ice, typically approaching a value of 1.0. Rime ice typically has greater adhesion properties than glaze ice but has a lower density. Adhesion properties increase with lower temperature up to a test point where no additional adhesion is gained with additional lower temperature. ③ Mixed or intermediate ice. A combination of glaze and rime ice which forms with rime patches slightly aft of the glaze ice portions. This ice forms at temperatures, liquid water content, and droplet sizes between those that produce rime and glaze ice.

## 结冰遭遇;Icing Encounter

暴露于连续或断续的结冰条件下,直到比一些预选的距离或持续时间长的间隙或中断出现。

An exposure to continuous or broken icing conditions until a gap or interruption longer than some preselected distance or duration occurs.

## 结构修理手册;Structural Repair Manual; SRM

定义对结构强度或寿命没有显著影响及有显著影响的损伤的手册。

SRM defines both damage that has and hasn't significant effect on the strength or life of the structure.

## 截获效应;Capture Effect

接收机仅对通频带内两个不同载波频率上的被接收电磁信号中的较强者做出响应(或"被其截获")的现象。

The phenomenon in which a receiver responds to (or is "captured by") only the stronger of two electromagnetic signals received which are on different carrier frequencies within the pass band of the receiver.

## 金属油箱;Metal Tank

各种类型的可从航空器上拆下来修理的金属材质燃油容器。

Various type fuel containers which are made from the metal material and can be removed from the aircraft for repair.

## 襟翼展开速度;Flap Extended Speed

襟翼在规定的放下位置时的最大允许速度。

The highest speed permissible with wing flaps in a prescribed extended position.

## 仅依据型号合格证书的生产;Type Certificate Only

当某种产品已获得型号合格证件,但尚未获得生产许可证(PC)时,型号合格证件(或其转让协议书)持有人进行该产品的生产活动。

The production activity done by the holder of TC (or agreement of assignment), when the product has obtained TC, but not yet gain PC.

## 紧凑模式;Compact Mode

在显示使用方面,通常指在不利或失效情况下使用数字形式的单个压缩的显示。

In display use, compact mode most frequently refers to a single, condensed display presented in numeric format that is used during reversionary or failure conditions.

## 紧急适航指令；Emergency Airworthiness Directive；EAD

具有紧急性，作为立即生效的法规而颁发的适航指令。

An AD provided with emergency nature and shall be issued as a legal regulation with immediate effectiveness.

## 近距导航；Short Range Navigation；SHORAIN

一种电子测位系统，其脉冲发射机触发两台位置已知的应答机以发射应答信号，用从发射触发信号到接收每台应答机应答信号之间经过的时间来确定定位。

An electronic position-finding system in which a pulse transmitter trigger responses from two transponders of known location, the elapsed time between transmittal of a triggering signal and the reception of a response from each transponder is used to establish a fix.

## (仪表着陆系统)近距指点标；ILS Inner Marker；IM

与Ⅱ类精密进近的仪表着陆系统一起使用的指点信标，位于中指点标与仪表着陆系统跑道终端之间。它以每秒六点的键调发射一个辐射图，从听觉和视觉上告诉驾驶员，他/她位于仪表着陆系统Ⅱ类进近的指定决断高(DH)，通常距降落地面 100 英尺以上。

A marker beacon used with an ILS (CAT Ⅱ) precision approach located between the middle marker and the end of the ILS runway, transmitting a radiation pattern keyed at six dots per second and indicating to the pilot, both aurally and visually, that he/she is at the designated decision height (DH), normally 100 ft above the touchdown zone elevation, on the ILS CAT Ⅱ approach.

## 进口全新民用航空器；Exotic Brand New Civil Aircraft

由中华人民共和国之外的主权国家制造，通过合法方式被中华人民共和国公民或机构拥有或使用，并申请(或已经)在中华人民共和国注册登记的全新民用航空器。

The brand new civil aircraft designed and manufactured by a exotic country，and owned by citizens or institutions of the People's Republic of China through legitimate possession，which intends to apply for registration in the People's Republic of China.

### 经批准；Approved

除了使用时注明为其他人员外，均指经局方或获局方对相关事务授权的任何人的批准，或依据国家与某外国或外部辖区之间的双边协议条款获取的批准。

Unless used with reference to another person，this term means approved by the Authority or any person to whom the Authority has delegated its authority in the matter concerned，or approved under the provisions of a bilateral agreement between the country and a foreign country or jurisdiction.

### 经批准的设计资料；Approved Design Data

CAAC 已批准的相关适用设计资料（如 TC、STC、TSO 授权、PMA 或等效证书）。

The relevant applicable design documents approved by CAAC（such as TC，STC，TSO authorization，PMA or equivalent certificate）.

### 经批准的资料；CAAC-approved Data

CAAC 或 CAAC 委任代表批准的文件，包括其中的引用文件。这些资料可包括下述方面：设计图纸、手册、程序和规范。

Data specifically approved by the CAAC or CAAC – delegated representatives，including any document referenced therein. These data may include design drawings，manuals，procedures，and specifications.

### 经停站；Transit Station

执管航空器的维修基地以外的国内外和地区的民用航空港。

The civil aviation airport which is domestic and overseas except maintenance base.

**经营人所在国；State of the Operator**

经营人的主要业务地点所在的国家，如没有这种业务地点，则为经营人的永久居住地点所在国。

The state in which the operator's principal place of business is located or, if there is no such place of business, the operator's permanent residence

**精密进近；Precision Approach**

利用精密进近程序的标准仪表进近。

A standard instrument approach using a precision approach procedure.

**精密进近程序；Precision Approach Procedure**

配备有诸如仪表着陆系统和精密进近雷达等电子下滑信标台的标准仪表进近程序。

A standard instrument approach procedure in which an electronic glide slope is provided，such as ILS and PAR.

**精密进近和着陆运行；Precision Approach and Landing Operation**

使用精密横向和垂直引导，按相应运行种类所确定的实施最低标准的仪表进近和着陆。横向和垂直引导指由以下方式提供的引导：① 地基助航设备；② 计算机生成的导航数据。

An instrument approach and landing using precision lateral and vertical guidance with minima as determined by the category of operation. Lateral and vertical guidance refers to the guidance provided either by：① a ground-based navigation aid；② computer generated navigation data.

**精密进近雷达；Precision Approach Radar；PAR**

终端区空中交通管制系统中的一种雷达设施，用以准确地探测和显示朝跑道做最终进近的航空器的方位角、距离和标高，使空中交通管制专家能向驾驶员提供咨询服务。

A radar facility in the terminal air traffic control system used to detect and display with a high degree of accuracy，azimuth，range，and elevation of an aircraft on the final approach to a runway，enabling the air traffic control specialist to provide advisory service to the pilot.

### 精密仪表进近；Precision Instrument Approach

使用仪表着陆系统(ILS)或者精密仪表进近雷达(PAR)提供方位和下滑引导的进近。

An approach in which azimuth guidance and glide guidance are provided by means of ILS or PAR.

### 警告；Alert

视觉、听觉或触觉信号，用于引起注意或表达系统状态信息。

A visual, aural, or tactile stimulus presented either to attract attention or to convey information regarding system status or condition, or both.

### 警告功能；Alerting Function

向机组提供有关非正常运行或飞机系统状态告警的飞机功能。

The airplane function that provides alerts to the flightcrew for non-normal operational or airplane system conditions.

### 警告规则；Alerting Philosophy

在驾驶舱实施警告功能的原则、指南和规则。

The principles, guidance, and rules for implementing alerting functions within a flight deck.

### 警告旗(旗标警告)；Flag (Flag Alarm)

某些机载导航设备中的一种警告装置，当信号强度或接收到的信号质量低于可接受值时，向驾驶员显示警告旗。

A warning device in certain airborne navigation equipment which is displayed to the pilot when signal strength or the quality of the received signal falls below acceptable values.

### 警告信息；Alert Messages；AM

由机载系统或地面设备提供的信息，以警示驾驶员注意 34 秒内的潜在碰撞危险，可以是视频或音频信息，如一个闪烁信号或耳机声音。

The messages that are provided by either the airborne system or ground equipment to alert the pilot to identify a potential collision hazard within 34 seconds. This alert may be visual and/or audible, such as a flashing display symbol or a headset tone.

### 警告抑制；Alert Inhibit

使用特定逻辑来防止告警出现。告警可通过告警系统自动抑制或机组人工抑制。

Application of specific logic to prevent the presentation of an alert. Alerts can be inhibited automatically by the alerting system or manually by the flightcrew.

### 警戒级警告；Caution Alert

需要机组立即知晓的警告，后续通常需要采取纠正措施。

An alert requiring immediate flightcrew awareness, and subsequent corrective action normally will be necessary.

### 净距；Net Clearance

两物体最近两点间的水平距离。

The horizontal distance between two nearest points of the two objects.

### 静电释放敏感材料；Electrostatic Discharge Sensitive Service

静电释放时通过或穿过材料表面，从而改变其物理特性或电子特性的材料。

A kind of material which physical properties can be changed due to electrostatic release pass or penetrate its surface.

### 静压源误差；Static Source Error

由静压孔测量到的气压与未受扰动的环境气压之间的差异。

Differences of air pressure between the air pressure measured by the static ports and undisturbed environment pressure.

### 静压源误差修正；Static Source Error Correction；SSEC

对静压源误差的修正。

The correction of static source error.

### 纠正措施；Rectifying Measures

申请人针对审查发现的问题提交的改正方法和行动措施。

The correction method and action for the problem of examine submitted by the applicants.

### 纠正决断咨询；Corrective Resolution Advisory

建议驾驶员偏离当前垂直速度的决断咨询，例如当飞机处于水平时的"爬升"。

A resolution advisory that advises the pilot to deviate from current vertical speed，such as CLIMB when the aircraft is level.

### 救生应急定位发射器；Survival ELT；ELT(S)

可从航空器上取下，在紧急情况下便于随时取用，由遇险者手工启动的应急定位发射器。

An ELT which is removable from an aircraft，stowed so as to facilitate its ready use in an emergency，and manually activated by survivors.

### 局部腐蚀；Local Corrosion

腐蚀范围不超过一个隔框、桁条或加强杆的蒙皮或腹板格子（机翼、机身、尾翼或吊架），一般局限于单个隔框、桁条或加强杆或者一个以上的隔框、桁条或加强杆，但腐蚀构件的相邻部位不存在腐蚀。

The corrosion area is not more than a frame，stringer，or the skin of strengthening rod，or web grid（wing，fuselage，empennage or pylon），generally limited in the single frame，stranger or strengthening rod；or more than one of the frame，stranger or strengthening rod，but the corrosion not appeared on the all adjacent parts.

### 局方；Administrator

各国的适航主管部门，对中国，指中国民用航空局、民航地区管理局。

The competent airworthiness authorities of various countries，for example，in the case of China，refer to the Civil Aviation Administration of China and the Regional Administration of Civil Aviation.

### 局方飞行试验；Official Flight Tests

在颁发了 TIA 或 LOA 之后才能开始进行正式的审定飞行试验。

Formal certification flight tests initiating after TIA or LOA has been issued.

### 距离-角度位置；Rho-theta Position

用距离和角度表示的坐标位置，通常为距某一设施的距离（海里）和方位。

A coordinate position described by distance and angle，usually referring to the distance in nautical miles and bearing from a facility.

### 距离-角度系统；Rho-theta System

依据极坐标的一种电子导航系统，在此系统中，位置用相对于某一发射台的距离或半径（$p$）和方位或角度（$e$）来划定。

Any electronic navigation system based on polar-coordinates in which position is defined in terms of distance or radius（rho）and bearing or angle（theta）with respect to a transmitting station.

### 决断高；Decision Height；DH

在精密进近中，如果不能建立继续进近所必需的目视参考，则应当开始复飞的特定高（海拔高度）。

An altitude used during a precision approach to trigger the decision of either landing or go-around.

### 决断高度（DA）或决断高（DH）；Decision Altitude Decision Height；DA/DH

精密进近或使用垂直引导的进近中的一个特定高度或高，在该高度上，假如尚未建立继续进近所需的目视参考，则必须开始复飞。① 决断高度（DA）以平均海平面为基准，决断高（DH）以跑道入口标高为基准。② 所需的目视参考是指为了使驾驶员能够判断航空器相对于所需飞行航径的位置及位置变化率，应能够看见(并保持一段足够的时间)的目视助航设施或进近区的一部分。在带决断高的Ⅲ类运行中,所需的目视参考是指特定程序和运行所规定的目视参考。③ 在同时使用决断高度和决断高时,为方便起见,可写成"决断高度/高",缩写为"DA/H"。

A specified altitude or height in the precision approach or approach with vertical guidance at which a missed approach must be initiated if the required visual reference to continue the approach has not been established. ① Decision altitude (DA) is referenced to the sea level and decision height (DH) is referenced to the threshold elevation. ② The required visual reference means that section of the visual aids or of the approach area which should have been in view for sufficient time for the pilot to have made an assessment of the aircraft position and rate of change of position, in relation to the desired flight path. In Category Ⅲ operations with a decision height, the required visual reference is that specified for the particular procedure and operation. ③ For convenience where both expressions are used they may be written in the form "decision altitude/height" and abbreviated "DA/H".

### 决断咨询;Resolution Advisory;RA

TCAS Ⅱ 向机组发出的声音和显示信息,指示执行或者不执行某些特定的机动动作以获得或保持与一个障碍物最低的垂直安全间隔。

Aural voice and display information provided by TCAS Ⅱ to a flight crew, advising that a particular maneuver should, or should not, be performed to attain or maintain minimum safe vertical separation from an intruder.

### 绝对压力;Absolute Pressure

相对于真空(无压力)测得的液体或气体的压力。

Pressure of a liquid or gas measured relative to a vacuum.

### 绝缘闪络;Insulation Flashover

电弧径迹的一种结果,绝缘电线在瞬间烧穿,并可能继续燃烧至周围的电线。绝缘体在经历电弧期间引起的高温退化后,可能扩展到整个组件的电线束。

A result of arc tracking, an instantaneous burn-through of the insulated wire with the possibility of continuing the burn into surrounding wires. Insulation flashover is a result of the high temperature degradation of the insulation experienced during arcing，and may propagate through a complete wire bundle severing the entire grouping.

# K

**抗扰度;Immunity**

在存在射频场的情况下,系统或设备部件以一种可接受的方式继续完成其预定功能的能力。

The capacity of a system or piece of equipment to continue to perform its intended function, in an acceptable manner, in the presence of RF fields.

**抗闪燃;Flash Resistant**

点火后不易猛烈燃烧。

Not susceptible to burning violently when ignited.

**科目飞行员;Subject Pilot**

被要求或批准作为机组人员参与具体飞行研究项目的局方或非局方工作人员。

Authority or non-Authority personnel who are required and approved as flightcrew members for specific research project flights.

**可达的;Accessible**

为了符合 FAR - 25.1357(d),可达的电路保护装置(CPD)是指某位飞行机组不需要离开他们的座位就可以容易地复位或更换 CPD。尽管 FAR - 25.1357(d)并不使用"可达的"这个词,但是 FAR - 25.1357(d)要求断路器或保险丝的"位置和标识必须清晰易读,以便在空中快速复位或更换"体现了"可达的"要求。CPD 不能接触到身体,但是可以对其进行远距离控制,如果按照本定义,则该 CPD 的控制装置对应飞行机组是可达的,该 CPD 也被认为是可达的。

For the purpose of compliance with § 25.1357(d), an accessible CPD is one that can be readily reset or replaced by a member of the flightcrew without leaving their seat. Although § 25. 1357(d) does not use the term "accessible" which is what § 25. 1357(d) means when it says that the circuit breaker or fuse "... must be located and identified so that it can be readily reset or replaced in flight." A physically inaccessible but remotely controllable CPD is considered accessible if its control device is accessible to the flightcrew in accordance with this definition.

## 可复用软件组件;Reusable Software Component; RSC

一种软件,其支持表明对 RTCA/DO - 178B 符合性的软件生命周期数据以及其他支持性文件是被认为可以重复使用的。

RSC is a kind of software. It supports RTCA/DO - 178B software life cycle data, and other supporting documentation being considered for reuse.

## 可接受的符合性方法;Acceptable Means of Compliance; AMC

表明规章条款符合性的可接受的方法。

Acceptable means of compliance for Airworthiness Regulation.

## 可靠性;Reliability

系统、子系统、组件或部件在一定时间内及一定条件下执行指定功能的能力或可能性。

The capacity and possibility that a system, subsystem, unit, or part performs its intended function for a specified interval under certain operational and environmental conditions.

## 可靠性评估方案;Reliability Assessment Plan

记录评估飞机和发动机及其电子电气系统(包括机电元器件和设备)可靠性的可控的、可重复的工序文件。

A document recording the controlled, repeatable processes for assessing the reliability of aircraft and engine electronic and electrical systems, including their electromechanical elements and equipment.

### 可控撞地飞行；Controlled Flight into Terrain；CFIT

飞机在机组控制下不经意地撞向地面、障碍物或水面的事故或事件，机组未能充分或及时察觉危险并阻止其发生。

An accident or incident in which the airplane，under the flightcrew's control，is inadvertently flown into terrain，obstacles，or water without either sufficient or timely flightcrew awareness to prevent the event，or both.

### 可能的失效状态；Probable Failure Condition

一架飞机在其整个运行寿命期间预计出现一次或多次的失效状态。

One or more failure conditions expected in service life circle of an aircraft.

### 可能失效；Probable Failures

在每架飞机的全生命周期内可能出现几次的失效，发生概率为 $1 \times 10^{-5}$ 或更大（见 AC 25.1309 - 1 附录 A）。失效的后果或要求的纠正措施不会严重影响飞机安全性或机组人员应对不利使用条件的能力。

Probable failures may be expected to occur several times during the operational life of each airplane. The probability of occurrence is on the order of $1 \times 10^{-5}$ or greater (see Advisory Circular 25.1309 - 1A). The consequence of the failure or the required corrective action may not significantly impact the safety of the airplane or the ability of the crew to cope with adverse operating conditions.

### 可燃的；Flammable

对液体或气体，可燃的意味着易于点燃或爆炸。

Flammable，with respect to a fluid or gas，means susceptible to igniting readily or to exploding.

### 可燃性包线；Flammability Envelope

压力/温度范围，在该范围内油/气混合体是可燃的。

Flammability envelope is the pressure/temperature domain where the fuel vapor/air mixture is flammable.

### 可燃性暴露时间;Flammability Exposure Evaluation Time; FEET

从飞机准备飞行开始,经过飞行和着陆,直到所有负载荷被卸除且所有乘客和机组已下机的这段时间。

The time from the start of preparing the airplane for flight, through the flight and landing, until all payload is unloaded and all passengers and crew have disembarked.

### 可燃性液体;Flammable Liquid

闪点为 38 摄氏度以下的液体。

The liquid of which flash point is under 38℃.

### (软件)可设置参数;Software Settable Parameters

在部件执行前设定的软件参数。

Software component data that are set before execution of the component.

### 可生存事件;Survivable Accident

驾驶舱和/或结构保持相对完整且乘客所经受的力不超过或不应该超过人类可承受的过载的事件。在这样的事件中,即使一些或所有的乘客都受到了致命伤,也仍将其归为可生存的(调查员记录机上乘客受致命伤的原因,为航空安全做出最大贡献)。

An accident in which the cockpit and/or structure remains relatively intact and the forces experienced by the occupants did not exceed or should not have exceeded the survivable limits of human G-tolerance. Such an accident is classified as survivable even if some or all occupants were fatally injured. (the investigator makes his or her greatest contribution to air safety by documenting the reasons why aircraft occupants were fatally or seriously injured in survivable accidents).

### 可修件;Repairable Part

技术上可以修复但缺乏厂家正式发布的技术文件的航空产品。

The product which is repairable technically, but lacks public technical data published by OEM.

## 可疑非批准件;Suspected Unapproved Part; SUP

被怀疑没有满足批准件要求的零件、部件或材料。

A part, component, or material that is suspected of not meeting the requirements of an approved part.

## 可用备降机场;Adequate Alternate Aerodrome

着陆性能要求能够得到满足并且预计在需要时可使用的备降机场。该机场应有必要的设备和服务,如空中交通管制、灯光、通信、气象服务、导航设备、救援和消防服务以及适用的仪表进近程序。

An alternate aerodrome at which the landing performance requirements can be met and which is expected to be available, if required. It should have the necessary facilities and services, such as air traffic control, lighting, communications, meteorological services, navigation aids, rescue and fire-fighting services and one suitable instrument approach procedure.

## 可用件;Serviceable Part

使用过的,并经合格厂家恢复至适航状态的航空设备。

The used aviation equipment which is recovered by qualified manufacturer to airworthiness condition.

## 可预见的条件;Foreseeable Conditions

显示或显示系统假定在其给定的功能范围内运行的全部环境。

The full environment that the display or the display system is assumed to operate within, given its intended function.

## 可追溯性;Traceability

在过程中两个或多个要素之间建立的可记录的联系。

The recorded relationship established between two or more elements of the development process.

## 客舱乘务员;Cabin Crew Member

为了乘客的安全,受运营人或机长指派执行值勤任务,但不得作为飞行机组成员的机组成员。

A crew member who performs, in the interest of safety of passengers, duties assigned by the operator or the pilot-in-command of the aircraft, but shall not act as a flight crew member.

### 客/货舱；Passenger/Cargo Compartments

专门设计的可以在飞机处于所有工作模式时，为人和动物提供一个合适的生活配套环境的舱室。这些区域可能是或可能不是增压的。这些区域包括但不限制于驾驶舱、客舱、厨房、所有级别货舱和行李舱。

All compartments specifically designed to provide a suitable life support environment for people and animals during all operating modes of the airplane. These areas may or may not be pressurized. These areas include, but are not limited to, the cockpit, passenger compartments, galleys, and all classes of cargo and baggage compartments.

### 客货混用飞机；Combi Aircraft

设计/配置为在机身同一水平内同时运输乘客和货物的飞机。

Aircraft designed/configured to transport both passengers and cargo on the same level within the fuselage.

### 空地飞行时间；Ground and In-flight Hours

航空器每次飞行自撤轮挡起至着陆后滑回机坪挡轮挡所经历的时间总和。

The sum of the time that the aircraft spends in each flight from removing the wheel block to sliding back to the apron wheel back after landing.

### 空速；Airspeed

航空器在空中相对于周围空气运动的速度，空速与地面上的飞行距离无关。

The speed at which an aircraft is moving through the air relative to the surrounding air. Airspeed is entirely independent of any distance covered on the surface of the earth.

### 空中飞行时间；In-flight Hours

航空器每次飞行自起飞时机轮离地到着陆后机轮触地所经历的时间总和。

The sum of time that the aircraft spends in each flight from wheels off the ground at the takeoff time to wheels touching down after landing.

### 空中告警与防撞系统；Traffic Alert and Collision Avoidance System; TCAS

一种机载系统，它执行所有必要的功能，以使其输出在适当时刻指示或开始采取相应规避机动动作的信号。

An airborne system that performs all necessary functions such that its output is a signal indicating or initiating an appropriate avoidance maneuver at a suitable time.

### 空中交通状况显示；Traffic Display

针对装备 TCAS 的飞机，应答机水平位置的显示。

A display of the horizontal position of transponder equipped aircraft relative to the TCAS equipped aircraft.

### 空中交通组织；Air Traffic Organization; ATO

主要致力于推动空中交通的安全性和有效性的组织。参与者包括商业和私人航空和军队。

The primary service of the Air Traffic Organization is to move air traffic safely and efficiently. Its stakeholders are commercial and private aviation and the military.

### 空中停车；In-flight Shut Down; IFSD

在空中的飞机一台发动机停止工作并且关车，无论是由发动机本身引起的，还是由飞行机组启动或外部影响所致的。

When an engine ceases to function (when the airplane is airborne) and is shutdown, whether self-induced, flightcrew initiated or caused by an external influence.

### 空中位置指示器；Air Position Indicator；API

机载位置指示器，靠连续计算航空器航向、空速和经过时间来进行工作。

Airborne position indicator operating by continuous computation of aircraft heading, airspeed, and elapsed time.

### 控制项目风险系数；Control Program Risk Factor

在控制项目期间整个全球机队（或适用的相关受影响子机队）预期发生的未来事件预测数量。

The forecasted number of future events expected to occur in the entire worldwide fleet (or, if applicable, the relevant affected subfleet) during the control program.

### 控制项目风险指南；Control Program Risk Guideline；CPRG

可接受风险的上限，以协助航空安全工程师决定为降低风险所提出的候选纠正措施是否足够。该指南同时按机队风险和单机风险进行描述。

The upper limit of acceptable risk which assists the Aviation Safety Engineer (ASE) in determining the adequacy, in terms of risk reduction, of a proposed candidate corrective action. These guidelines are characterized in terms of both fleet risk and individual risk.

### 库存寿命；Shelf Life

航空器材出厂时，厂家给定的库存时间期限。

The storage time limit which the manufacturer provides when leaving factory.

### 扩编飞行机组；Augmented Crew

配备三名驾驶员的机组。

The crew equipped with three pilots.

# L

**拉飘;Ballooning**

着陆前拉平过程中的高度瞬时增加。

A momentary increase in altitude during flareout prior to landing.

**拉平;Flareout**

着陆机动的一部分,此时降低下降率以减少着陆撞击。

The portion of a landing maneuver in which the rate of descent is reduced to lessen the impact of landing.

**雷达/导航基线;Radar/Navigation Baseline**

① 在雷达显示中指代表雷达扫描波束轨迹的可见线;② 在导航中指用于确定导航坐标的两点之间的测地线。

① In radar displays, the visual line representing a track of the radar scanning beam; ② in navigation, the geodetic line between two points used to determine navigational coordinates.

**雷击电流冲击;Lightning Current Stroke**

雷击过程中,电能传导至地表或其他电荷载体所产生的电流冲击。

Current surge occurring when lightning makes contact with the ground or another charge center.

**雷击区域;Lightning Strike Zones**

飞机表面区域和结构中,容易被雷电附着的区域,并且具有持续停驻和电流导通的特征。

Aircraft surface areas and structures that are susceptible to lightning attachment, dwell time, and current conduction.

### 雷击事件;Lightning Event

雷击的全过程,包括一次或者多次雷击,伴随间断或者持续的电流。

The total lightning event consisting of one or more lightning strokes, plus intermediate or continuing currents.

### 离岸机场;Offshore Airport

与城市完全隔离、四面环水、完全离开海岸的机场。

An airport entirely offshore and completely isolated from urban development and surrounded by water on all sides.

### 离地;Liftoff

航空器在起飞滑跑中在跑道上脱离接触地面的点。

The point on the runway at which the aircraft breaks contact with the ground on the takeoff roll.

### 连续起落次数;Continuous Take-offs

航空器接地后连续起飞未使用刹车的次数(几次接地算几次)。

The number of times for continuous takeoff after the aircraft is grounded without using brakes (several times are counted as its number).

### 连续最大结冰;Continuous Maximum Icing; CMI

大气结冰条件下的最大连续强度,由云中液态水含量、云中水滴平均有效直径、周围空气温度,以及上述三个变量间的相互关系定义。

The maximum continuous intensity of atmospheric icing conditions which is defined by the variables of the liquid water content in cloud, the mean effective diameter of the cloud droplets, the ambient air temperature, and the inter-relationship of these three variables.

### 联邦航空条例(美国);Federal Aviation Regulations; FAR

由 FAA 颁布的法规,用于管理在美国境内运行的所有航空活动。

Rules prescribed by the Federal Aviation Administration (FAA) governing all aviation activities in the United States.

### 联合维修委员会;Joint Maintenance Coordination Board; JMCB

负责实施及更改 MAG 以符合协议要求的委员会。

The JMCB is responsible for the implementation of, and any changes to, the MAG for proper application of the agreement.

### 列入持续适航文件的可修件;Defined Repaired Part in Continuous Airworthiness Document

原制造厂家正式出版、公布的持续适航文件(如翻修手册、零部件修理手册等)中已经列出并具体说明的翻修和修理方法、规范、标准和流程的零部件。

The component which OEM has officially published continuous airworthiness document (such as overhaul manual, component maintenance manual) has listed and specified overhaul and repair methods, specifications, standards and procedures.

### 临界发动机;Critical Engine

失效对航空器的性能或操纵品质影响最大的发动机。

The engine whose failure would most adversely affect the performance or handling qualities of an aircraft.

### 临界发动机失效速度;Critical Engine Failure Speed

用以确定所需起飞距离的飞机速度,假设临界发动机在此速度下失效。

The airplane speed used in the determination of the takeoff distance required at which the critical engine is assumed to fail.

### 临界高度;Critical Altitude

在标准大气和规定转速下,能保持规定功率或规定进气压力的最大高度。除非另有说明,临界高度指在最大连续转速下能保持下述两者之一的最大高度:

① 最大连续功率,指发动机在海平面和额定高度上的功率额定值都相同的情况;② 最大连续额定进气压力,指发动机的最大连续功率受恒定的进气压力控制的情况。

The maximum altitude at which, in standard atmosphere, at a specified rotational speed, is possible to maintain a specified power or a specified manifold pressure. Unless otherwise stated, the critical altitude is the maximum altitude at which it is possible to maintain, at the maximum continuous rotational speed, one of the following: ① the maximum continuous power, in the case of engines for which this power rating is the same at sea level and at the rated altitude; ② the maximum continuous rated manifold pressure, in the case of engines, the maximum continuous power of which is governed by a constant manifold pressure.

### 临界航空器;Critical Aircraft

在机场设计中,就某具体机场而言,控制一个或几个设计项目参数(如跑道长度、铺面强度、横向间隔等)的航空器。对所有的设计项目,临界航空器不一定是同一种航空器。

In airport design, the aircraft which controls one or more design items such as runway length, pavement strength, lateral separation, for a particular airport. The same aircraft may not be critical to all design items.

### 临时型号合格证书;Provisional Type Certificate

局方基于要求颁发的有时间和操作限制的设计批准。尽管局方没有完全审核颁发 TC 的问题项,当申请人完成必要的测试和分析时,局方仍会颁发的一个临时的设计批准。

A time and operationally limited design approval that the Authority issues, upon request. Even though the Authority has not completed its findings of compliance to issue a TC, the Authority issues a provisional type certificate after the applicant has completed the necessary tests, analyses, and computations to show that the product complies with the applicable regulations.

### 零部件；Components and Parts

任何用于民用航空产品或者拟在民用航空产品上使用和安装的材料、零件、部件、机载设备或者软件。

Any materials, parts, components, airborne equipment or software used or intended for use and installation on civil aviation products.

### 零部件制造人批准书；Parts Manufacturer Approval；PMA

除下列之外的已经获得型号合格证、型号设计批准书的民用航空产品上的加改装或者更换用的零部件的一种批准书：① 根据型号合格证、型号设计批准书或者生产许可证生产的零部件；② 根据局方颁发的技术标准规定项目批准书生产的项目；③ 符合局方认为适用的行业技术标准或者国家技术标准的标准件，如螺栓、螺母等；④ 航空器所有人或者占有人按照局方批准的其他方式为维修或者改装自己的航空器而生产的零部件。

An approval document for the modification or replacement of parts and components in the production of civil aviation products that have obtained type certificate and type design approval documents except for the following: ① components produced according to the type certificate, type design approval, or production certification; ② projects produced according to CTSOA issued by the Authority; ③ standard parts, such as bolts, nuts, conforming to the industry technical standards or national technical standards deemed applicable by the Authority; ④ parts and components produced by the owner or possessor of the aircraft for the purpose of repairing or refitting his own aircraft in accordance with other methods approved by the Authority.

### 零燃油重量；Zero Fuel Weight

通常，运输类航空器将油箱设计在机翼中。除其他优点外，将燃料储存在机翼中可以减轻机翼的弯曲压力，并且与将燃料存储在机身上相比，机翼可以承受更大的重力。对此类航空器，零燃油重量是确保最大机翼弯曲压力高于将机翼中的燃料换作同等重量载荷作用于机身时的重力。当机身配有辅助油箱时，现有的零燃油重量不再适用，因为该辅助油箱中的燃料不会减轻机翼的弯曲压力。因此，有必要在零燃油重量中扣除该辅助油箱的最大燃油容量。或者，将零燃油重量定义为最大零燃油重量。此时，辅助油箱中的燃油将被视为飞机的有效载荷,无论采取哪种方式计算,AFM都必须明确说明航空器的重量及其含义。

Typically, transport category airplanes are designed to carry the fuel supply in the wings. In addition to any other advantages, locating the fuel in the wings relieves wing bending stresses and allows a higher maximum weight than the quantity of fuel located within the fuselage. For such airplanes, zero fuel weight is established as a limit to ensure that maximum wing bending stresses are not exceeded by replacing fuel in the wings with an equal weight of payload carried in the fuselage. When an auxiliary fuel tank is installed within the fuselage, the existing zero fuel weight limit is no longer directly applicable because the fuel contained in that tank does not relieve wing bending stresses. It is, therefore, necessary to reduce the zero fuel weight limit by the maximum usable fuel capacity of the auxiliary tank. Alternatively, the zero fuel weight limit may be redefined as the maximum zero wing fuel weight limit. Any fuel contained in the auxiliary tank would then be treated as payload from a weight and balance standpoint. Regardless of which procedure is used, the AFM must clearly state the limit and its meaning.

### 陆基增强系统;Ground-based Augmentation System; GBAS

陆基增强系统对全球定位系统来说是一个局部覆盖的增强系统,它计算完整的和校正过的数据,并且直接从一个地面发射机为全球定位系统/地面增强系统用户传播数据。

GBAS is a localized coverage augmentation system to GPS that calculates integrity and correction data, and broadcasts the data to GPS/GBAS users directly from a ground-based transmitter.

### 旅客便利项目;Passenger Convenience Items

在 MMEL 中所说的旅客便利项目,是那些与旅客方便性、舒适性或是娱乐性有关的项目,如(但不限于)厨房设施、电影放映设备、烟灰缸、音响设备、头顶阅读灯。厕所门外部的烟灰缸不属于旅客便利项目。

The passenger convenience items which are mentioned in the MMEL are related to passenger convenience, comfort or entertainment system, such as, but not limited to, kitchen facilities, film projection equipment, ashtrays, stereo equipment, overhead reading lights. The ashtray outside the toilet door is not included in the passenger convenience items.

### 履历文件;Log Book

记载航空器及其部件和工具设备使用和维修历史的记录性文件。

Record files of a plane that records its components, and the use of tools, equipment, maintenance history.

### 轮胎层级;Tire Ply Rating

用于确定指定轮胎的最大推荐额定载荷和充气压力。它是对轮胎强度的量度。

This term is used to identify the maximum recommended load rating and inflation pressure for a specified tire. It is an index of tire strength.

### 轮胎/机轮安全箱;Tire/Wheel Safety Cage

一种加强结构,专门设计用于保护操作人员免受在轮胎/机轮组件增压过程中可能发生的任何爆炸的影响。

A reinforced structure that has been specifically designed to protect service personnel from the effects of any explosive event that may occur during the pressurization of a tire/wheel assembly.

### 轮胎压力指示系统;Tire Pressure Indicating System

安装在飞机上用于检测轮胎压力的系统。

A system installed on an aircraft used for tire pressure checks.

# M

**马赫数；Mach Number**

飞机速度与当地声速的比值。

The ratio of the speed of the aircraft to the local speed of sound.

**卖方采购设备；Seller Purchase Equipment**

由航空器制造厂协助买方订购的买方提供设备。

Buyer furnished equipment which is ordered by buyer through the assistance of aircraft manufacturer.

**卖方提供设备；Seller Furnished Equipment**

由买方选择、卖方提供的设备。

The equipment selected by buyer and provided by seller.

**慢车推力；Idle Thrust**

把发动机推力操纵杆置于能放置的最小推力位置止动点时获得的喷气推力。

The jet thrust obtained with the engine power control lever set at the stop for the least thrust position at which it can be placed.

**盲降系统；Navaglide**

用于仪表着陆系统和自动着陆系统；通过测距系统从地面向航空器提供上、下、左、右的信号以及距离数据；按时间分享原则同时为数架航空器服务；是一种微波系统。

Navaglide is used in ILS and automatic landing systems；provides left， right, up, down signals to aircraft from the ground, also the distance data through DME system; operates on time-sharing principle to serve several aircraft simultaneously; and is a microwave system.

### 没有归组的航空器；Non-group Aircraft

运营人申请批准是根据特定航空器的特性而非一个组别的特性航空器。对没有归组的航空器，适航批准基于数据采集，同组别航空器需收集相同数量的数据来证明导航系统的持续完整性和准确性。

An aircraft for which the operator applies for approval on the characteristics of the unique airframe and navigation system used, rather than on a group basis. For non-group aircraft where airworthiness approval has been based on data collection, the continuing integrity and accuracy of the navigation system is demonstrated by the same amount of data collection as required for group aircraft.

### 美国材料与试验协会；American Society for Testing and Materials; ASTM

包括生产者、消费者和监管者的国际机构，负责制定、修订和管理材料标准。

An international body of producers, consumers, and regulators that develops, maintains, and administers material standards.

### 美国国家标准协会；American National Standards Institute; ANSI

非营利性的民间标准化组织。通过它，政府和民间相关系统相互配合，起到了连接联邦政府和民间标准化系统的作用。它协调并指导全国标准化活动，制定标准，给标准化研究和使用单位提供帮助，提供国内外标准化信息，同时扮演标准化管理机关的角色。

ANSI is nonprofit qualitative folk standardization organizations. Relevant government system and folk system can cooperate with each other through ANSI, which builds a bridge between federal government and folk standardization system. It coordinates and guides the national standardization activities, standard formulating, research and use the unit to help, providing standardized information at home and abroad. It plays the role of administrative authority.

**美国航空航天协会;American Institute of Aeronautics and Astronautics; AIAA**

世界上最大的致力于全球航空航天事业的技术协会。

The world's largest technical society dedicated to the global aerospace profession.

**美国航空人员信息手册;U. S. Airman's Information Manual**

一套驾驶员的操作手册,内含在国家空域系统(仅美国边界以内的)内计划与实施飞行所需的信息。

A set of pilot's operational manuals containing information needed for the planning and conduct of flight in the National Airspace System (conterminous the United States only).

**美国航空运输协会;Air Transport Association of America; ATA**

美国定期航空工业主要的贸易和服务组织,美国的成员航空公司占每年美国定期航空公司承运的客运和货物交通运输的 95%。

The principal trade and service organization of the United States scheduled airline industry, and the United States airlines account for 95% of the passenger and cargo traffic carried annually by the United States scheduled airlines.

**美国交通运输部;U.S. Department of Transportation; DOT**

成立于 1966 年 10 月 15 日,协调美国交通运输的各种需要和计划。

DOT was founded in October 15, 1966, coordinating the transportation needs and plans of the United States.

**美国联邦航空交互式报告系统;Federal Aviation Interactive Reporting System; FAIRS**

通用服务管理处按照当前版本的办公管理和预算通告 A-126,收集和维护政府飞机数据的综合数据系统。

The comprehensive data system used by General Services Administration (GSA) to collect and maintain government aircraft data in accordance with Office of Management and Budget (OMB) Circular A-126, current edition.

## 密度高度；Density Altitude

由压力高度和当前环境温度决定的高度。在标准大气下，密度高度等于压力高度。在给定压力高度上，温度越高，密度高度也越高。

The altitude which is determined by pressure altitude and existing ambient temperature. In standard atmosphere, density altitude equals to pressure altitude. For a given pressure altitude, the higher the temperature, the higher the density altitude.

## 描述性数据；Descriptive Data

必须完整地定义设计修改或安装的数据。

Data necessary to completely define the design of the modification or installation.

## 民航局批准；Approved by the CAAC

民航局、民航地区管理局或者民航局授权的机构或者个人所进行的批准。

The approval approved by the CAAC Headquarter Office, the CAAC RA, or organizations or individuals authorized by the CAAC.

## 民航局批准的生产制造系统；Manufacture System Approved by CAAC

根据 CCAR - 21 部批准的生产系统，包括：① 零部件制造人批准书(PMA)持有人；② 技术标准规定项目批准书(CTSOA)持有人；③ 仅依据型号合格证生产的型号合格证持有人；④ 生产许可证(PC)持有人。

Production system approved according to CCAR - 21, include: ① holder of PMA; ② holder of CTSOA; ③ TC only; ④ holder of PC.

## 民航局批准的维修培训机构；Maintenance Training Organization Approved by CAAC

按照 CCAR - 147 部获得了民航局颁发的民用航空器维修培训机构合格证的培训机构。

Organization which obtained civil aircraft maintenance training organization certificate issued by CAAC accordance with CCAR - 147.

**民航局认可的部件；Approved Part**

下述认可的装于型号审定产品的零部件：① 指根据 CCAR - 21 部及双边适航协议，装于经型号认可的外国航空产品上的零部件；② 根据 CCAR - 145 部及有关维修合作安排或协议认可的维修单位维修的零部件；③ 按照航空器及其部件制造厂家指定方式进行的因设计或制造导致的索赔修理或执行强制性改装的零部件；④ 航空器制造厂家确定的标准件（如螺母和螺栓）；⑤ 航空运营人根据民航局批准的程序制造的用于自身维修目的的零部件；⑥ 由民航局授权的人员确定符合批准的型号设计数据的零部件；⑦ 其他民航局规定的情况。

The following part installed in TC product：① according to CCAR - 21 and BAA，installed in foreign aircraft obtained VTC；② maintenance by MRO according to CCAR - 145 and its related arrange or agreement；③ claim repair or forced modification due to defect of design or manufacture；④ standard component confirmed by aircraft maintenance；⑤ operator according to the CAAC approved procedures for the manufacture of self repair purpose parts；⑥ by the CAAC authorized personnel to determine compliance with approved models of design data of parts；⑦ other according to CAAC specified.

**民航局认可的培训机构；Training Agency Recognized by CAAC**

未按照 CCAR - 147 获得民航局颁发的民用航空器维修培训机构合格证，但民航局按照规定的程序认可其培训结论的机构。

The agency failed to gain the Civil Aircraft Maintenance Training Agency Certificate according to CCAR - 147，but is the recognized training agency according to relative program.

**民航无损检测人员资格鉴定与认证委员会；Qualification and Certification of Personnel for Nondestructive Testing Board of Civil Aviation；CANDTB**

由民用航空主管部门批准的，主要管理无损检测人员资格鉴定与认证的组织（简称民航 NDT 委员会）。该组织按本标准提供或支持 NDT 资格鉴定、考试和资格认证服务。

Management of NDT personnel qualification and certification organization approved by the civil aviation authority (referred to the Civil Aviation NDT Committee abbreviations：CANDTB). The organization provides or supports

NDT qualification, examination and certification services in accordance with this standard.

### 民用对应航空器型别;Civil Counterpart

已通过标准类别型号合格审定,并与军用航空器型别相同或相似的型别。

The model of an aircraft that has been type-certificated in a standard category, and that is the same or similar to a military model aircraft.

### 民用航空标准件;Civil Aviation Standard Part

制造符合确定的工业或国家标准或规范的零件,包括设计、制造和统一标志要求。这些标准或规范应是公开发布并在航空器或其部件制造厂家的持续适航性资料中明确的。

A part manufactured in compliance with an established industry or government standard or specification which includes design, manufacturing and uniform identification requirements. This standard or specification must be publicly available and specified in the continuing airworthiness data of the aircraft or component manufacturers.

### 民用航空测试设备;Civil Aviation Test Equipment

对航空产品、材料、零部件或结构的成分、特性或性能进行测量、检测、试验和鉴定的设备。

Aviation products, materials, parts, or the structure of the composition, characteristics or performance measurement, testing, testing, and identification equipment.

### 民用航空产品;Production/Product

民用航空器、航空发动机和螺旋桨。

Civil aircraft, aero engines and propellers.

### 民用航空规章;Civil Air Regulations; CAR

美国联邦法规(CFR)的前身。

The predecessors of the current CFR (Code of Federal Regulations).

**民用航空器;Civil Aircraft**

除用于执行军事、海关、警务飞行外的航空器。

Aircraft other than those used in flight missions of military, customs and police services.

**民用航空器国籍;Nationality of Civil Aircraft**

民用航空器属于某个国家的法律资格。

The legal qualification of a civil aircraft belonging to a country.

**民用航空器国籍登记变更申请人;Applicant of Civil Aircraft Certificate of Change Nationality Registration**

民用航空器国籍登记变更后的所有人或占有人,通常为该民用航空器在国籍登记变更后的实际运营人。

Owner or occupier of aircraft after change nationality registration, also the operator of aircraft after change nationality.

**民用航空器国籍登记证申请人;Applicant of Civil Aircraft Certificate of Nationality Registration**

符合《民用航空器国籍登记规定》(CCAR‐45‐R1)规定的民用航空器所有人或者占有人,通常为民用航空器的实际运营人。

Owner or occupier of aircraft conforms with the regulations of CCAR‐45‐R1, also the operator of aircraft.

**民用航空器临时登记证申请人;Applicant of Civil Aircraft Certificate of Temporary Nationality Registration**

民用航空器的制造人、销售人或中国民用航空局认可的其他人。

Manufacturers, salers or others authorized by CAAC.

**民用航空器适航管理;Airworthiness Management of Civil Aircraft**

以保障民用航空器的安全性为目标的技术管理,是政府适航管理部门在制定了各种最低安全标准的基础上,对民用航空器的设计、制造、使用和维修等环节进行科学统一的审查、鉴定、监督和管理。适航管理分为初始适航管理和持续适航管理。

The target of the civil aircraft airworthiness management is to ensure the security of aircraft, which is the scientific review, appraisal, supervision and management to the design, manufacture, usage and maintenance of civil aircraft by government, on the basis of minimum safe standard. Airworthiness management can be divided into initial airworthiness management and continued airworthiness management.

### 民用航空委员会;Civil Aeronautics Board; CAB

政府委任的管理航空运输经济方面的机构,对航空公司、通用销售代理、货物销售代理和航空货运代理及其财产、产权、设备、设施和特许经营权具有全面的监督和控制权限。

The Civil Aeronautics Board (CAB) is the agency of the government mandated to regulate the economic aspect of air transportation, and shall have the general supervision, control and jurisdiction over air carriers, general sales agents, cargo sales agents, and air freight forwarders as well as their property, property rights, equipment, facilities, and franchise.

### 民用衍生类飞机;Civil-derived Aircraft

限制类别的特殊用途飞机。

Special-purpose aircrafts type-certificated in restricted category.

### 模拟冰型;Simulated Ice Shapes

由木头、环氧树脂或其他材料,通过制造技术制成的冰型。

Ice shapes made of wood, epoxy, or other materials by any construction technique.

### 模拟机;Simulator

模拟飞行状态、空中交通管制或其他复杂问题的装置,用于培训有关人员解决这些问题。

A device which simulates the conditions of flight, air traffic control or other complex problems for the purpose of training personnel in their solutions.

### 模拟机飞行检查员;Check Airman (Simulator)

有资格针对特定型别的航空器,在飞行模拟机或者飞行训练器上实施飞行检查的人员。

A person who is qualified to conduct flight checks, but only in a flight simulator or a flight training device for a particular type aircraft.

### 模拟结冰;Simulated Icing

制造模拟冰块的过程,如使用喷雾在飞机或飞机表面的结冰道或结冰罐后积聚冰块。

The process of creating simulated ice, for example, accumulating ice on an aircraft or aircraft surface by using a spray array, in an icing tunnel or behind an icing tanker.

### 模拟质量管理系统;Simulation Quality Management System; SQMS

FSTD 持续资格认证的质量管理体系的一部分。

The elements of a quality management system for FSTD continuing qualification.

### 模式;Mode

显示控制系统的功能状态。

The functional state of a display and/or control system(s).

### 磨损;Attrition

飞机在使用寿命中的普通摩擦、磨损。

The general wear and tear of an aircraft during its service life.

### 目视参考;Visual References

驾驶员使用观察机舱外部环境获取的视觉信息,作为飞机控制和飞行航路评估的主要参考。

Visual information the pilot derives from the observation of real-world cues, out the flight deck window, used as a primary reference for aircraft control or flight path assessment.

### 目视飞行;Contact Flight/Visual Flight

驾驶员运用对地平线和地标的目视参考,来确认航空器的姿态并寻找从一地到他地的路线的飞行。

Flying in which a pilot ascertains the attitude of his aircraft, and finds his way from place to place by visual reference to the horizon and landmarks.

### 目视飞行规则;Visual Flight Rules; VFR

在目视条件下进行飞行的程序的控制规则。

Rules that govern the procedures for conducting flight under visual conditions.

### 目视检查;Visual Inspection; VI

直接用肉眼或者借助多种辅助设备,对被检对象的状态进行判断的过程。

The process of using the eye, alone or in conjunction with various aids, as the sensing mechanism from which judgments may be made about the condition of a unit to be inspected.

### 目视警告;Visual Alert

使用显示信息来呈现条件、情况或事件。

The projected or displayed information to present a condition, situation, or event.

### 目视可见碰撞损伤;Visible Impact Damage; VID

可以被及时发现,并可很快修理或更换的受损部件。

The damaged parts that could be checked in time, and repaired or changed.

### 目视下降点;Visual Descent Point; VDP

进近路线上的一点,依据不精确直线进场程序,从 MDA 正常下降到跑道接地点。

The VDP is a defined point on the final approach course of a nonprecision straight，in approach procedure from which normal descent from the MDA to the runway touchdown point may be commenced.

# N

**耐火;Fire Resistant**

应用于安装动力装置(如流体输送导管、易燃流体系统零部件、导线、空气导管、接头和动力装置)时,耐火指材料或零部件处于特定位置,在热和可能发生的其他情况下完成其预定功能并经受 2 000 华氏度(±150 华氏度)的火焰至少 5 分钟的能力。例如,耐火软管可经受 2 000 华氏度火焰,历时至少5 分钟。

When applied to powerplant installations such as fluid-carrying lines, flammable fluid system components, wiring, air ducts, fittings and powerplant controls, fire resistant means the capability of a material or component to perform its intended functions under the heat and other conditions likely to occur at the particular location and to withstand a 2 000℉ flame(±150℉) for 5 minutes minimum. For example, a fire resistant hose can withstand a 2 000℉ flame for 5 minutes.

**耐久性;Durability**

航空设备或部件的性能,在经过经济性维护后,能在现场运行并保持工作到指定时间。

Ability of avionics equipment or components to function and sustain stress in field service for a specific period of time with economical maintenance.

**南极区域;South Polar Area**

南纬 60 度以南的整个区域。

The entire area south of 60° latitude.

### 内部 HIRF(高强度辐射场)环境;Internal HIRF Environment

机身内、设备内或空腔内的辐射场环境。

The RF environment inside an airframe, equipment enclosure, or cavity.

### 内部程序;Internal Procedure

生产批准书持有人或其相关机构不包括在局方批准的资料中的程序。

A PAH's or associate facility's procedures that are not included as part of the Authority-approved data.

### 内部评估;Internal Evaluation

对飞行项目内部的组织政策、程序和系统进行有效性确定的独立审查和分析。评估采取审计和检查的方式,但不严格限于这两种方法。

An independent review and analysis, internal to the flight program, of organizational policies, procedures, and systems to determine their effectiveness. Evaluations make use of auditing and inspection techniques, but not strictly limited to the two methods.

### 内部审核;Internal Audit

由被审核组织或其代表实施的审核。

An audit conducted by, or on behalf of, the organization being audited.

### 能力;Capability

局方使用"能力"一词来描述正确地完成所分配的任务需要的知识和技能。

The Authority uses the word "capability" to describe the knowledge and skills required to properly accomplish the assigned tasks.

# O

**ODA 持有人;ODA Holder**

获得指定 ODA 信函的组织。

ODA Holder is the organization that obtained the ODA letter of designation.

**ODA 任命办公室;ODA Appointing Office**

负责选择、任命和监督机构委任授权(ODA)持有者的办公室。

Appointing office is the lead office，as appropriate，responsible for selection，appointment and oversight of ODA holders.

**ODA 申请人;ODA Applicant**

至少包括两个人申请指定的机构委任授权(ODA)的组织。

An organization that consists of at least two individuals applying for an ODA designation.

**OMT 领导;OMT Lead**

来自组织管理团队(OMT)办公室的成员,作为 OMT 主要联络人。

A member of the Organization Management Team（OMT）office who serves as the primary contact for OMT.

**欧洲航空安全局;European Aviation Safety Agency; EASA**

欧盟的一个机构,它负责在航空安全和环境保护领域的具体管理和执行任务。因此,该机构成为了欧盟建立和维持民用高航空安全性水平和环境保护水平的策略的关键一部分。

An agency of the European Union, which has been given specific regulatory and executive tasks in the field of aviation safety and environmental protection. This agency therefore constitutes a key part of the European Union's strategy to establish and maintain a high level of civil aviation safety and environmental protection in Europe.

## 欧洲航空局;European Aviation Authority; AA

欧盟成员国的责任政府机构或实体,它代表欧盟对被监管的实体行使法律监督,并决定他们对标准、规章和其他欧盟权限内要求的符合性。

A responsible government agency or entity of an EU member state that exercises legal oversight on behalf of the EU over regulated entities, and determines their compliance with applicable standards, regulations, and other requirements within the jurisdiction of the EU.

# P

**PMA 件;PMA article**

依据零部件制造人批准书生产的零部件。

Parts manufactured in accordance with PMA.

**爬升率;Rate of Climb; R/C**

航空器单位时间内增加的高度。

The increased height of the aircraft unit time.

**跑道接地区的海拔标高;Touch Down Zone Elevation; TDZE**

着陆表面前 3 000 英尺的最高点。

The highest elevation in the first 3 000 ft of the landing surface.

**跑道入口跨越高度;Threshold Crossing Height; TCH**

目视或电子下滑波束直线延伸段高出跑道入口的高度。

The height of the straight line extension of the visual or electronic glide slope above the runway threshold.

**培训大纲;Training Syllabus**

局方要求的最低培训要求和培训内容。

The lowest requirements and training contents prescribed by CAAC.

**配平;Trim**

航空器想要达到的稳定姿态。

A stable condition of desired aircraft attitude.

## 喷气式飞机；Jet Airplane

所有由涡喷或涡扇发动机提供动力的固定翼飞机，无论是按 CCAR－23（小型或通用类飞机）还是按 CCAR－25（运输类航空器）审定的飞机。

A fixed-wing aircraft powered by turbojet or turbofan engine, regardless of the certification within CCAR－23 (aviette or general aviation airplane) or CCAR－25 (transport airplane).

## 喷气助推起飞；Jet Assisted Takeoff；JATO

利用辅助的喷气或火箭发动机协助航空器在少于正常需要的距离上或在主动力源失效的情况下起飞。

The use of supplemental jet or rocket engines to assist an aircraft to takeoff in less than the normally required distance or in the event of failure of the primary power source.

## 批次；Batch

同一类型和组成并且同一批生产的或同一批交付的包装产品的总体。

The same type and composition and with a batch of production or the same delivery packaging products overall.

## 批准；Approval

经局方或由局方在有关事项上授予其权力的任何人批准，或根据本国与（或）外国或司法管辖区之间的双边协议的规定批准。

The approval of the authority or any person authorized by the authority in relevant matters, or in accordance with the provisions of bilateral agreements between the country and/or foreign countries or jurisdictions.

## 批准的维修单位；Approved Maintenance Organization；AMO

经 145 部所认证的维修服务提供机构。

The organization providing aircraft maintenance service that is certificated by Part－145.

**批准放行证书/适航批准标签; Authorized Release Certificate/Airworthiness Approval Tag; ARC/AAT**

制造符合性检查代表或委任制造检查代表签发的,用于证实试验产品已经过制造符合性检查,符合型号资料的标签。

A tag issued by manufacturing conformity inspection representative or designated manufacturing inspection representative, used to confirm that test products have been conformity inspected and comply with design data.

**批准生产检查系统; Approved Production Inspection System; APIS**

确保每一个产品与型号设计一致并且处于安全运行状态。

APIS is used to insure that each product conforms to the type design and is in condition for safe operation.

**批准手册; Approved Manuals**

作为审定计划的一部分,需民航管理当局的手册或其章节,包括飞行手册的批准部分、持续适航说明的适航性限制部分、发动机和螺旋桨的安装和操作说明书、审定维护要求等。

Manuals or sections of manuals, requiring approval by the aviation authorities as part of the certification program. These include the approved sections of the flight manual, the airworthiness limitation section of the instructions for continued airworthiness (ICA), the engine and propeller installation and operating instruction manuals, and the certification maintenance requirements, where applicable.

**(维修或更改)批准数据; (Maintenance and Alteration) Approved Data**

被局方批准的数据,用于产品维修更改。

Data that has been approved by civil aviation authority and is used to perform maintenance and alterations on products.

**批准睡眠区; Authorized Sleeping Area**

经局方批准,为使机组成员获得良好睡眠而指定的场所。

The designated place approved by the Administrator to enable crewmembers to sleep soundly.

## 批准型号清单；Approved Model List；AML

多个补充型号合格证使用同一个批准模型清单来控制安装资格的情况。

Special case of multiple STC using an approved model list（AML）to control installation eligibility.

## 疲劳；Fatigue

在经过循环应力或载荷应力或低于其抗拉强度的载荷作用下，材料以一种易碎的方式破裂的趋势。

The tendency of a material to break in a brittle manner under repeated stresses or load stresses at or below its tensile strength.

## 疲劳评定；Fatigue Evaluation

对强度、细节设计和制造的评定，必须表明飞机在整个使用寿命期间可避免由疲劳引起的灾难性破坏。

An evaluation of the strength, detail design, and fabrication which must show that catastrophic failure due to fatigue will be avoided throughout the operational life of the airplane.

## 疲劳损伤；Fatigue Damage；FD

材料承受高于疲劳极限的交变应力时，每一次循环都使材料产生一定量的损伤，导致疲劳强度下降的现象。

Material under alternating stress higher than fatigue limit，each cycle made material produce a certain amount of damage，leading to the phenomenon that fatigue strength drops.

## 偏差；Discrepancy

任何设施的工作参数，其允差在飞行检查手册中已给出，但根据飞行检查的测量结果，该参数不在给出的允差之内。

Any facility operating parameter for which a tolerance is given in the Flight Inspection Manual which is not within given tolerance values，as determined by flight inspection measurements.

### 偏离；Deviation

对规章中明确允许偏离的条款，合格证持有人在提出恰当理由和证明能够达到同等安全水平的情况下，经局方批准，可以不遵守相应条款的规定或者遵守替代的规定、条件或者限制。

The act that a certificate holder is allowed not to comply with the requirements or may comply with the substituted requirements, conditions, or limitations of applicable provisions from which deviation is clearly allowed by applicable regulations, if the certificate holder is able to give appropriate reasons and prove that it can reach the equivalent safety level and is authorized by the administrator.

### 偏流角；Drift Angle; DA

航空器纵轴与其相对于地面的航迹之间的水平夹角。

Horizontal angle between the longitudinal axis of the aircraft and its path relative to the ground.

### 偏流修正；Crabbing

为补偿风引起的偏流，将航向保持在航迹偏左或偏右。

Holding a heading to left or right of course to compensate for wind drift.

### 偏流修正角；Crab Angle

风的修正角，由侧风影响造成的航迹与航空器航向之间的角度差。

A wind correction angle referring to the angular difference between the course and the heading of an aircraft due to the effects of a crosswind.

### 飘降着陆；Pancake Landing

在此种着陆中，飞机在离地面相当高的空中拉平并失速，因而沿陡的航径迅速下沉并猛烈撞地。

A landing in which the airplane is leveled off and stalled rather high above the surface, as a result of which the airplane settles rapidly on a steep flight path and strikes the surface forcefully.

### 平衡尾翼载荷；Balancing Tail Load

俯仰加速度为零时使航空器平衡所需的载荷。

A load necessary to place the airplane in equilibrium with zero pitch acceleration.

### 平衡载荷；Balancing Loads

在规定加载条件的外力作用下，使航空器处于平衡状态的载荷。由此所获得的平衡状态可以是真实的，也可以是假设的。平衡载荷可以代表空气载荷或惯性载荷，或两者兼之。

Loads by which the airplane is placed in a state of equilibrium under the action of external forces resulting from specified loading conditions. The state of equilibrium thus obtained may be either real or assumed. Balancing loads may represent air loads, inertia loads, or both.

### 平均非计划拆换间隔时间；Mean Time between Unscheduled Removals

通过数学方法计算、用可靠性估计得到的数值。

The value calculated by mathematical methods and estimated by the reliability.

### 平均无故障时间；Mean Time Between Failure；MTBF

一定时间内，多次故障间隔时间的平均值。

The average of multiple failure intervals within a certain time.

### 平视显示器；Head Up Display；HUD

采用透明屏幕（合成器），用来显示主要飞行信息（例如姿态、大气数据、导航等）的显示系统，在驾驶员的前向视野，驾驶员和挡风玻璃之间。

A display system that projects primary flight information (such as attitude, air data, guidance) on a transparent screen (combiner), in the pilot's forward field of view, between the pilot and the windshield.

## 平稳运行；Steady Operation

在结冰试验中，发动机在设定的测试点以及在油门瞬时变换时，应表现出平稳、可靠、稳定运行等特点。

During icing testing, the engine should demonstrate steady, reliable, and smooth operation while sitting on test point, as well as during throttle transients.

## 平行仪表着陆系统程序；Parallel ILS Procedures

为同一机场的平行跑道服务的两套仪表着陆系统装置，供在批准的最低气象条件下同时平行进近使用。

A dual ILS installation which serves parallel runways at the same airport and provides for simultaneous parallel approaching to authorized minimums.

## 评估；Evaluation

项目审查的术语，它包括：① 对室或处的负责人或部门负责人进行的审查；② 对被审查的组织机构的某一部门进行的审查；③ 根据局方的技术行政目标、政策和标准，着重对受检查的局方业绩情况进行的衡量。

The term applied to program review which is performed：① for the head of an office or service, or a division chief；② by an element of the organization being reviewed；③ emphasizing the measurement of agency performance of the functions under scrutiny, in terms of agency technical administrative objectives, policies, and standards.

## (Ⅱ类运行)评估程序；(Category Ⅱ Operations) Evaluation Program

通过一系列飞行验证确定飞行控制引导系统的性能，在该性能下，系统可满意地执行Ⅱ类运行。

An operational program based on a series of demonstration flights to establish the capability of the flight control guidance system to perform satisfactorily in Category Ⅱ operations.

## 评估员；Evaluator

评定或评价机组成员、教员、其他评估员、飞行签派员或其他运行人员表现的人员。

A person who assesses or judges the performance of crewmembers, instructors, other evaluators, aircraft dispatchers, or other operations personnel.

### 评审委员会；Evaluation Panel；EP

局方指定的评审申请的团队。该团队包括申请单位管理小组、委任办公室代表以及适当的总部政策代表。

The group the Authority assigns to review the application. The evaluation panel consists of the prospective organization management team (OMT), a representative from the appointing office, and the appropriate permanent headquarters policy representatives.

### 评审子系统；Review Subsystem

能影响保持适航部门批准的设计资料或质控资料的一种特殊的活动或功能，如设计资料控制，特种制造工艺和适航性审定，这些活动或功能要接受对批准的程序的适宜性及贯彻情况的评审。

A specific activity or function that may affect the maintenance of authority-approved design or quality data, such as design data control, manufacturing control, and supplier control. Such activities are subject to evaluation of the adequacy and implementation of approved procedures.

### 坡度角；Bank Angle

航空器相对于纵轴的姿态或绕纵轴滚转的角度。

Airplane attitude with respect to or around the longitudinal axis, or roll angle (degrees).

### 迫近的地形碰撞；Imminent Terrain Impact

当飞机的飞行路径预计将发生地形碰撞时，通知飞行机组的一个警告。

An alert that will notify the flightcrew when the flight path of the airplane is projected to impact terrain.

# Q

### 起飞；Take Off；T/O

航空器从停止状态加速到飞行状态,开始飞行的动作。

The act of beginning flight in which an aircraft is accelerated from a state of rest to flight.

### 起飞安全速度；Takeoff Safety Speed；$V_2$

离地后获得的参考空速,在该速度下可以达到规定的单发不工作爬升性能。

A referenced airspeed obtained after lift-off at which the required one-engine-inoperative climb performance can be achieved.

### 起飞额定功率；Takeoff Rating

发动机和螺旋桨的规定额定值,指发动机或螺旋桨经合格审定,用于起飞的适用功率、转速或其他限制。

A rating assigned to engines and propellers which means the applicable power, rotation speed, or other limits for which the engine or propeller is certificated for takeoff operation.

### 起飞功率；Takeoff Power

① 对活塞式发动机,指在标准海平面条件下和批准供正常起飞的最大曲轴转速与发动机进气压力条件下产生的制动马力,其连续使用时间限于经批准的发动机技术说明书中的规定;② 对涡轮发动机,指在特定高度和大气温度的静止条件下和批准供正常起飞的最大转子转速与燃气温度条件下产生的制动马力,其连续使用时间限于经批准的发动机技术说明书中的规定。

① With respect to reciprocating engines, it means the brake horsepower that is developed under standard sea level conditions, and under the maximum conditions of crankshaft rotational speed and engine manifold pressure approved for the normal takeoff, and limited in continuous use to the period of time shown in the approved engine specification; ② with respect to turbine engines, it means the brake horsepower that is developed under static conditions at a specified altitude and atmospheric temperature, and under the maximum conditions of rotorshaft rotational speed and gas temperature approved for the normal takeoff, and limited in continuous use to the period of time shown in the approved engine specification.

### 起飞功率和/或推力;Take-off Power and/or Thrust Rating

在发动机型号合格证数据单上规定,当在指定的条件下和适当的认可限制范围内运行时,系列和新翻修的发动机的最小测试台验收功率和/或推力。

The minimum test bed acceptance power and/or thrust as stated in the engine type certificate data sheet, of series and newly overhauled engines when running at the specified conditions and within the appropriate acceptance limitations.

### 起飞警告系统;Takeoff Warning System

在起飞滑跑的开始阶段,如果飞机处于任何一种不允许安全起飞的形态,则警告系统必须自动向驾驶员发出音响警告。

In the beginning of takeoff, if the plane is under the condition when safe take-off is not allowed, the warning system must send out an audible warning to the pilot automatically.

### 起飞决断速度;Takeoff Decision Speed; $V_1$

地面校正空速,在此速度下,出于发动机失效或其他原因,驾驶员必须做出继续起飞或中断起飞的决断。

The calibrate airspeed on the ground at which, as the result of engine failure or other reasons, the pilot is assumed to have made a decision to continue or reject the takeoff.

## 起飞抬轮速度;Takeoff Rotation Speed; $V_R$

飞机起飞滑跑加速到机组可以开始抬轮的安全速度。在该速度下飞机开始抬轮,以便在跑道上空 35 英尺时达到 $V_2$ 速度。

The speed used by the pilot to begin raising the nose wheel off the runway during the acceleration to takeoff safety speed. The speed at which aircraft rotation begins in order to achieve a speed of $V_2$ at 35 ft above the runway.

## 起飞推力;Takeoff Thrust

对涡轮发动机,指在特定高度和大气温度的静止条件下和批准供正常起飞的最大转子转速与燃气温度条件下产生的喷气推力,其连续使用时间限于经批准的发动机技术说明书中的规定。

With respect to turbine engines, it means the jet thrust that is developed under static conditions at a specific altitude and atmospheric temperature under the maximum conditions of rotorshaft rotational speed and gas temperature approved for the normal takeoff, and limited in continuous use to the period of time shown in the approved engine specification.

## 起落次数(循环);Flight Cycle

航空器从滑跑至离地、最后落地的完整过程。

A complete process which the aircraft taxis and takes off and finally lands on the ground.

## 起落架放下速度;Landing Gear Extended Speed

航空器起落架放下后能安全飞行的最大速度。

The maximum speed at which an aircraft can safely fly with the landing gear extended.

## 起落架收放速度;Landing Gear Operating Speed

起落架安全放下或收上时的最大速度。

The maximum speed at which the landing gear can be safely extended or retracted.

## 气动力系数；Aerodynamic Coefficients

空气动力和力矩的无量纲系数。

Non-dimensional coefficients for aerodynamic forces and moments.

## 气流分离；Burbling

机翼或机身周围流线型气流的分离。它特别指越过机翼上表面的流线型气流分离，这会造成升力损失和阻力增加。

A separation and breakdown of the streamline flow about an airfoil or body. It is especially concerned with the breakdown of the streamline flow of air across the upper surface of an airfoil which results in a loss of lift and an increase in drag.

## 气象雷达；Weather Radar

适用于或可用于探测降水或足够密度的云层的雷达。

Radar which is suitable for or can be used for the detection of precipitation or clouds of sufficient density to be detected.

## 前掠；Sweepforward

翼面或翼面的后缘或其他基准线从根部到梢部的前斜。

The forward slant from root to tip of an airfoil or the trailing edge or other reference line of an airfoil.

## 前三点式起落架；Tricycle Landing Gear

三轮式起落架，主起落架位于重心之后，第三个机轮位于航空器机头前侧的下部。

A three-wheel landing gear with the main gear located to the rear of the center of gravity and the third wheel located forward under the nose of the aircraft.

## 前缘；Leading Edge

翼面或螺旋桨叶片首先接触空气的边缘。

The edge of an airfoil or propeller blade which first meets the air.

**潜在的结冰条件；Potential Icing Conditions**

飞机制造商就在地面或飞行中可能会导致飞机积冰的温度和可见湿气，定义的大气结冰条件。

Atmospheric icing conditions that airframe manufacturers typically define in terms of temperature and visible moisture that may result in aircraft ice accretion on the ground or in flight.

**强制持续适航资料；Mandatory Continuing Airworthiness Information；MCAI**

对改装、换件或检查航空器的强制要求以及对使用限制和程序的修改。上述资料还包括缔约国以适航指令形式发布的资料。

Mandatory continuing airworthiness information is intended to include mandatory requirements for modification, replacement of parts or inspection of aircraft and amendment of operating limitations and procedures, also including the information issued by contracting states in the form of airworthiness directives.

**强制性标准；Mandatory Standard**

具有法律属性，在一定范围内通过法律、行政法规等强制手段加以实施的标准。

Mandatory standards having legal property, and are implemented within a certain range through the coercive means, such as the laws, administrative regulations.

**勤务车辆；Service Vehicle**

为航空器提供各种勤务保障的机动车辆。

Vehicles which provide various of service support for aircraft.

**倾斜角；Angle of Bank**

航空器横轴与水平面的夹角，亦称为滚转角。

The angle between the lateral axis of an aircraft and the horizontal, also called angle of roll.

**晴空湍流；Clear-air Turbulence；CAT**

飞机在无云的空中遭遇的湍流，通常指与风切变相关的高度湍流，在急流附近经常会遇到。

Turbulence encountered by aircrafts in air where no clouds are present; more popularly applied to high-level turbulence associated with wind shear; often encountered in the vicinity of the jet stream.

**区域安全性分析；Zonal Safety Analysis**

影响系统安全的安装、系统界面以及与潜在维修差错相关的安全分析标准。

The safety analysis standard with respect to installation, interference between systems, and potential maintenance errors that can affect system safety.

**取消；Cancel**

航空器取消预定的飞行计划。

Aircraft cancels a scheduled flight plan.

**全动飞行模拟器；Full Flight Simulator；FFS**

特定类别、制造厂商、型号和系列的航空器驾驶舱的复制件，包括航空器在地面和飞行条件下各种操作必需的设备和计算机程序、一套能够提供驾驶舱外视景的目视系统、一套能提供至少等同于三自由度运动系统指示的系统，以及 FAR - 60 所规定的该装置内安装的具有满量程能力的各类系统，并达到相应于具体 FFS 鉴定等级的鉴定性能标准(QPS)。

A duplicate of a aircraft cockpit with the specific category, manufacturer, model and series, including the required equipments and computer programs for aircraft operation on the ground and flight conditions, a set of the visual system that can can provide the view outside the cockpit, a set of system that can provide the indication of at least equivalent to three degrees of freedom motion system, and all kinds of systems with the full scale ability installed in this device stated in FAR - 60, and achieve the identification performance standards corresponding to the specific FFS qualification level.

### 全货运承运人;All-cargo Carrier

一种持有民航局所发的临时"公众便利需求证"的航空承运人。获准在规定航线上从事定期航空货运、快递和邮件运输,以及进行包括客运在内的非定期运营。

One of a class of air carriers holding temporary Certificates of Public Convenience and Necessity, issued by the Civil Aeronautics Board, authorizing the performance of scheduled air freight, express, and mail transportation over specified routes, as well as the conduct of non-scheduled operations, which may include passengers.

### 全球导航卫星系统;Global Navigation Satellite Systems; GNSS

一个星基导航专用术语,包括 GPS、GPS/SBAS、GPS/GBAS、GLONASS、伽利略(Galileo)和任何其他卫星导航或在航空无线电导航服务(ARNS)频带中适合航空用的增强系统。

GNSS is a generic term for satellite-based navigation, including GPS, GPS/SBAS, GPS/GBAS, GLONASS, Galileo and any other satellite navigation or augmentation system suitable for aviation use within the aeronautical radio navigation service (ARNS) frequency band.

### 全球定位系统;Global Positioning System; GPS

利用卫星传输信号确定飞机位置的导航系统。

A navigation system that employs satellite transmitted signals to determine the aircraft's location.

### (发动机)全权数字电子控制;Full Authority Digital Electronic Control; FADEC

由数字计算机组成的系统,称为电子发动机控制器(EEC)或发动机控制单元(ECU),及其相关配件,控制航空发动机性能。

FADEC is a system consisting of a digital computer, called electronic engine controller (EEC) or engine control unit (ECU), and its related accessories that control all aspects of aircraft engine performance.

**全时间显示;Full-time Display**

专用的连续信息显示。

A dedicated continuous information display.

**全向信标航路;Victor Airway**

VOR 航路的语音代号。

Phonetic designation of VOR airways.

**全新民用航空器;Brand New Civil Aircraft**

符合下列条件的民用航空器：① 航空器所有权被制造厂商或专门的交易商或租机公司所拥有且使用未超过 100 小时或 1 年（以两者中先实现者为准）；② 航空器所有权未曾被私人拥有、出租或安排短暂使用；③ 航空器未曾被任何培训学校专门用作培训驾驶员或参加空中出租业务。

The following civil aircraft is considered a brand new civil aircraft: ① the aircraft is owned by manufacturer, specialized dealer or lease companies, and has been used less than 100 hours or 1 year (whichever is come first); ② the aircraft has not privately owned, leased or temporary used; ③ the aircraft is not used by any training school to train driver or to lease.

**全新型号航空器;New Type Aircraft**

首次申请中国民用航空器局型号合格证（或认可证）的航空器型号。对进口航空器，即使在申请中国民用航空局型号认可证之前已研制并已获得所在国/地区民航当局的型号合格证，仍属于全新型号航空器。

The type of aircraft for which the type certificate (or validation of TC) of the Civil Aviation Administration of China is first applied for. For imported aircraft, even if they have been developed before applying for the type certification of the Civil Aviation Administration of China and have obtained the type certificate of the civil aviation authority of the country/region where they are located, they are still new type aircrafts.

**全重;Gross Weight; GW**

整架飞机的重量，包括飞机、燃油、滑油、货物和乘客等。

Total weight of the entire airplane, including fuel, oil, cargo, passenger, etc.

## 权益转让协议;licensing Agreement

设计批准持有人与生产批准持有人或者申请人之间签署的、以确定双方为生产民用航空产品或者零部件使用所需的设计资料的权利及责任的合同或者安排。

A contract or arrangement between a design approval holder and a production approval holder (or applicant) formalizing the rights and duties of both parties to use the design data for the purpose of manufacturing the product or article.

## 确认;Validation

确定需求被正确完成。

Determining that requirements are both correct and complete.

## 确认与验证;Validation and Verification; V&V

确认是确定对产品的要求是否正确和完整;验证是对要求的实施过程进行评估,以确定是否满足这些要求。

Validation is to determine whether the requirements for a product are correct or complete; verification is the evaluation of an implementation of requirements to determine that they have been met.

# R

**燃油系统限制；Fuel System Limitation；FSL**

在持续适航文件中的适航限制部分确定的任何检查，以确保作为关键设计布局控制限制项目的完整性。

Any inspection that is identified in the limitations section of the instructions for continued airworthiness as required to assure integrity of items identified as critical design configuration control limitations.

**染色渗透检查；Dye Penetrant Inspection**

检查表面裂纹的一种方法，该方法使渗透性染料进入已有裂纹，并用吸收性显影剂从裂纹里吸收。

An inspection method for surface cracks in which a penetrating dye is allowed to enter any cracks present and is pulled out of the crack by an absorbent developer.

**热电偶；Thermocouples**

应采用裸露接点的、1/16～1/8 英寸、金属铠装的、陶瓷封装的、铬镍-镍铝合金，具有名义尺寸 22 到 30 AWG(美国线规)或等效导线的热电偶。不应该采用抽气的、带防护罩的热电偶。

The thermocouples to be used should be bare junction，1/16 to 1/8 - in，metal sheathed，ceramic packed，chromel-alumel，thermocouples with nominal 22 to 30 AWG (American Wire Gage) size conductors or equivalent. An air aspirated，shielded，thermocouple should not be used.

## 人为因素;Human Factors

一个多学科领域,致力于优化人为性能并减少人为差错,包含行为和社会科学、工程学和生理学的方法和原则。

Human factors is a multidisciplinary field devoted to optimizing human performance and reducing human error, incorporating the methods and principles of the behavioral and social sciences, engineering, and physiology.

## 认可审查局方审定基础;Validation Authority Certification Basis

包含适当的由审定当局确认的适航标准、豁免、专用条件,以及审定当局发布的等效安全性,以接受进口产品的设计或设计更改。

The applicable airworthiness standards identified by the VA plus any exemptions, special conditions, and equivalent level of safety findings declared by VA to establish design acceptance of an imported product or to certify the design change.

## 认可审查小组;Validation Team

审定当局召集技术专家,实行认可审查任务的小组。

Team of technical specialists assembled by VA to conduct the validation program.

## 认可审查遗留项目;Validation Action Item

认可审查的某个阶段结束时,尚待确定的认可审定要求和/或符合性方法,以及尚待完成的符合性证明活动等。

An item for which the validation requirements and/or means of compliance is not designated, and the substantiation is not closed at the end of a validation phase.

## 认可审查资料;Data of Validation

与申请、颁发型号认可证、补充型号认可证以及进口材料、零部件、机载设备的设计批准认可有关的资料,包括申请人提供的资料和局方完成的认可审查文件。

Data related with application and issue of VTC, VSTC and design approval validation for imported civil aviation materials, parts and appliances including data from applicants and the authority for validation.

### 认可审定项目；Validation Project Item

某认可审定的独特审查项目，如独特的设计、方法、符合性方法。

Certification item unique to a particular validation project, for example, unique design, usage, or methods of compliance (MOC).

### 认可适航证书；Rendering a Certificate of Airworthiness Valid

一个缔约国为承认任何其他缔约国颁发的适航证替代等效于本国颁发的适航证所采取的行动。

The action taken by a contracting state, as an alternative to issuing its own Certificate of Airworthiness, in accepting a Certificate of Airworthiness issued by any other contracting state as the equivalent of its own Certificate of Airworthiness.

### 认可项目；Validation Item; VI

审定当局特别关注的审定项目或适航标准，如标准显著差异、特定审定项目、一般审定项目。

Certification item or airworthiness standard of particular interest to the VA, such as an SSD, a project validation item, and generic validation item.

### 日历日；Calendar Day

按照世界协调时或者当地时间划分的一个时间段，从当日 0 点到次日 0 点之间的 24 小时。

The period of elapsed time, using coordinated universal time or local time, that begins at 0 o'clock at midnight and ends 24 hours later at the next midnight.

### 日历月；Calendar Month

按照世界协调时或者当地时间划分，从本月 1 日 0 点到下个月 1 日 0 点之间的时间段。

The time period between 0 o'clock of the first day in this month and 0 o'clock of the first day in the next month in accordance with GMT or local time.

## 冗余;Redundancy

存在多种独立的方式实现一项既定的功能或飞行运行。每种方式没有必要是相同的。

The presence of more than one independent means for accomplishing a given function or flight operation. Each means need not necessarily to be identical.

## 冗余性分析;Redundancy Analysis

如果被选部件或系统的用途或功能能够被其他设备项目所代替,那么该部件或系统可以被认为是冗余项目,条件是能够证实该设备项目的替代设备正常工作。如果航空器型号审定基础要求具有两项(或多于两项)功能或信息来源,那么冗余就不能被视为将该设备项目归于 MMEL 的充分理由。对这种情况,可以用另一种验证方法,例如安全分析。

The component or system may be considered to be redundant project if component or system can be replaced by other equipment on condition of the confirmation of the alternative equipment working properly. If the basic certification of the aircraft requires two (or more) function or information sources, then redundancy can not be treated as the sufficient reason for equipment attributed to MMEL. In this case, other authentication method can be used, such as safety analysis.

## 软件采样;Software Sampling

选择一组有代表性的软件生命周期数据进行检查或分析。其目的是确定到项目的某一时间点,开发的所有软件生命周期数据的符合性。抽样是评估软件过程和数据符合性的主要手段。

Selecting a representative set of software life cycle data for inspection or analysis. The purpose is to determine the compliance of all software life cycle data developed up to that point in time in the project. Sampling is the primary means of assessing the compliance of the software processes and data.

### 软件工具鉴定；Software Tool Qualification

在特定机载系统范围内获得软件工具合格审定信用的必要过程。

The process necessary to obtain certification credit for a software tool within the context of a specific airborne system.

### 软件构型索引；Software Configuration Index (SCI)

用于明确软件产品的构型。它可以包含一个构型项或一组构型项。

Software configuration index (SCI) identifies the configuration of the software product. It can contain one configuration item or a set of configuration items.

### 软件后续审定项目；Software Subsequent Certification Project

后续的项目，首次认证项目的软件生命周期数据被重复使用。

The follow-on project in which software life cycle data from the original certification project is reused.

### 软件计划和标准；Software Plans and Standards

指导软件开发过程和整体过程的一组数据。

A set of data that directs the software development processes and integral processes.

### 软件库；Software Library

软件及相关数据和文档的受控存储库，旨在帮助软件开发、使用或修改。

Software library is a controlled repository of software and related data and documents designed to aid in software development, use, or modification.

### 软件评审；Software Review

检查或检验软件生命周期数据、软件项目进度和记录以及其他证据是否符合 RTCA/DO‑178B/C 的行为。评审是一个包罗万象的术语，可能包括审阅文件、面谈项目人员、目击活动、采样数据和参加情况介绍会。评审可以在自己的办公桌、申请人的工厂或申请人供应商的工厂进行。

Software review is the act of inspecting or examining software life cycle data, software project progress and records, and other evidence to assess compliance with RTCA/DO－178B/C objectives. Review is an encompassing term and may consist of a combination of reading documents, interviewing project personnel, witnessing activities, sampling data, and participating in briefings. A review may be conducted at your own desk, at an applicant's facility, or at an applicant's supplier's facility.

### 软件生命周期数据;Software Life Cycle Data

在软件生命周期过程中产生的数据,用于计划、指导、解释、定义、记录或提供活动圆满完成的证据。

Data produced during the software life cycle to plan, direct, explain, define, record, or provide evidence of successful completion of activities.

### 软件原始审定项目;Software Original Certification Project

第一次使用的已完成的审定项目中的软件生命周期数据。

The first use of the software life cycle data in a completed certification project.

### 软控制;Soft Control

显示元素,用以操作、选择或取消选择信息(如菜单和软键)。

Display elements used to manipulate, select, or de-select information (for example, menus and soft keys).

# S

### 三级腐蚀;Level 3 Corrosion

通过第一次及其后续检查发现的腐蚀现象,运营人判断为影响适航并且需要及时、迅速地处理。

Damage found during the first or subsequent inspection(s) which is determined by the operator to be a potential airworthiness concern requiring expeditious action.

### 三角翼;Delta Wing

平面形状很像希腊字母"Δ"或等腰三角形的机翼,其底边构成机翼的后缘。

A wing which is shaped in planform substantially like the Greek letter delta, or is like an isoceles triangle, with the base forming the trailing edge.

### 闪点;Flash Point

易燃液体的最低温度,在该温度下使火焰靠近已加热的样品,可使蒸气瞬间被点燃或"闪光"。

The flash point of a flammable fluid defined as the lowest temperature at which the application of a flame to a heated sample causes the vapor to ignite momentarily, or "flash."

### (燃油系统)闪电防护;Lightning Protection

燃油系统的设计和布置,必须能防止在下列情况下点燃该系统内的燃油蒸气:① 在雷击附着概率高的区域发生直接雷击;② 在极可能受扫掠雷击影响的区域发生扫掠雷击;③ 在燃油通气口处产生电晕放电和流光。

The design and arrangement for the fuel system must be able to prevent the igniting of fuel within the system in the following cases: ① a direct lightning strike occurs in a region with a high probability of lightning attached; ② may be affected in a very sweeping swept lightning region occurs sweep lightning; ③ at the vent in the fuel to produce corona discharge and streamer.

## 闪电接触点;Lightening Attachment Point

闪电和飞机接触的点。

The point where the lightning flash contacts the aircraft.

## 商业航空;Air Commerce

洲际、海外或与外国间的商业航空,或用航空器运输邮件,或航空器在任何国内的任何运营或航行,或直接影响或可能危及洲际、海外或与外国间商业航空安全的航空器的任何运营或航行。

Interstate, overseas, or foreign air commerce or the transportation of mail by aircraft or any operation or navigation of aircraft within the limits of any domestic airway or any operation or navigation of aircraft which directly affects, or which may endanger safety in, interstate, overseas, or foreign air commerce.

## 商业航空安全小组;Commercial Aviation Safety Team; CAST

商业航空安全小组成立于 1998 年,建立了一套集成的、以数据驱动的策略,以减少美国商业航空死亡风险,并在全世界推动新的政府和行业安全举措。

CAST was founded in 1998, it has developed an integrated, data-driven strategy to reduce the commercial aviation fatality risk in the United States, and promotes new government and industry safety initiatives throughout the world.

## 商业航空运输;Commercial Air Transportation

以取酬或出租为目的,运送乘客、货物或邮件的航空运输。

The transportation by air of passengers, cargo or mail for remuneration or hire.

**商业和政府机构代码(美国);Commercial and Government Entity Code; CAGE**

标志与联邦政府、北约成员国以及其他国外政府有生意往来的承包商的五位代码。

A five-position code that identifies contractors doing business with the Federal government, NATO member nations, and other foreign governments.

**商业货架产品;Commercial Off the Shelf; COTS**

通过正常商业渠道可获得的产品、零部件或软件,而不是实现相同功能的客户定制产品。

Products, components, or software that are readily available through normal commercial channels, as opposed to custom-built units that would achieve the same functionality.

**商业运行飞行时间;Commercial Flight Time**

有经济收入的空中飞行时间。

The flight time in air for a flight which has economic income.

**商业运营起落次数;Commercial Flight Circle**

有经济收入的正常起落次数。

The flight cycles for a flight which has economic income.

**商业运营人;Commercial Operator**

为取酬或出租而用航空器在商业航空中从事人员或财物运载的法人,航空承运人或外国航空承运人或依据美国联邦法规 14 卷 375 部授权的法人除外。对某项运行是否为了"取酬或出租"有疑问时,应根据该项航空运输业务只是该法人其他业务中的附属部分,或者该项业务本身就是其主要谋利业务来判断。

A person who, for compensation or hire, engages in the carriage by aircraft in air commerce of persons or property, other than as an air carrier or foreign air carrier or under the authority of Part 375 of 14 CFR. Where it is doubtful that an operation is for "compensation or hire", the test applied is whether the carriage by air is merely incidental to the person's other business or is, in itself, a major enterprise for profit.

## 商用件;Commercial Part

非专门为飞机上的应用而设计或生产的零件。如果局方认为某件商品符合以下条件,则设计批准持有人可以将其指定为"商用件":① 不是为飞机上的应用而专门设计或生产的;② 仅根据商用件制造商的规范进行生产,并且仅以商用件制造商的标记进行标记。

An article that is not specifically designed or produced for applications on aircraft. A design approval holder may designate an article as a commercial part if the Authority finds the part:① is not specifically designed or produced for applications on aircraft;② is produced only under the commercial part manufacturer's specification and marked only with the commercial part manufacturer's markings.

## 商用衍生飞机;Commercial Derivative Aircraft; CDA

被转为政府机构运行用途的通过商用型号合格审定的飞机,相关的改装和设备通过合格审定程序以满足民航适航标准。

A commercial type-certificated aircraft converted for operational use by government agencies, with associated mission modifications or equipment approved to civil airworthiness standards through the type certification process.

## 上诉委员会;Appeal Panel

包括至少三个人,其资历相当于顾问水平,或者比那些不涉及行使最初否决或最终决策的人高一层次的组织。

The appeal panel consists of at least three persons equivalent to the advisor level or above who were not involved in the original denial or termination decision.

## 设备类型清单;Kind of Equipment List; KOEL

所有已安装的影响飞机操作限制的设备列表,该列表中的设备是根据其操作功能来确定的。

A list of all installed equipment affecting the airplane operation limitations. The list also identifies this equipment according to its operational function.

**设备清单;Equipment List**

由制造人或运营人安装于特定飞机上的设备的清单。

An inventory of equipment installed by the manufacturer or operator on a particular aircraft.

**设备旋转方向;Rotational Direction of Equipment**

面对设备旋转面方向看到的设备的旋转方向。

The direction of rotation as observed when looking at the drive face of the equipment.

**设计;Design**

说明部件构型及所有必要的定义部件特征的尺寸、公差、材料、工艺和程序的图纸和规范。

All drawings and specifications that show the part's configuration and all information on dimensions, tolerances, materials, processes, and procedures necessary to define all part characteristics.

**设计保证;Design Assurance**

型号合格证或型号设计批准书申请人为了充分表明其具有以下能力,所应具备的所有有计划的系统性措施:① 设计的产品应符合适用的适航规章和环境保护要求;② 应表明并证实对适航规章和环境保护要求的符合性;③ 向型号合格审定委员会(TCB)和型号合格审定审查组表明符合性。

The TC OR TDA applicant should have necessary planned systematic measures in order to show that it has the following capabilities: ① designed products should comply with applied airworthiness regulations and environment protection requirement; ② should indicate and prove the compliance to airworthiness regulation and environment protection requirement; ③ show compliance to TCB and Type Certificate Review Team.

**设计保证系统;Design Assurance System; DAS**

按照 CCAR - 21 要求,设计批准申请人或持证人建立的、使得其具备所要求的设计职能、适航职能和独立监督职能的体系。

In accordance with the requirements of CCAR – 21, the system established by the design approval applicant or holder enables it to have the required design functions, airworthiness functions and independent supervision functions.

### 设计大改;Major Change

显著影响重量、平衡、结构强度、可靠性、操纵特性、适航特性、动力及噪声特性或排污等的飞机、发动机、螺旋桨的更改。

The change to an aircraft, engine, or propeller which appreciably affects weight, balance, strength, reliability, operational characteristics, airworthiness characteristics, power, noise characteristics, or emissions.

### 设计符合性;Design Compliance

民用航空产品和零部件的设计符合规定的适航规章和要求。

The design of aircraft and part conforms with the CCAR and its requests.

### 设计国;State of Design

对型号设计机构具有管辖权的国家。

The state having jurisdiction over the organization responsible for the type design.

### (硬件)设计过程;(Hardware) Design Process

使用如下过程,从一组要求中研发出的硬件项目:捕获需求、概念设计、详细设计、应用和生产过渡。

Creating a hardware item from a set of requirements using the following processes: requirements capture, conceptual design, detailed design, implementation and production transition.

### 设计滑行质量;Design Taxiing Mass

航空器的最大质量,在这个质量下,对航空器在地面开始起飞前使用时可能发生的载荷做了结构规定。

The maximum mass of the aircraft at which structural provision is made for load liable to occur during use of the aircraft on the ground prior to the start of take-off.

### 设计批准；Design Approval

局方颁发的用以表明该航空产品或者零部件设计符合相关适航规章和要求的证件，其形式可以是型号合格证、型号认可证、型号合格证更改、型号认可证更改、补充型号合格证、改装设计批准书、补充型号认可证、零部件设计批准认可证，或者零部件制造人批准书、技术标准规定项目批准书对设计部分的批准，或者其他方式对设计的批准。

The certificates issued by the Authority to show that the design of the aviation product or parts meets the relevant airworthiness regulations and requirements may be in the form of TC, VTC, amended TC or VTC, STC, MDA, VSTC, VDA, PMA, CTSOA or other approved design.

### 设计批准持有人；Design Approval Holder；DAH

设计批准的持有人，包括型号合格证、型号合格证更改、补充型号合格证、补充型号合格证修订、PMA、TSO 授权、TSO 设计批准信函和现场批准。

Holder of any design approval, including TC, amended TC, STC, amended STC, PMA, TSO authorization, letter of TSO design approval, and field approvals.

### 设计批准函；Letter of Design Approval；LODA

局方颁发给符合一个特殊技术标准规定的国外制造商的产品的设计批准信函。设计批准函是一个生产许可，并不是安装许可。地区适航审定办公室负责向申请人所在民航当局颁发设计批准函。

A design approval that Authority issues only to a foreign manufacturer of an article that meets a specific TSO. A LODA is a production approval and is not an installation approval. The geographic ACO is responsible for issuing the LODA to the applicant's civil aviation authority (CAA).

### 设计起飞重量；Design Take-off Weight

出于结构设计的目的设计的航空器最大重量，假定飞机按这个重量开始起飞滑跑。

The maximum weight at which the aircraft, for structural design purposes, is assumed to be planned to be at the start of the take-off run.

## 设计使用寿命;Design Service Goal; DSG

航空器设计中飞行周期、飞行小时和日历时间的目标值。在航空器制造中称为设计服务目标(DSO)和设计寿命目标(DLG)

The flight cycle, flight hours, and calendar time goals used in the design of the airplane. Other common terms used in the aircraft industry are design service objective (DSO) and design life goal (DLG).

## 设计水平速度;Design Level Speed

选定用于确定有关结构载荷情况的指示空速。这个数值将列入航空器合格证内,作为水平和爬升飞行的最大运行限制。

The indicated airspeed chosen for use in determining the pertinent structural loading conditions. This value will be incorporated in the aircraft certificate as a maximum operational limitation in level and climbing flight.

## 设计小改;Minor Change

民用航空产品的重量、平衡、结构强度、可靠性、使用特性以及对民用航空产品适航性没有显著影响的更改。

Change to an aircraft which does not appreciably affect weight, balance, strength, reliability, operational characteristics, airworthiness characteristics.

## 设计眼位;Design Eye Position; DEP

每个驾驶员站位的某个位置,坐在此位置的驾驶员获得要求的外部可见度和仪表扫视的组合。

The position at each pilot's station from which a seated pilot achieves the required combination of outside visibility and instrument scan.

## 设计值;Design Values

为保证整个结构的完整性具有高置信度,由试验数据确定及被选用的材料、结构元件和结构细节的性能。这些值通常基于为考虑实际结构状态而修正过的许用值,并用于分析、计算安全裕度。

Material, structural elements, and structural detail properties that have been determined from test data and chosen to assure a high degree of confidence in the integrity of the completed structure. These values are most often based on allowable adjusted to account for actual structural conditions, and used in analysis to compute margins-of-safety.

## 设计着陆重量;Design Landing Weight

在以最大下降速度的着陆情况的结构设计中使用的最大飞机重量。

The maximum airplane weight used in structural design for landing conditions at the maximum velocity of descent.

## 设计最大重量;Design Maximum Weight

表明符合每个适用的结构载荷情况的最大重量。

The maximum weight at which compliance with each applicable structural loading condition is shown.

## 设计最小重量;Design Minimum Weight

表明符合每个结构载荷情况的最小重量。

The minimum weight at which compliance with each structural loading condition is shown.

## 射频;Radio Frequency; RF

无线电传输的频率。

Frequency useful for radio transmission.

## 射频干扰;Radio Frequency Interference; RFI

电磁辐射的频谱范围从亚声速频率至 X 射线。

Frequency spectrum of electromagnetic radiation extending from subsonic frequency to X-rays.

### 申请人；Applicant

个人、商行、合伙企业、集团、公司、协会、联合证券协会或政府实体，包括受托人、接收人、受让人，或他们中的任何相似代表。

Individual, firm, partnership, corporation, company, association, joint stock association, or governmental entity, including a trustee, receiver, assignee, or similar representative of any of them.

### 申请人的飞行试验；Applicant Flight Test

为了符合《民用航空产品和零部件合格审定规定》(CCAR - 21)第 21. 35 条的规定，申请人所进行的飞行试验。申请人实施试验是为了表明提交给审查方进行地面和飞行试验的产品满足最低的质量要求，符合型号设计，对计划的试验是安全的，表明符合规章要求。申请人的飞行试验可简称为申请人试飞或公司试飞。申请人的飞行试验按飞行试验的目的可分为研发试飞和表明符合性试飞两类。

Flight tests carried out by the applicant in order to comply with the provisions of Article 21. 35 of the Civil Aviation Products and Components Certification Regulations (CCAR - 21). The applicant performs the tests to demonstrate that the product submitted to the Authority for ground and flight testing meets the minimum quality requirements, conforms to the type design, is safe for the planned tests, and demonstrates compliance with regulatory requirements. The applicant's flight test may be referred to as the applicant's test flight or the company's test flight. According to the purpose of flight test, the applicant's flight test can be divided into two categories: research and development flight test and conformity test.

### 审查；Audit

定期、正式的评审、查证，以评价某机构是否符合规章、政策、标准和合同要求。审查从机构的管理和运行着手，并扩展至活动、产品和服务。

Scheduled, formal reviews and verifications that evaluate whether an organization has complied with policy, standards, and/or contract requirements. An audit starts with the management and operations of the organization and then moves to the organization's activities and products/services.

**审查员；Auditor**

满足规定的经历条件、通过规定的培训胜任执行审查任务的人员。

A person that meets specified experience condition, through a prescribed training up to perform audits.

**审定当局；Validating Authority；VA**

审定标准或补充型号合格证的局方，也可以被称为进口当局。

Aviation authority responsible for validating the type certificate or supplemental type certificate, may also be called the importing authority.

**审定飞行试验；Certification Flight Test**

用于核查申请人所提交的飞行试验数据。对与申请人共同开展的并行飞行试验，则从该飞行试验中获取符合性验证的数据资料。审定飞行试验用来评估航空器的性能、飞行操纵、操纵品质和设备的工作情况，并确定使用限制、操作程序和提供给驾驶员的信息。

Certification flight tests are used to verify the flight test data submitted by the applicant. For combined flight tests conducted with the applicant, data for compliance verification is obtained from the flight test. Certification flight tests are used to evaluate aircraft performance, flight controls, handling qualities, and equipment performance, and to determine operational limits, operating procedures, and information to be provided to the pilot.

**审定飞行试验风险管理；Flight Test Risk Management Process**

申请人负责所有飞行试验安全管理工作。对审定飞行试验，审查组还应按照民航局适航审定部门规定的飞行试验风险管理政策和本单位的程序要求进行试飞风险管理，确保审定飞行试验风险的可接受性。

The applicant is responsible for all flight test safety management. For the certification flight test, the review team shall also carry out flight test risk management according to the flight test risk management policy specified by the airworthiness certification department of Civil Aviation Administration and the procedure requirements of the organization, so as to ensure the acceptability of the risk of certification flight test.

## 审定计划；Certification Plan，CP

申请人用来表明产品符合相关规章的预期方式。

The intended means by which the applicant demonstrates that the product complies with the relevant regulations.

## 审定维修要求；Certification Maintenance Requirements；CMR

在航空器设计合格审定期间，作为型号合格审定运行限制要求的周期性任务，适用于设备、系统和动力装置安装。

Required periodic tasks established during the design certification of the airplane as an operating limitation of the TC，applying to equipment，systems，and powerplant installations.

## 甚低频；Very Low Frequency；VLF

低于 30 千赫的频带。

The frequency band below 30 kHz.

## 甚高频全向；VHF Omnirange；VOR

电子导航系统，为特定的地面站提供精确的方向信息。

An electronic air navigation system that provides accurate direction information in relation to a certain ground station.

## 渗漏检测；Leak Testing

检查是否渗漏或对渗漏进行定位、定量的方法。

Methods for checking，targeting or quantifying the leak.

## 渗漏检测灵敏度；Sensitivity of Leak Test

在规定条件下，仪器、方法或系统能够检测出的最小渗漏率。

Under specified conditions，the minimal leakage rate detected by equipment or testing method.

## 升力不对称；Dissymmetry of Lift

升力不平衡或非对称分布，对旋翼航空器，指横穿旋翼航空器旋翼桨盘的不平衡升力,这是由前进桨叶和后退桨叶的风速差别造成的。

The unequal or asymmetric distribution of lift. In rotorcraft, the unequal lift across the rotor disc of a rotary-wing aircraft, owing to the difference in wind velocity over the advancing and the retreating blades.

### 升压;Boost Pressure

相对于标准海平面大气压力计量的歧管压力。

The manifold pressure measured relative to standard sea level atmospheric pressure.

### 生产检验委任代表;Designated Manufacturing Inspection Representative; DMIR

由生产许可证书持有人推荐,经适航部门审查后授权其代表适航部门从事《民用航空器适航委任代表和委任单位代表的规定》中第十条工作的专业技术人员。

The professional and technical personnel recommended by the PC holder and authorized by the airworthiness department to work on behalf of the airworthiness department in Article 10 of the Regulations on the Airworthiness Designated Representative and Designated Organization Representative for Civil Aircraft.

### 生产模型(首件模型);Production (First Article) Model

用生产工具、型架、夹具设计的最终机械和电气结构的模型。其主要目的是确认生产工艺并标志供训练用或先行试验用的首件,不是为了修改生产过程。此种模型是实际生产过程的首件。

A model in its final mechanical and electrical form of final production design made by production tools, jigs, fixtures and methods. Its primary purpose is to confirm the production process and identify first articles for training purposes or advanced testing not intended to alter the production run. This model is the first article of the actual production run.

### 生产批准书;Production Approval

局方颁发的用以表明允许按照经批准的设计和经批准的质量系统生产民用

航空产品或者零部件的证书,其形式可以是生产许可证(**PC**)、零部件制造人批准书(PMA)或技术标准规定项目批准书(CTSOA)。

The certificate issued by the Authority to indicate that it is allowed to produce civil aviation products or parts in accordance with the approved design and approved quality system may be in the form of PC，PMA or CTSOA.

### 生产批准书持有人；Production Approval Holder，PAH

生产许可证(PC)、零部件制造人批准书(PMA)、技术标准规定项目批准书(CTSOA)持有人的统称。

PAH is collectively referred to as the holders of PC，PMA and CTSOA.

### 生产许可审查委员会；Production Certification Board；PCB

由局方组织成立的,代表局方负责某一项目生产许可审定工作的最高评审机构。

The highest evaluation function in charge of a project of production certification on behalf of the Authority.

### 生产许可审查委员会专家；PCB Specialist

针对生产许可审查委员会(PCB)的设施审查功能,专家指局方制造检查员/管理员或飞行试验、结构、系统和/或设备工程人员。

As related to the facility audit function of PCB, it refers to Authority manufacturing inspectors/supervisors or flight test，structures，systems，and/or equipment engineering personnel.

### 生产许可审定；Production Certification

局方对已获得民用航空产品型号设计批准并欲重复生产该民用航空产品的制造人所进行的资格性审定,以保证该民用航空产品符合经批准的设计。生产许可审定的最终批准形式是颁发生产许可证。

The qualification certification of the manufacturer who has obtained the approval of the civil aviation product type design and wants to repeat the production of the civil aviation product，so as to ensure that the civil aviation products conform to the approved design. The final approval form of production certification is to issue production certificate.

## 生产许可委任单位代表；Designated Production Certification Organization Representative；DPCOR

根据 AP‐183‐AA‐2018‐11《适航委任单位代表管理程序》的规定，获得民航行政机关的授权，代表民航行政机关执行部分或全部生产审查工作的单位或机构。生产许可委任单位代表是委任单位代表的一种。

The organization or institution authorized by the civil aviation administrative organ to carry out part or all of the production reviews on behalf of the civil aviation administrative organ in accordance with AP‐183‐AA‐2018‐11 Airworthiness Designated Organization Representative Management Procedure. DPCOR is one of the designated organization representatives.

## 生产制造地区监督办公室；Manufacturing Inspection District Offices；MIDO

生产制造监督办公室的次级地区办公室。该办公室负责监督区域内的产品证书、适航证书、许可持有人（生产制造厂家）和设计者。在型号证书的取证过程中，生产制造地区监督办公室要协助航空器审定办公室的工作，负责调查和提交规章的不符合报告。生产制造地区监督办公室负责调查和确认服务困难项目的纠正措施，以及其在质量系统中的执行情况。

Subordinate offices to the MIO in its geographical area. This office oversees production certification, airworthiness certification, approval holders (manufacturing facilities), and designees, in its geographical area. MIDO supports ACO during type certification programs, investigating and submitting enforcement reports on noncompliance with regulations. MIDO investigates and ensures corrective measures, for service difficulties, implemented as identified in the quality system.

## 生产制造卫星监督办公室；Manufacturing Inspection Satellite Office；MISO

一个下属办公机构，地理位置相距遥远，该机构向生产制造地区监督办公室（MIDO）负责，并履行与其同样的职责。

A subordinate, geographically remote, office that reports to a MIDO and is responsible for the same activities as the MIDO.

## 声爆；Sonic Boom

航空器或其他物体以声速或超声速运动时产生的激波所引起的噪声。激波是一种压力扰动，耳朵听来像噪声或轰声。

A noise caused by a shock wave that emanates from an aircraft or other object traveling at or above the speed of sound. A shock wave is a pressure disturbance, and is received by the ear as a noise or clap.

## 声级；Sound Level

以分贝为单位的量，用符合美国国家标准协会声级计标准 51.4 - 1971 的声级计测量。

A quantity, in decibels, measured by a sound-level meter satisfying the requirements of the American National Standards Institute Specifications for Sound-Level Meters, 51.4 - 1971.

## 声疲劳强度；Acoustic Fatigue Strength

下列两者之一必须用有试验依据的分析，或者用具有类似结构设计和声激励环境的飞机的服役历史表明：① 承受声激励的飞行结构的任何部分都不可能产生声疲劳裂纹；② 载荷作用在所有受疲劳裂纹影响的部位，声疲劳裂纹不可能引起灾难性破坏。

One of the following two circumstance must be indicated by analysis with test basis or aircraft history service experience which has similar structure design and acoustic excitation environment: ① any part of flight structure which withstands acoustic excitation should not produce acoustic fatigue flaw; ② acoustic fatigue flaw should not lead to catastrophic damage if load act on fatigue flaw affected part.

## 声学更改；Acoustical Change

可能增加航空器噪声级的型号设计更改。声学更改应当符合航空器噪声标准。

A type design change that may increase the aircraft noise. An acoustic change should comply with aircraft noise standards.

### 声压级;Sound Pressure Level

声压与基准声压之比以 10 为底的对数的 20 倍,以分贝为单位。机上基准声压为 20 微牛顿/平方米(或 0.000 2 微巴)。

20 times the logarithm to the base ten of the ratio of a sound pressure to a reference sound pressure, expressed in decibels. The reference pressure for airborne sound is 20 $\mu$N/m$^2$(or, 0.000 2 mbar).

### 剩余的静压源误差;Residual Static Source Error

经过 SSE 修正后,剩余的未经修正或修正过多的静压源误差。

The amount by which static source error(SSE) remains undercorrected or overcorrected after the application of SSE correction.

### 失速;Stall

一种飞行机动或状态,飞机处于这种状态时,流经机翼上下方的空气停止提供足以保持航空器高度的升力,这是由于飞行速度降低或机翼迎风角度大于提供所需升力的角度。

The flight maneuver or condition in which the air passing over and under the wings that stops providing sufficient lift to hold the altitude of the aircraft, caused by a reduction in speed or a wing angle into the wind greater than the angle which provides proper lift.

### 失速警告速度;Stall Warning Speed; $V_{sw}$

发出自然或人工失速警告的速度。

Speed at which onset of natural or artificial stall warning occurs.

### 失效;Failure

影响部件、零件或元件的工作,使其不再能够按预期工作的事件,包括功能的丧失和故障。

An occurrence that affects the operation of a component, part, or element that it can no longer function as intended, including both loss of function and malfunction.

### 失效结冰；Failure Ice

防冰系统（IPS）或其部件失效后，飞机上积累的冰。

Aircraft ice accretion following failure of the ice protection system (IPS), or its components.

### 失效影响；Failure Effect

对系统或组件在失效状态下运行情况的描述。

A description of the operation of a system or item as the result of a failure.

### 失效状态；Failure Condition

在考虑飞行阶段、相关的不利运行或环境条件，或外部事件的情况下，由一个或多个失效错误引起的，对飞机、乘员具有直接或相继影响的状态。

Considering the adverse conditions of use or environmental conditions，by one or more failure cause or contribute to direct and indirect effects of the plane and crew.

### 湿滑跑道；Wet Runway

跑道表面潮湿、反光，但没有明显的积水。

A runway considered as wet when there is sufficient moisture on the surface causing reflection，but without significant areas of standing water.

### 湿租；Wet Lease

按照租赁协议，承租人租赁航空器时携带一名或者多名机组成员的租赁。

A lease in which the lessee leases aircraft including one or more crewmember of the lessor in accordance with a lease agreement.

### 时寿件；Life Limited Part

按照一定的飞行时间、起落次数（循环）、日历时间、APU 小时、APU 循环或其组合进行控制，到期需送到车间进行检测、翻修或拆下报废的部件。

A part needs inspection，overhaul or replacement according to flight hours，flights(circles)，APU hours，APU circles or a combination of them.

### 时寿件挂签;Time Limited Parts Tag

在时寿件上挂附标签或其他记录。标签或记录应包括时寿件的件号、序号和现行的寿命状况。

Hanging labels or other records attached on the time limited parts. Labels or records should include part number, serial number and current life situation of time limited parts.

### 实施程序方案;Schedule of Implementation Procedures; SIP

程序性文件,类似于有些国家适航当局签署的 IPA。主要是为了加快美国与另一个国家之间航空器和其他航空产品的进出口批准流程。

The procedural document, similar to an IPA that is associated with some BAA. It facilitates the approval process for aircraft and other aeronautical products being imported or exported between the United States and a foreign country.

### 实施机龄;Implementation Age; IA

腐蚀预防与控制大纲应该在受影响的区域被执行的飞机的机龄(从最初制造商交付开始)。对每个编号腐蚀任务,原始大纲都指定了实施机龄。

The airplane age (since initial manufacturer's delivery) at which the CPCP should be implemented in the affected area. The baseline program specifies the implementation age for each numbered corrosion task.

### 实质性更改;Substantive Change

有必要进行实质性全面调查,并由此需按 CCAR - 21 重新申请型号合格证书或型号认可证书的一类设计更改。

A type of design change that is sufficient to conduct a substantive and comprehensive investigation, and therefore requires reapplication for TC or VTC in accordance with CCAR - 21.

### 使用过;Used

不是全新的航空器、发动机或螺旋桨。

An aircraft, engine, or propeller that is not new.

## 使用过的航空发动机；Used Aircraft Engine

有累计运行时间的航空发动机，包括翻修后的发动机。

An aircraft engine that has accumulated time in service，including a rebuilt engine.

## 使用过的航空器；Used Aircraft

对某航空器而言，只要满足下列条件之一即为使用过的航空器：① 航空器的所有权曾经被除制造厂或专门的租机公司之外的第三方所持有；② 曾经被私人拥有、出租或短暂使用过；③ 曾经专门用作培训驾驶员或参与空中出租业务；④ 航空器所有权虽然一直被制造厂或专门的租机公司所持有，但累计使用超过100 飞行小时或 1 日历年（以两者中先实现者为准）。

As to aircraft，only meet one of the following conditions shall be regarded as used aircraft：① ownership of the aircraft had been held by third parties in addition to the manufacturer or specialized aircraft rental companies；② has been privately owned，leased，or arranged for short-term use；③ has specifically been used for training pilots involved in aerial rental business；④ ownership has been the manufacturer or specifically for aircraft rental company holds，but the cumulative use more than 100 flying hours/calendar year (whichever comes first).

## 使用仪表进近程序实施的进近和着陆运行；Approach and Landing Operations Using Instrument Approach Procedures

仪表进近和着陆运行分类如下：非精密进近和着陆运行；使用侧向引导（不使用垂直引导）的仪器进近和着陆运行。

Instrument approach and landing operations are classified as follows：non-precision approach and landing operations；an instrument approach and landing which utilizes lateral guidance but does not utilize vertical guidance.

## 事故；Accident

在民用航空器运行阶段或者在机场活动区内发生的与航空器有关的下列事件：① 人员死亡或者重伤；② 航空器严重损坏；③ 航空器失踪或者处于无法接近的地方。

The following incidents relating to civil aircraft that occur during the operation of civil aircraft or in the activity area of airport: ① death or serious personal injury; ② serious damage to the aircraft; ③ the aircraft is missing or in an inaccessible place.

### 事故调查和预防;Accident Investigation and Prevention; AIP

涉及飞机事故调查和所有相关美国国家运输安全委员会(NTSB)的活动。

Respect to aircraft accident investigation and all activities related to the National Transportation Safety Board (NTSB).

### 事故/事件数据系统;Accident/Incident Data System; AIDS

收集在美国注册飞机的事故和事件信息。系统用户可以在系统数据库中查询相关类型的飞机和其他特定因素,如维护操作。

Collecting information on aircraft accidents and incidents involving U. S. - registered aircraft. Users of the system can query the database for data concerning certain kinds of aircraft and specific causal factors, such as maintenance actions.

### 事故征候;Incident

按照《民用航空器飞行事故征候》确定的事故征候的事件或满足下列标准的事件。① 飞行控制系统故障或失效;② 因受伤或生病导致飞行机组人员不能够执行其正常的飞行职责;③ 涡轮发动机的结构部件失效,不包括压气机、涡轮叶片和导流片;④ 飞行中着火;⑤ 飞行冲突;⑥ 修理(包括材料和劳动力),在事件中财产或等值市场价值损失超过 200 000 元人民币(不包括本架飞机的自身损失);⑦ 对大型多发飞机(最大合格审定起飞重量超过 5 700 千克):(a) 飞行中的电子系统失效,该系统要求使用应急汇流条供电(比如电池、辅助电源或冲压发电机),以保持飞行操纵或基本仪表供电;(b) 飞行中的液压系统失效,这种故障导致持续依赖唯一保留的液压或机械系统,使飞行操纵面运动;(c) 两台或多台发动机产生的动力和推力持续损失;(d) 使用应急撤离系统进行的飞机撤离。

The event of incidents determined per to Civil Aircraft incidents or the events that meet the following criteria. ① Flight control system malfunction or failure; ② inability of any required flight crewmember to perform his normal flight duties as a result of injury or illness; ③ failure of structural components of a turbine engine excluding compressor and turbine blades and vanes; ④ in-flight fire; ⑤ aircraft collides in flight; ⑥ damage to property, other than the aircraft, estimated to exceed 200 000 RMB for repair (including material and labor) or fair market value in the event of total loss; ⑦ for large multiengine aircraft (more than 5 700 kg maximum certificated takeoff weight): (a) in-flight failure of electrical system which requires the sustained use of an emergency bus powered by a back-up source such as a battery, auxiliary power unit, or air-driven generator to retain flight control or essential instruments; (b) in-flight failure of hydraulic systems that results in sustained reliance on the sole remaining hydraulic or mechanical system for movement of flight control surfaces; (c) sustained loss of the power or thrust produced by two or more engines; (d) an evacuation of an aircraft in which an emergency egress system is utilized.

## 事件(飞行标准);Occurrence (Flight Standards)

除了要求飞行标准处进行调查的,对安全有潜在影响的事故或事故征候以外的事件。当不涉及损伤、损坏或 49 CFR 830.5 的报告要求时,包括: ① 中止起飞不包括跑道偏移;② 飞机返航,回到出发地机场并且没有事故征候,安全着陆;③ 飞机改航,出于天气条件之外的原因转移到不同的目的地。

An event, other than an accident or incident that requires investigation by the Flight Standards Service for its potential impact on safety. Includes the following when no injury, damage, or 49 CFR 830.5 reporting requirements are involved: ① aborted takeoffs not involving a runway excursion; ② air turnbacks where the aircraft returns to the departure airport and lands without incident; ③ air diversions where the aircraft diverts to a different destination for reasons other than weather conditions.

## 事件调查安全建议;Incident Investigation Safety Recommendation

调查机构根据调查材料提出预防事故或事件的建议。

The advise for preventing the incidents or accidents which is raised by any investigation organizations base on the investigating materials.

### 视觉包线;Viewing Envelope

显示器所能达到的最小视觉空间(例如,亮度、对比度、色度)。

Total volume of space where the minimum optical performance of the display is met (for example, luminance, contrast, or chromaticity).

### 视觉提示;Visual Cue

在参考面上积冰,已经证实测试在监视面上积冰。

Ice accretion on a reference surface that has been demonstrated by testing to correlate with ice accretion on a monitored surface.

### 试车台;Test Cell

在进行大修或其他维修作业时,为固定的仪器设备提供可控环境,来评估涡轮发动机性能的设备。

An enclosed facility that provides a controlled environment of fixed instrumentation designed to evaluate turbine engine performance following overhaul or other maintenance.

### 试飞;Flight Test

为测量航空器或航空器部件的运行/飞行特性而进行的飞行。

Flight for the purpose of investigating the operation/flight characteristics of an aircraft or aircraft component.

### 试飞安全评审委员会;Safety Review Board; SRB

当试飞小组确认试验已经准备就绪时,审查飞行试验大纲的委员会。

A board which provides an opportunity for review of flight test programs after test teams have determined that they ready for the tests.

### 试飞方案;Test Flight Program

维修单位和运营人根据航空器试飞任务的要求,编制的用于测试航空器飞行性能的有关技术文件。

Technical documents used by maintenance units and operators to test aircraft flight performance according to the requirements of aircraft test missions.

### 试验(可行性)模型;Breadboard (Feasibility) Model

以概略方式或试验方式证实一种装置、线路、设备、系统或原理的可行性使用的模型。它没有必要是完整的、最终的形式,不需要包括生产型部件,甚至不需要应用于实验室之外。它是验证可行性所需的最低限度模型。

A model used to prove the feasibility of a device, circuit, equipment, system or principle in rough or breadboard form. It is not necessarily complete, in final form, composed of production type parts or even usable outside of the laboratory. It is a minimum model required to demonstrate feasibility.

### 试验产品;Test Product

型号合格审定中用于各种验证试验的试验件、原型机及其零部件。

Parts, prototype and subparts used for verification test in type certification.

### 试验件;Test Article

专门用于审定试验的部件、航空器、发动机或螺旋桨。

Components, aircraft, engines, or propellers used specifically for certification testing.

### 试验类适航证书;Experimental Airworthiness Certificate

在 STC 颁发之前,针对改装飞机运行颁发的证件,目的是研究、开发试验和审定飞行验证。

A certificate operating the aircraft with the modification installed before STC approval is granted in order to perform research and development testing and certification flight testing.

### 试验模型;Experimental Model

用以验证基本概念的技术可靠性的模型。它是完整的并可用于实验室之外,尽管不一定是最终的形式,也不一定要由生产型零部件组成。

A model used to demonstrate the technical soundness of the basic idea. It is complete and usable outside the laboratory though not necessarily in final form or composed of production type parts.

### 适度可能；Reasonably Probable

在每类航空器的每次运行中不可能经常发生，但在每类航空器的运行全寿命周期内可能发生很多次。

Reasonably probable means unlikely to occur often during the operation of each aircraft of the type but which may occur several times during the total operational life of each aircraft of the types in which the engine may be installed.

### 适航；Airworthiness

按适用适航要求进行设计、制造、使用和维护的航空器，在规定的限制使用条件下所具有的固有安全特性。

The inherent safety characteristics of aircraft designed，manufactured，operated and maintained in accordance with applicable airworthiness requirements under specified limited service conditions.

### 适航标准；Airworthiness Standard；AS

满足民用航空器适航性的最低安全标准。

Airworthiness standard is the basic safe standard to confirm the aircraft airworthiness.

### 适航当局；Airworthiness Authority

民航局航空器适航司、航空器适航中心、各地区管理局适航处及各航空器审定中心。

The Aircraft Airworthiness Department of CAAC，the Aircraft Airworthiness Center of CAAC，the Airworthiness Divisions of Regional Administrations of CAAC，and the Aircraft Certification Centers.

### 适航放行；Airworthiness Release

CCAR - 121 第 379 条规定的飞机放行，或者 CCAR - 135 第 435 条规定的

航空器放行。

Aircraft release according to CCAR - 121. 379 or CCAR - 135. 435.

### 适航管理;Airworthiness Management

适航当局通过对航空器进行相关管理,确保航空器固有的安全特性处于高于最低标准的状态。

The Authority performs appropriate management to keep the inherent safety features of an aircraft higher than the lowest standards.

### 适航管理程序;Airworthiness Procedure, AP

中国民用航空局颁发的有关民用航空规章的实施办法或具体管理程序,是政府适航部门工作人员从事适航管理活动的依据,也是民用航空器设计、制造、使用和维修的单位或个人从事民用航空活动时应当遵守的行为规则。

The implementation measures or specific management procedures of relevant civil aviation regulations issued by the Civil Aviation Administration of China, which are the basis for the staff of the airworthiness department of the government to engage in airworthiness management activities, and also the rules of conduct that the organizations or individuals engaging in civil aviation activities in the design, manufacture, use and maintenance of civil aircraft should abide by.

### 适航规章;Airworthiness Regulations

某类航空产品的标准规范。

The regulations contain standards for specific types of aviation products.

### 适航规章标准差异表;Airworthiness Regulations and Standards Differences Lists

不同局方汇编的重大标准差异表和非重大标准差异表。标准差异表是由两个局方联合制定、出版和予以公布的。

Lists of significant standards differences and non-significant standards differences assembled for different authorities amendment pair. Standards differences lists are jointly developed, published, and made public by both authorities.

## 适 航 技 术 实 施 程 序；Technical Implementation Procedures （TIP） for Airworthiness

有关局方之间的设计批准、出口适航批准、证后活动、技术援助的程序文件。

The procedural document for design approval，export airworthiness approvals，post-design activities，and technical assistance between Authorities.

## 适航监察员；Airworthiness Inspector

从事适航审定的局方人员，包括局方的制造检查人员、工程人员、试飞工程师和试飞员等。

The person who engages in airworthiness inspection，including authority inspector for manufacturing，engineering，and flight test.

## 适航检查委任单位代表；Airworthiness Inspections Designated Organization Representative；AIDPR

根据 AP‐183‐AA‐2018‐11《适航委任单位代表管理程序》的规定，获得民航行政机关的授权，代表民航行政机关执行航空器适航检查工作的单位或机构。适航检查委任单位代表是委任单位代表的一种。

The organization or institution authorized by the civil aviation administrative organ to carry out aircraft airworthiness inspection on behalf of the civil aviation administrative organ in accordance with the provisions of AP‐183‐AA‐2018‐11 Airworthiness Designated Organization Representative Management Procedure. AIDPR is a kind of appointed organization representative.

## 适航批准；Airworthiness Approval

局方为某一航空器、航空发动机、螺旋桨或者零部件颁发的证件，表明该航空器、航空发动机、螺旋桨或者零部件符合经批准的设计并且处于安全可用状态。

A certificate issued by the Authority for an aircraft，aircraft engine，propeller，or article which certifies that the aircraft，aircraft engine，propeller，or article conforms to its approved design and is in a condition for safe operation.

### 适航审定部门;Airworthiness Certification Department

中国民用航空局航空器适航审定司(适航司)、各地区管理局适航审定处、各航空器适航审定中心(审定中心)等。

The Aircraft Airworthiness Department of CAAC (AAD), Airworthiness Certification Divisions of Regional Administrations of CAAC, Aircraft Airworthiness Certification Centers of Regional Administrations of CAAC (AACC), etc.

### 适航审定处;Airworthiness Certification Division

适航处业务受适航司领导,是地区性职能机构,在适航司授权下,负责所辖地区航空器的持续适航管理。

The regional functional body whose airworthiness operations ordered by the Airworthiness Division. Under the Airworthiness Division's authorization, it is responsible for the continuing airworthiness of the aircraft under management.

### 适航实施程序(美国);Implementation Procedures for Airworthiness; IPA

由双边航空安全协议行政协定授权的程序文件,该程序文件为设计批准、生产、出口适航审批、后期设计审批活动和局方之间的技术援助而制定。这个文件规定了适合进口到美国和 BASA 签约国的民用航空产品和零部件。该文件界定了接口要求,以及为了这些民用航空产品的进口和持续支持,局方之间的活动。

The procedural document authorized by the BASA Executive Agreement for design approval, production activities, export airworthiness approvals, post-design approval activities and technical assistance between authorities. This document defines the civil aeronautical products and parts eligible for import into the United States and the counterpart BASA signatory country. It defines the interface requirements and activities between the Authorities for the import and continued support of those civil aeronautical products.

### 适航限制;Airworthiness Limitation; AL

在型号审定过程中规定的某些结构项目(包括机体、发动机、螺旋桨)的使用限制。

Restrictions on the use of the certain structure items (including the airframe, engines, propellers) in the process of model validation.

### 适航限制部分;Airworthiness Limitation Section; ALS

任何"被批准的"由制造商或型号合格审定办公室确认的适航性限制。

Any "approved" airworthiness limitations identified by the manufacturer or Type Certificate Holding Office.

### 适航限制项目;Airworthiness Limitation Items; ALI

申请人通过型号合格审定过程制订并获审查部门批准的型号设计组成部分文件,通常列入飞机飞行手册或持续适航文件的适航限制部分,内容涉及对相关飞机结构和系统的强制性维修、检查、更换和评定的要求。

A document developed by the applicant through the type certification process and approved by the certification department as a component of the type design, usually included in the airworthiness limitations section of the aircraft flight manual or continuing airworthiness document, covering requirements for mandatory maintenance, inspection, replacement, and evaluation of related aircraft structures and systems.

### 适航证件;Airworthiness Certificate

型号审定、生产审定和适航审定相关证件。

Relevant certificates for type certification, production certification and airworthiness certification.

### 适航指令;Airworthiness Directive; AD

针对任何航空器上发现的、很可能存在于或发生于相同/相似型号设计的其他航空器中的不安全状态,所制定的强制性检查要求、改正措施或使用限制。

Mandatory inspection requirements, corrective actions or restrictions on the use of unsafe conditions found on any aircraft that are likely to exist or occur on other aircrafts of the same/similar type design.

### 适用的设计数据；Applicable Design Data

申请人为了申请设计机构批准、TC、STC、技术标准指令（TSO）授权、零部件制造批准（PMA）等，需要提供的所有图纸、规范和其他技术资料，出于生产目的发布时处于受控状态。

All necessary drawings, specifications, and other technical information provided by the applicant for, or the holder of, a design organization approval, TC, STC, technical standard order (TSO) authorization, parts manufacturer approval (PMA), or equivalent, and released in a controlled manner for production purposes.

### 适用机场；Adequate Airport

达到第 121.197 条规定的着陆限制要求且被局方批准的合格证持有人使用的机场，它可能是下列两种机场之一：① 经审定适合大型飞机公共航空运输承运人所用飞机运行的，或符合其运行所需等效安全要求的机场，但不包括只能为飞机提供救援和消防服务的机场；② 对民用开放的、可用的军用机场。

An airport that an airplane operator may list with approval from the authority because the airport meets the landing limitations of section 121.197 and is either: ① an airport that has been approved to be suitable for the operation of aircraft used by large aircraft public air transport carriers, or meets the equivalent safety requirements for its operation, excluding those apply to aircraft rescue and firefighting service; ② a military airport that is active and operational.

### 首次补充型号合格证；First-of-type STC

为特定构造和型号的飞机颁发的初始补充型号合格证（设计大改）。

An initial supplemental type certificate (major design change) issued for a particular make and model of aircraft.

### 寿命(负载)增强因子；Life (or Load) Enhancement Factor

应用于结构交变载荷测试中的附加载荷系数和/或测试周期，与设计载荷和寿命值有关，用于计算材料的易变性。

An additional load factor and/or test duration applied to structural repeated load tests, relative to the intended design load and life values, used to account for material variability.

### 寿命件;Life-limited Part

根据 CCAR 第 21.50、第 23.1529、第 25.1529、第 27.1529、第 29.1529、第 31.82、第 33.4 以及第 35.4 条款,在持续适航文件的适航性限制章节中规定了更换时间、检查间隔或相关程序的零部件。

Life-limited parts are those parts for which replacement times, inspection intervals or related procedures are specified in the airworthiness limitation section of the continuing airworthiness document in accordance with CCAR – 21.50, 23.1529, 25.1529, 27.1529, 29.1529, 31.82, 33.4, 35.4.

### 授权;Authorizations

由管理部门向个体授予的完全或者部分接入管理信息系统的许可。

Permission granted by management to individuals authorized full or partial admission to restricted access information management systems.

### 授权放行文件;Authorized Release Document

PAH 为飞机发动机、螺旋桨或零部件发放的,证明其符合批准的设计并且处于安全运行状态的文件。

Document issued by the PAH certifying that an aircraft engine, propeller, or component conforms to an approved design and is in a safe operating condition.

### 授权区域;Authorized Area

对于委任工程代表,授权区域适用于飞机的特定部分或系统,或发动机和螺旋桨型号,或委托适用的专业领域。

For DER, an authorized area applies to the specific portion or system of the aircraft, or the type of engine or propeller, or specialized area to which a delegated function applies.

## 授权任命者;Authorized Designee

申请人为完成申请而授予局方执行特定任务,例如目击一次实验或实施一次符合性检查。

Authority designee grants Authority approval to perform a specific task for the applicant, such as witnessing a certain test or conducting a certain conformity inspection.

## 属地管理局;Local Administration

航空运营人各分基地所在地区管理局。针对主基地和同一地区内的分基地,合格证管理局同样视为属地管理局。

A regional administration in charge of the satellite bases of air operator. For the main base and satellite bases within the same region, the certificate administration is also considered as local administration.

## 数字地形立体数据;Digital Terrain Elevation Data; DTED

地球表面间隔均匀的网格点,表明三维坐标:经度、纬度和海拔。

An evenly spaced grid of points on the Earth's surface illustrating three-dimensional coordinates, latitude, longitude and elevation.

## 数字签名;Digital Signature

在电子文档上使用安全数字签名与个人手写签名具有等效的意义,打印时也许会或者不会包含与原手写签名一模一样的拷贝。

A secure digital means conveying the same meaning as an individual's handwritten signature in an electronic document, which when printed may or may not contain an exact copy of the originating handwritten signature.

## 数字式飞行数据记录器;Digital Flight Data Recorder; DFDR

使用数字方法将数据记录和储存在存储介质中,并能将数据从介质中恢复的记录设备。DFDR 可以是包括 DFDAU 和 FDAU 的记录系统中的存储设备,也可以是单个设备,使用内部的数据采集系统以将飞机的模拟和离散信号转换为数字形式。

A recording device using a digital method to record and store data onto a storage medium and to retrieve the data from the medium. A DFDR may be the storage device in a recording system that includes a DFDAF, DFDAU, or an FDAU. Or, it may be a stand-alone device using an internal data collection system to convert aircraft analog and discrete signals to digital form.

### 数字证书；Digital Certificate

以数字形式签署的声明，将用户、计算机或对公众服务的识别信息绑定到私人密钥。

A digitally signed statement that binds the identifying information of a user, computer, or service to a public/private key pair.

### 双边航空安全协议；Bilateral Aviation Safety Agreement；BASA

两国之间签订的以提高航空安全为目的的行政协议。

An executive agreement concluded between two countries for the purpose of promoting aviation safety.

### 双边监督委员会；Bilateral Oversight Board；BOB

负责保证（双边）协议有效运作的联合执行团队。

A joint executive-level group responsible for ensuring the effective functioning of the agreement.

### 双边适航协议；Bilateral Airworthiness Agreement；BAA

由两国政府代表签署或通过交换外交照会生效的关于适航的技术性协定或协议。

The technical implementation conventions or agreement about airworthiness signed by both government representative or become effective by exchange diplomatic note.

### 水平品质因数；Horizontal Figure of Merit；HFOM

圆心位于水平面上的在真正位置上的圆的半径，描述了该区域在适当时、无故障条件下包含的至少具有 95% 概率的指示水平位置。

The radius of a circle in the horizontal plane with its center being at the true position，that describes the region assured to contain the indicated horizontal position with at least 95% probability under fault-free conditions at the time of applicability.

## 水上迫降；Ditching；DITCH

水面上的迫降或预防性降落。

A forced or precautionary landing on water.

## 水下定位装置；Underwater Locator Device；ULD

能在水下发出位置信号的仪器。

The device transmitting the position signal under the water.

## 睡眠铺位；Sleeping Surface

符合 AC 121 - 31 要求的水平铺位，如床、卧铺等。

Any horizontal surface such as a bed or bunk that meets the guidelines referenced in this AC 121 - 31.

## 顺序区旅客容量；Sequential Zone Passenger Capacity

2 个或多个邻近区域的乘客最大数量。

The maximum number of passengers which may be seated in two or more adjacent zones.

## 瞬态抑制装置；Transient Suppression Device；TSD

将连接到系统的燃油量、燃油温度传感器或燃油水平开关等电路的瞬态电压或电流限制到预定水平的装置。

A device that limits transient voltages or currents on wiring to systems such as the fuel tank quantity, fuel temperature sensors, and fuel level switches, to a predetermined level.

## 说明性资料；Indicative Material

完整、准确地定义或描述重要改装的制造、装配、安装、使用和维护的一类设计资料。这些资料通常包括图纸、示意图、带标记的照片、工艺规范、安装说明

（如工程指令和零部件安装手册等）、软件设计及配置文档、飞行手册补充、主最低设备清单补充、重量平衡手册补充、持续适航文件补充、服务通告等。

Design information which can completely and accurately define or describe the important modification of the manufacture, assembly, installation, use and maintenance, including drawings, diagrams, tagged photos, production engineering specification(PES), installation instructions (such as engineering order and parts installation manual), software design and configuration documentation, supplemental flight manual, supplemental master minimum equipment list, supplemental weight and balance manual, supplemental continuing airworthiness documents, service bulletin.

### 私用大型航空器运营人；Private Large Aircraft Operator

经局方认证，根据运营规章获得授权，开展私人飞行的航空器运营人。

Aircraft operators those are certificated by the authority in accordance with the rules and get the authorized operation regulation to implement the private flight.

### 搜索救援；Search and Rescue；SAR

寻找失踪的航空器并对需得到协助的航空器给予支援的一项工作。

A service to seek missing aircrafts and to assist the aircrafts found to be in need of assistance.

### 随机审定管理；Random Certificate Management；RCM

根据需要完成的型号审定任务。

The performance of CM tasks that may be accomplished on an as-needed basis.

### 碎片扩散角；Fragment Spread Angle

从单个转子旋转平面中心的前部和后部测量，始于发动机或辅助动力装置轴向中心线的角度。

The angle measured, fore and aft from the center of the plane of rotation of an individual rotor stage, initiating at the engine or APU shaft centerline.

## 损伤；Damage

制造（加工、生产、装配或处理）或使用引起的结构异常。

A structural anomaly caused by manufacturing (processing, fabrication, assembly or handling) or service usage.

## 损伤（离散源）；Discrete Source Damage

鸟击、风扇叶片非包容性故障、发动机非包容性破坏、高能旋转机械非包含性破坏或类似原因可能对飞机造成的结构性损伤。

Structural damage of the aeroplane that is likely to result from: impact with a bird, uncontained fan blade failure, uncontained engine failure, uncontained high-energy rotating machinery failure or similar causes.

## 损伤标准；Injury Criteria

描述对人体组成部分的损伤限定，更普遍适用于各种冲击损伤的保护系统设计。为了提供防止严重人身伤害或死亡的数据，在测试中使用生物替代物来代替人类受试者；然而，建立生物替代物与人类之间的数据相互关系是困难的。为评估保护系统的性能，可以使用仿人测试设备来替代生物代替物，并且仿人测试设备仅是人体的一个最基本的表现形式。冲击损失标准应用仿人测试设备测量得到的参数来表示。

Injury criteria describes the trauma limits of individual human body components, which is more generally applicable to a variety of impact injury protection system designs. To provide data for protection against serious injury or death, biological surrogates are used instead of human subjects in tests; however, correlation of data between the biological surrogates and living humans is difficult. Moreover, for evaluating the performance of a protection system, an anthropomorphic test device (ATD) may be used to instead a biological surrogate, and the ATD is only a rudimentary representation of the human body. Impact injury criteria should be expressed in parameters which can be measured on an ATD.

## 损伤类别；Category of Damage

根据 AC 20 - 107B，基于剩余强度能力、要求的载荷水平、可检性、检测间隔、损伤威胁，以及产生损伤的事件是否明显可见，定义的 5 类损伤。

According to AC 20 - 107B, five categories of damage have been defined based on residual strength capability, required load level, delectability, inspection interval, damage threat and whether the event creating damage is obvious.

### 损伤容限；Damage Tolerance

结构承受一定水平的疲劳、腐蚀、偶然或离散源损伤后，能在一定时间内保持必要的剩余强度的属性。

The attribute of the structure that permits it to retain its required residual strength for a period of use after the structure has sustained a given level of damage from fatigue, corrosion, accident, or a discrete source.

### 损伤容限评定；Damage Tolerance Evaluation；DTE

确定是否要开展维修活动以检测或排除可能导致灾难性破坏的疲劳裂纹的过程。当损伤容限评定应用于修理和改装时，它包括对修理或改装的评估以及对受修理或改装影响的疲劳关键结构的评定。

A process that leads to a determination of maintenance actions necessary to detect or preclude fatigue cracking that could contribute to a catastrophic failure. As applied to repairs and alterations, DTE includes the evaluation both of the repair or alteration and of the fatigue critical structure affected by the repair or alteration.

### 所需导航性能；Required Navigation Performance；RNP

在一划定空域内实施运行所必需的导航性能的说明。对某一特定所需导航性能类型和/或其应用，规定了导航性能和要求。

Statement of the navigation performance necessary for operation within a defined airspace. Navigation performance and requirements are defined for a particular RNP type and/or application.

### 所需地形高度；Required Terrain Clearance；RTC

按照终端区仪表程序（TERPS）和航空信息手册定义的最低障碍地形高度

要求。

The minimum requirements for obstacle terrain clearance as defined by the terminal instrument procedures (TERPS) and the Aeronautical Information Manual(AIM).

### 所有人;Owner

拥有飞机、气球、飞机发动机或螺旋桨的人。

A person who owns an aircraft, balloon, aircraft engine, or propeller.

# T

**TAWS 干扰警告;TAWS Nuisance Alert**

在正常安全程序期间,因地形识别和警告系统(TAWS)设计性能限制而出现的一个不适当的警告。

An inappropriate alert, occurring during normal safe procedures, that occurs as a result of a design performance limitation of TAWS.

**TAWS 虚假警告;TAWS False Alert**

未超过系统设计的地形警告或警示阈值时发出的警告或警戒信息。

A warning or caution that occurs when the design terrain warning or caution threshold of the system is not exceeded.

**TCCA 批准的资料;Data Approved by TCCA**

通过加拿大运输部、基础设施和社会部部长或部长代表批准的数据。

Data approved by the Canadian Minister of Transport, Infrastructure and Communities (Minister) or a delegate of the Minister.

**TF 类;Class TF**

用在亚声速航空器上的所有涡扇和涡喷发动机。

All turbofan or turbojet aircraft engines employed for propulsion of aircraft designed to operate at subsonic flight speeds.

**TP 类;Class TP**

所有航空涡桨发动机。

All aircraft turboprop engines.

**胎压指示系统；Tire Pressure Indicating System（Installed on an Aircraft)**

安装在飞机上的一种用于检查轮胎压力的系统。

A system installed on an aircraft used for tire pressure checks.


**特别繁忙运输机场；Very Busy Transportation Airport**

由局方指定的交通流量较大的国际机场，包括北京首都机场、上海虹桥机场、上海浦东机场和广州白云机场。

The international airports specified by Authority with heavy traffic flow, including Beijing Capital International Airport, Shanghai Hongqiao International Airport, Shanghai Pudong International Airport and Guangzhou Baiyun International Airport.


**特大维修事故；Jumbo Maintenance Accident**

由于维修造成下列情况之一者为特大维修事故：① 航空器及部件在地面发生损坏，直接经济损失超过事故当时同型或同类可比新航空器（最大起飞质量小于或等于 5.7 吨的航空器除外）整机价格的 3% 或超过 500 万元（含），以低者为准；② 在地面发生事故人员死亡四人（含）以上；③ 重大飞行事故。

One of the following situations caused by maintenance: ① the aircraft and components are damaged on the ground, and the direct economic loss exceeds 3% of the whole aircraft price of the same type or comparable new aircraft (except aircraft with maximum take-off mass less than or equal to 5.7 t) at the time of the accident or exceeds 5 million yuan (inclusive), whichever is lower; ② four or more persons died in the accident on the ground; ③ major aviation accidents.


**特定风险；Particular Risk**

在飞机外部，或在飞机内部但在被分析的系统和项目外部，可能违反失效独立性的事件或影响。

Particular risk is defined as those events or influences which are external to the aircraft or within the aircraft but external to the system(s) and item(s) being analyzed, which may violate failure independence claims.

## 特定用途集成电路；Application Specific Integrated Circuit；ASIC

结合硬件和软件进行综合设计的产物，在未来航空电子系统中需要解决软/硬件的协同设计过程问题。

Incorporate design characteristics and techniques from both hardware and software disciplines. Issues relating to the hardware/software co-design process will need to be addressed in future avionics systems.

## 特高频；Ultra High Frequency；UHF

300～3 000 兆赫的频带。

A frequency band between 300 and 3 000 MHz.

## 特技飞行；Acrobatic Flight

含有正常飞行不需要的航空器姿态突变、非正常姿态或非正常加速的有意机动飞行。

An intentional maneuver involving an abrupt change in the attitude of the aircraft, an abnormal attitude or abnormal acceleration which is not necessary for normal flight.

## 特殊适航证；Special Airworthiness Certificate；SAC

对标准适航证以外的取得 CCAR‑21 第 21.24 条、第 21.25 条或者第 21.26 条的型号合格证或者按照第 21.29 条取得型号认可证的初级类、限用类、轻型运动类航空器，以及局方同意的其他情况，颁发特殊适航证。特殊适航证分为初级类、限用类和轻型运动类三类。

The special airworthiness certificate can be issued for primary category, restricted category or light sport category aircraft entitled in type certificate under CCAR‑21.24, 21.25 or 21.26, or validation of type certificate under 21.29, or for the aircraft agreed by the CAAC-AAD. The special airworthiness certificates are classified into three types: primary category, restricted category and light sport category.

## 特殊适航指令；Special Airworthiness Directive

带有敏感技术内容，涉及航空安保的适航指令。

An AD furnished with sensitive technical content and involved with defense and security of aviation.

### 特殊维修项目；Special Maintenance Project；SMP

经专门论证并有专项基金的项目，适用于设施设备部件、道路、建筑物和公用系统组成部分的改型、修理、改进或更换。

A specifically justified and specially funded project for the modification, repair, improvement or replacement of facility equipment components, roads, buildings, and utility system components.

### 特殊详细检查；Special Detailed Inspection；SDI

为了检测损伤、失效或不正常现象而对一个特定的项目、安装件和组件进行详细检查。

Detailed examination performed on a specific project, installation parts and components, in order to detect damage, failure or abnormal signs.

### 特殊项目；Special Project

除 TC、STC、TC 更改、STC 更改或 PMA 项目外的审定项目。

A certification project that is not a TC, STC, amended TC, amended STC, or PMA project.

### 特许飞行证；Special Flight Permit

民用航空器尚未取得有效适航证或目前可能不符合有关适航要求，但在一定限制条件下能安全飞行时进行飞行需要的一种证件，可分为第一类特许飞行证和第二类特许飞行证。① 从事下列飞行之一的、尚未取得有效适航证的民用航空器，应取得第一类特许飞行证：(a) 为试验航空器的新设计构思、新设备、新安装、新操作技术及新用途而进行的飞行；(b) 为证明符合适航标准而进行的试验飞行，包括证明符合型号合格证、型号设计批准书、补充型号合格证和改装设计批准书的飞行、证实重要设计更改的飞行，证明符合标准的功能和可靠性要求的飞行；(c) 新飞机的生产试飞；(d) 制造人为交付或出口航空器而进行的调机飞行；(e) 制造人为训练机组而进行的飞行；(f) 为航空比赛或展示航空器的飞行能力、性能和不寻常特性而进行的飞行，包括飞往和飞离比赛、展览、拍摄场所的飞行；(g) 为航空器市场调查和销售而进行的表演飞行；(h) 局方同意的其他飞行。② 从事以下飞行之一的、尚未取得有效适航证或目前可能不符合有关适航要求，但在一定限制条件下能安全飞行的航空器，应当取得第二类特许飞行证：(a) 为改装、修理航空器而进行的调机飞行；(b) 运营人为交付或出口航空器而进行的调机飞行；(c) 为撤离发生危险的地区而进行的飞行；(d) 局方认为必要的其他飞行。

A permit when civil aircraft has not yet obtained an airworthiness certificate or may not meet the relevant airworthiness requirements at present, but necessary to fly under certain limited conditions, which can be divided into Class Ⅰ special flight permits and Class Ⅱ special flight permits. ① A Class Ⅰ special permit must be obtained for a civil aircraft without an airworthiness certificate, for the following purposes: (a) flying the aircraft for testing new aircraft design concepts, new aircraft equipments, new aircraft installations, new aircraft operational techniques, or new uses for aircraft; (b) flying the aircraft for showing compliance with applicable airworthiness standards, including demonstration flights to show compliance for issuance of type certificate, supplemental type certificate, and modification design approval, flights to substantiate major design changes, and flights to show compliance with function and reliability requirements of the applicable regulations; (c) production flight testing for new production aircraft; (d) flight conducted by manufacturer for delivering or exporting the aircraft; (e) flying for flight crew training; (f) flying for air competitions or exhibiting an aircraft's flight capabilities, performance, or unusual characteristics, including flying to and from competitions, air shows and productions; (g) flying the aircraft for market surveys and sales promotion; (h) other flights authorized by the authority. ② A Class Ⅱ special flight permit that must be obtained for an aircraft without an airworthiness certificate or an aircraft that currently may not meet applicable airworthiness requirements, but capable of safe flight under special limitations for the following purposes: (a) ferry flights for aircraft repairs or alterations; (b) ferry flights conducted by operator for aircraft delivery or export; (c) evacuating aircraft from areas of impending danger; (d) other flights considered necessary by the Authority.

**特种设备;Special Equipment**

国家认定的,受设备本身和外在因素的影响而容易发生事故,并且一旦发生事故会造成人身伤亡及重大经济损失的危险性较大的设备。

Dangerous equipment which is certified by state and is easy to cause accidents if it is affected by itself together with external factors and which may lead to casualties and significant economic loss.

**梯度;Gradient**

在给定的距离内,获得或失去的高度与所经过距离之比。

The ratio of height gained or lost in a given distance to the distance traveled.

**替代类适航指令;Superseding Airworthiness Directive**

对 AD 进行实质性更改后,以替代版来颁发。

A substantive change to an AD should be issued as a supersedure.

**替换件;Replacement Article**

能够直接替换某一已经获得型号合格审定或型号认可审定批准的件(被替换件)的零部件。设计上,替换件能够从装配、外形和功能方面替代被替换件。

A part that can directly replace another part that has been approved with type certification or type approval (the part to be replaced). In design, the replacement article can replace the replaced part in terms of assembly, shape and function.

**调制效应;Modulation Effect**

与飞行检查有关的,指除产生正常航线信息的调制之外的任何一种调制所造成的航线偏离。

Course deviation caused by any type of modulation other than that producing normal course line information pertaining to flight inspection.

**停机坪;Ramp**

陆上机场的一个划定区域,用于停放航空器以便乘客登机、离机、装卸货物、加油、停机或维修。

A defined area on a land airport, intended to accommodate aircraft for purposes of loading or unloading passengers or cargo, refueling, parking or maintenance.

### 停机位;Gate Position

停机坪上的一个指定空间或位置,供乘客上、下和装卸货物时停放航空器用。

A designated space or position on an apron for an aircraft to remain parked while loading or unloading passengers and cargo.

### 通风舱;Ventilated Compartment

换气时间已知且不超过六分钟的舱室。

A compartment where the air change time is known and does not exceed 6 minutes.

### 通航空域;Navigable Airspace

最低飞行高度及以上的空域,包括安全起降所需的空域。

Airspace at and above the minimum flight altitudes, including airspace needed for safe takeoff and landing.

### 通勤航空承运人(美国);Air Carrier-commuter

从事任一下列业务的航空出租运营人:① 在两点或更多点之间每周至少进行 5 次往返飞行并发布飞行时刻表,说明上述往返飞行的时刻、日期和地点;② 根据与美国邮政局的现行合同空运邮件。

An air taxi operator which: ① performs at least five round trips per week between two or more points and publishes flight schedules which specify the times, days of the week and places between where such flights are performed; ② transports mail by air pursuant to a current contract with the United States Postal Service.

### 通勤类飞机;Commuter Category Airplanes

座位设置(不包括驾驶员)为 19 座或以下,最大审定起飞重量为 8 618 千克

(约 19 000 磅)或以下,用于非特技飞行的按照通勤类飞机适航标准进行合格审定的多发动机飞机。通勤类飞机要求表明对 CCAR – 23 的符合性。

The airplanes limited to propeller-driven, multiengine airplanes that have a seating configuration, excluding pilot seats, of 19 or less, and a maximum certificated takeoff weight of 8 618 kg (19 000 lb) or less, intended for non-acrobatic operations. Commuter category airplanes are required to show compliance with CCAR – 23.

### 通用工具/设备;General Tools/Equipment

制造厂家没有推荐,仅提出了尺寸、精度和范围要求的工具或设备。

Tools/equipment not specially suggested by manufacturers, but the size, accuracy and scale range of them are required by manufacturers.

### 通用航空;General Aviation

民用航空中包含所有航空业务的一个部分,但不包括持有民航局所发"便利需求证"的航空承运人和大型航空器商业运营人。

The portion of civil aviation which encompasses all facets of aviation except for air carriers holding a certificate of convenience and necessity from the Civil Aeronautics Board, and large aircraft commercial operators.

### 通用航空航空器;General Aviation Aircraft

除航空承运人使用的民用航空器之外的所有民用航空器。

All civil aircrafts except for those used by air carriers.

### 通用航空机场;General Airport

无公共航空运输定期航班到达的民用机场。

A civil airport where no public air transportation scheduled flights are operated.

### 通用航空运行;General Aviation Operation

除商业航空运输运行或航空作业运行以外的航空器运行。

An aircraft operation other than a commercial air transport operation or an aerial work operation.

### 通用类飞机;Utility Category Airplane

座位设置(不包括驾驶员)为 9 座或以下,最大审定起飞重量为 5 700 千克(约 12 500 磅)或以下,用于有限特技飞行的飞机。要求通用类飞机表明对 CCAR - 23 的符合性。

The utility category is limited to airplanes that have a seating configuration, excluding pilot seats, of nine or less, a maximum certificated take-off weight of 5 700 kg(12 500 pounds ) or less, and intended for limited acrobatic operation. Utility category airplanes are required to show compliance with CCAR - 23.

### 通用运输机场;General Transport Airport

适合于通用航空中使用总重不超过 79 379 千克(约 175 000 磅)的运输型飞机的机场。

An airport that accommodates transport type airplanes up to 79 379 kg (175 000 lb) gross weight used in general aviation.

### 同一性;Identicality

针对替换件的特有概念,指所申请零部件的设计与某一经批准的被替换件的设计是相同/本质上相同的。

A concept specific to replacement parts, meaning that the design of the requested part is identical/essentially identical to the design of an approved replacement part.

### 头戴式显示器;Head Mounted Display; HMD

安装在驾驶员头部的 HUD 显示器。

A special case of HUD mounted on the pilot's head.

### 突(阵)风包线;Gust Envelope

假定飞机在平飞时遇到对称的垂直突风,由此引起的限制载荷系数必须对应按突风速度确定的情况。

The airplane is assumed to be subjected to symmetrical vertical gusts in level flight. The resulting limit load factors must correspond to the conditions determined by the gust speed.

## 突风载荷；Gust Load

由风突变产生的载荷。

The load generated by the windshear.

## 涂刷液体检测；Liquid Application Test

将检测液体涂刷在检测区域的低压侧，观察液体中产生的气泡情况，以检出渗漏的一种检测方式。

A method brushing the detection liquid in the detection area of the low pressure side to observe the bubble generated in the liquid in order to detect leakage.

## 途中飞行咨询服务；En Route Flight Advisory Service；EFAS

向飞行中的驾驶员提供近乎实时的天气服务的专门化系统。

A specialized system providing near real-time weather service to pilots in flight.

## 湍流；Turbulence

① 围绕机翼或其他翼面的一种气流状态，此时在各层气流之间以不同的速度和压力横向混合；② 大气中由上升气流、下降气流和突风表征的一种局部状态。

① A condition in the airflow about a wing or other airfoil in which different velocities and pressures arc laterally mixed between layers of the airflow；② in the atmosphere，a local condition characterized by updrafts，downdrafts，and gusts.

## 推进系统；Propulsion System

由一个动力装置和随同装到机身上的所有相同设备共同组成的一个系统，用以提供支持、监测与控制任一个动力装置的功率或推力输出等必要功能。

A system composed of a power unit and all other necessary equipment installed on the fuselage to support, monitor, and control the power or thrust output of any power unit.

### 推算定位;Dead Reckoning

运用方向和速度数据向前推算已知的位置,估计或确定位置。

The estimating or determining of position by advancing an earlier known position by the application of direction and speed data.

### 脱离式自动应急定位发射器;Automatic Deployable ELT;ELT(AD)

紧固在航空器上,通过撞击,在某些情况下也通过液压传感器自动脱离和启动的应急定位发射器,也具备手动脱离功能。

An ELT which is rigidly attached to an aircraft and automatically deployed and activated by impact, and, in some cases, also by hydrostatic sensors. Manual deployment is also provided.

# V

**VFR 云上飞行条件；VFR Conditions-on-top**

应驾驶员请求发布的经批准的 IFR 许可，此时航空器可以在等于或高于 MEA 高度的任何相应 VFR 高度上飞行，此高度至少在云、烟、霾和雾层之上 1 000 英尺。

An authorized IFR clearance，issued upon pilot's request，in which the aircraft may be flown at any appropriate VFR altitude at or above the MEA which is at least 1 000 ft above a cloud，smoke，haze，or fog layer.

**VHF 全向信标；VHF Omnidirectional Range；VOR**

导航系统中的一种无线电发射设备，使用由两个信号调制的 VHF 无线电波，经兼容的机载接收机比较、解算和显示这两个信号的相对相位，从而向驾驶员提供相对于此设施方位的定向指示。

A radio transmitter facility in the navigation system radiating a VHF radio wave modulated by two signals，the relative phases of which are compared，resolved，and displayed by a compatible airborne receiver to give the pilot a direct indication of bearing relative to the facility.

**$V_{s0}$**

一种在着陆形态下的失速速度或最小稳定飞行速度。

A stalling speed or minimum steady flight speed in the landing configuration.

**$V_{s1}$**

一种失速速度或最小稳定飞行速度。

A stalling speed or minimum steady flight speed.

# W

**外部负载和连接装置;External Load and Attached Device**

连接外部负载与航空器的结构件,包括外载容器、连接支持结构以及用于抛投外部负载的任何快投装置。

Structure component which connects extras loading and aircraft including external load containers, connection support structure and any quick cast device used for cast extras load.

**外部高强度辐射领域环境;External High-intensity Radiated Fields Environment**

航空器外部射频辐射场环境。

RF fields at the exterior of an aircraft.

**外部审核;External Audit**

由被审核组织之外的组织执行的审核。

An audit conducted by an entity outside of the organization being audited.

**外部事件;External Event**

区别于飞机事件的起因,如大气状态,跑道状态、客舱和行李着火及鸟击等。

An occurrence which has its origin distinct from the aircraft or the system being examined, such as atmospheric conditions, operating environment, cabin and baggage fires, and bird-strike.

**外部载荷;External Load**

在飞机机身外部承载或延伸的载荷。

A load that is carried, or extends, outside the aircraft fuselage.

**外国民航局;Foreign Civil Aviation Administration; FCAA**

根据国家法律,授权对民用航空产品进行适航管理的外国国家政府机构或民间组织。

Foreign government agencies or non-governmental organizations responsible for airworthiness management of civil aviation products authorized by national law.

**外国适航证认可书;Validation of Foreign Airworthiness Certificate**

为在外国登记注册,持有外国适航当局颁发的现行有效适航证,且型号设计已经民航局认可,并由中国占有人或使用人运行的航空器颁发的认可书。

The certificate which is registered in foreign countries, owning the current available airworthiness certificate issued by foreign airworthiness authorities, its type design has been validated by CAAC, and operated by Chinese occupant or user.

**外胎;Casing**

轮胎结构的一部分。

The structural part of a tire.

**完全 RVSM 包线;Full RVSM Envelope**

航空器在 RVSM 空域中可使用的马赫数、$W/\delta$ 和高度值的整个范围。

The entire range of operational Mach numbers, $W/\delta$, and altitude values over which the aircraft is operated within RVSM airspace.

**完整性;Integrity**

系统或项目表明可依赖其正确地完成工作的定性或定量属性。它有时用不能满足正确工作标准的概率来表示。

Qualitative or quantitative attribute of a system or an item indicating that it can be relied upon to work correctly. It is sometimes expressed in terms of the probability of not meeting the work correctly criteria.

**危害;Hazard**

有可能导致人员受到伤害、疾病或死亡,或者系统、设备或财产受损,或者环境受到破坏的任何现有的或潜在的状况。

Any existing or potential condition that can lead to injury, illness, or death; damage to or loss of a system, equipment, or property; or damage to the environment.

### 危机响应;Crisis Response

针对紧急事件的响应水平,包括人力资源管理服务和/或家庭援助。

The level of response to a critical incident involving human resource management services and/or family assistance.

### 危急事件;Critical Incident

事故/事件,例如有人员死亡的空难、恐怖主义活动、炸弹威胁、有毒物质泄漏、被延迟的营救或康复手术,以及地震、飓风这类自然灾害。发生危急事件时,员工,包括管理层官员,会经历痛苦。重大危急事件后的伴随研究活动,以及过多的媒体报道,将给员工增加额外的要求。

Accidents/incidents, such as an aviation disaster with loss of life, acts of terrorism, bomb threats, exposure to toxic materials, prolonged rescue or recovery operations, and natural disasters such as earthquakes and hurricanes. When a critical incident occurs, employees, including management officials, may experience distress. The accompanying investigative activities following the major critical incident, as well as excessive media coverage may place additional demands on employees.

### 危险材料;Hazardous Material

国际上称作危险品,被发现在许多飞机零部件上使用,同时也基本被使用在航空工业中。

Hazardous materials (known internationally as dangerous goods) are found in many aircraft parts and also in substances used in the aviation industry.

### 危险等级;Hazard Level

根据对飞机、乘客和机组的影响确定的事故的后果等级。

Level of event outcome, as defined by its effect on the aircraft, passengers, and crew.

### 危险警报; Warning Alert

需要机组立即采取行动的撞地威胁告警。

An alert for a detected terrain threat that requires immediate flightcrew action.

### 危险性失效状态; Hazardous Failure Condition

此种失效状态可能降低飞机性能或机组处置不利工作条件的能力,以至于会有如下情况: ① 安全裕度或功能大大降低;② 影响人体正常功能或产生较重的工作负荷,使机组不能准确地或全面地完成任务; ③ 一些乘员受到严重或致命的伤害。

This kind of failure condition may reduce aircraft performance or the ability of the crew to cope with adverse operating conditions, so the below conditions can be resulted: ① the safety margin or function is reduced greatly; ② affecting the body's normal functions or excessive workload, making the crew not complete the task exactly or fully; ③ a few passengers receive serious or fatal injuries.

### 微波着陆系统; Microwave Landing System; MLS

以微波频率(5.0～5.25 GHz/15.4～15.7 GHz)工作的仪表进近与着陆系统,提供方位、标高和距离测量方面的精密导引。

An instrument approach and landing system operating in the microwave frequencies (5.0 ～ 5.25 GHz/15.4 ～ 15.7 GHz) that provides precision guidance in azimuth, elevation and distance measurement.

### 微动腐蚀; Fretting Corrosion

一种由两个滑动表面之间重复摩擦引起的损坏形式,并且因共同的腐蚀作用而加速。

A form of deterioration caused by repetitive friction between two sliding surfaces and accelerated by a common corrosive action.

### 维修；Maintenance

为确保航空器持续适航所需执行的任务，包括大修、检查、更换、缺陷纠正以及具体的改装或修理，或以上各项的组合。

The performance of tasks required to ensure the continuing airworthiness of an aircraft, including any one or combination of overhaul, inspection, replacement, defect rectification, and the embodiment of a modification or repair.

### 维修保留工作项目；Maintenance Work Item Deferred

在航空器计划维修工作中，因工具、设备、器材、工作条件等原因不能正常按计划完成的维修工作项目。

In the aircraft planned maintenance work, the maintenance work can not be finished according with the plan due to the reason of tools, equipment, material, working conditions, etc.

### 维修参考文件；Reference Documents for Maintenance

载明某一具体维修工作的实施方法和标准的技术文件，维修单位可视具体情况根据适航性资料自己制订或直接使用有关适航性资料中已核准其适用性的内容。在实际使用中，维修单位自己制订的文件可以称为实施指南、实施说明、工艺等。

Reference documents for maintenances giving detailed methods and standards concerning the maintenance task. The maintenance organization can choose to develop the airworthiness data or just use the applicable part (previously verified) of the airworthiness data as their reference documents for maintenance. Documents established by maintenance organization themselves can be named as accomplishment instructions, accomplishment guidance, work process, etc.

### 维修差错；Maintenance Errors

在维修活动中，由于维修责任造成的威胁飞行安全、违反适航规章或具有一定直接经济损失的航空器、航空器部件、车辆、设备、设施损坏和人员受伤，但其程度未构成维修事故征候的事件。

During maintenance activities, the event leading to potential flight safety hazard or violation of airworthiness regulation or economic losses caused by the maintenance responsibility such as aircraft, aircraft parts, vehicles, equipment, facilities damage and person injury, but not reach maintenance accidents.

### 维修大纲；Maintenance Programme

描述具体的定期维修任务及其完成频次和相关程序（如可靠性方案）的文件，这些程序对适用的航空器的安全运行是必要的。

A document which describes the specific scheduled maintenance tasks and their frequency of completion and related procedures, such as a reliability programme, necessary for the safe operation of those aircrafts to which it applies.

### 维修代码；Maintenance Code

供维修人员使用的，装载在机载计算机中为机载维修计算机提供接口的代码。

The code in an airborne computer-based system that interfaces with an onboard maintenance computer or computer used by maintenance personnel.

### 维修单位/维修培训机构责任经理；Maintenance Organization/Training Organization Accountable Manager

维修单位/维修培训机构中能对本单位满足本规定的要求负责，并有权为满足本规定的要求支配本单位的人员、财产和设备的人员。

The person of a maintenance organization/training organization who can be responsible for guaranteeing that the organization concerned meets the requirements specified by the present regulations and who can dispose the personnel, assets and equipment thereof with the purpose of guaranteeing such satisfaction.

### 维修单位质量经理；Maintenance Organization Quality Manager

维修单位中由责任经理授权对维修工作质量进行管理和监督，并直接向责任经理负责的人员。

The person appointed within a maintenance organization (or a training organization), who is authorized by the accountable manager to manage and supervise the quality of the maintenance work performed by the maintenance organization. The quality manager shall be directly responsible to the accountable manager.

### 维修放行文件;Maintenance Release

证明已按照批准的数据和维修机构的程序手册所述的程序或根据相当的制度,令人满意地完成有关维修工作的文件。

A document which contains a certification confirming that the maintenance work to which it relates has been completed in a satisfactory manner, either in accordance with the approved data and the procedures described in the maintenance organization's procedures manual or under an equivalent system.

### 维修改装必检项目;Maintenance and Alterations Required Inspection Items

需要由执行工作人员以外的人员进行检查的维修和改装项目。必检项目至少包括由实施不正确或零附件错装、材料错用引发危及飞行安全操作的失效、缺陷或故障的项目。

The items of maintenance and alterations requiring inspection by a person other than the one who performed the work. These items include at least those where incorrect performance or improper parts or materials use could result in a failure, malfunction, or defect endangering the safe operation of the aircraft.

### 维修机构的程序手册;Maintenance Organization's Procedures Manual

经维修机构的首脑核准的文件,其中详细载明了维修机构的结构和管理职责、工作范围、设施介绍、维修程序及质量保证制度或质量检查制度。

A document endorsed by the head of the maintenance organization which details the maintenance organization's structure and management responsibilities, scope of work, description of facilities, maintenance procedures and quality assurance or inspection systems.

### 维修机库;Maintenance Hanger

可容纳、停放航空器,并能满足对航空器进行检查、维修、部件更换等作业要求的建筑物。

The building or structure which can be used to accommodate and park aircraft，and meet the requirements of aircraft inspection，maintenance，parts replacement.

**维修机坞；Maintenance Dock**

用于航空器维护、维修的大型工作平台。

Huge working platform for airplane maintenance and repair.

**维修计划文件；Maintenance Planning Document；MPD**

由航空器制造厂家提供的该型航空器必需的维护信息和方案，航空器运营人可依据该方案制订适合自己机队情况的维护计划。该方案包含所有制造厂家推荐的、满足制造国当局的持续适航要求的维修任务和计划。

The necessary maintenance information and program provided by the aircraft manufacturer for each type of their aircraft. The operator can establish the maintenance planning applicable to the own fleet per to this program. The program contains all the maintenance tasks and plans which are recommended by the manufacturer and meet the requirements of the continued airworthiness of the manufacturer's Authority.

**维修记录；Maintenance Records**

对航空器及航空器部件所进行的任何检测、修理、排故、定期检修、翻修和改装等不同形式维修工作的记录。

Working records of maintenance work for aircraft and component including detection and maintenance trouble shooting，scheduled check，overhaul and modification.

**维修警告指令；Maintenance Alert Directive；MAD**

为向维修人员发出临时性或警告性信息而颁发的一种指令。

A directive issuance used to transmit temporary or alert information to maintenance personnel.

### 维修培训机构管理手册；Management Manual for the Maintenance Personnel Training Organization

由维修培训机构制订的，阐述本单位如何符合各项要求及各项培训管理制度和培训管理程序。

The manual developed by the maintenance personnel training organization to expound the organization how to comply with the requirements, training management system and training management program of this provisions.

### 维修人为因素；Maintenance Human Factor

在航空器维修工作过程中，应当考虑人的行为能力和局限性对航空器或航空器部件的维修有效性和安全性的影响，以及考虑人与其他因素的协调关系的基本原则。

The basic principle which applies to the maintenance of civil aircraft or aircraft components, with adequate consideration of the impact of human performance and limitations on the effectivity and safety of the maintenance work as well as the perfect harmonization of the human and other factors.

### 维修人员的工作时间；Working Time of the Maintenance Personnel

也称为维修人员的值勤时间，指维修人员在接受维修单位安排的工作任务后，从为了完成该次任务到指定地点报到开始（不包括从居住地或驻地到报到地点所用的时间），到工作任务完成或解除为止的连续时间段。

Is also termed as on-duty time of the maintenance personnel, it means the continuous period from the maintenance personnel reporting for his duty while accepting the maintenance task assigned by the maintenance organization till the maintenance task is completed or terminated. The period the maintenance personnel take from his inhabitation or residence to the reporting place is not included.

### 维修审查委员会；Maintenance Review Board；MRB

由有资格的局方人员组成，其任务是在制定 MRBR 及改版过程中，向工业指导委员会（ISC）和工作组（WG）提出审查意见。它受航空器评审组（AEG）指派的 MRB 主席的领导。

MRB is composed of the qualified staff from authority, whose task is to forward the comments to the industry steering committee (ISC) and the working group(WG) during MRBR development and revision. It is under the MRB chairman's leadership, and the MRB chairman is assigned by aircraft evaluation group(AEG).

### 维修审查委员会报告；Maintenance Review Board Report；MRBR

维修审查委员会报告是由制造国当局制定和批准的、针对衍生型号或新型号审定航空器的初始最低维护和检查要求，该报告包含了对航空器、翼上发动机维修方案的初始最低维护和检查要求，但并未包含对独立未装机发动机的维修方案。该报告将成为航空器运营人建立自己维修方案的基础，其中的要求对相同型号的航空器都是适用的。

MRBR is developed and approved by the manufacture's Authority for the initial minimum requirements of the maintenance/inspection for the variant or new type certification of aircraft. It contains the initial minimum maintenance/inspection requirements for the aircraft and on-wing engine program, but does not include the maintenance program for off-wing engine. MRBR is the basis for the aircraft operator to establish their own maintenance program, the requirements thereof is applicable to each aircraft of the same model.

### 维修实施依据文件；The Basic Maintenance Document for the Maintenance Work

说明某一具体维修工作实施方法和标准的技术文件。在维修单位已核准其适用性并且保证维修人员能够正确理解的情况下，维修工作实施依据文件可以直接使用适航性资料中的有关内容。

The technical document defining the methods and the standards to perform the specific maintenance work. The information of the relevant airworthiness data may be quoted directly in such document in the case that the information of the relevant airworthiness data has already been verified as applicable and can be understood correctly by the maintenance personnel.

### 维修事故；Maintenance Accidents

在维修活动中，由维修责任造成的具有重大直接经济损失的航空器、航空器部件、车辆、设备、设施损坏和人员重伤或人员死亡的事件。

During maintenance activities, the accident leading to huge economic losses caused by the maintenance responsibility such as aircraft, aircraft parts, vehicles, equipment, facilities damage and severe injury or fatality.

### 维修事故征候;Maintenance Incidents

在维修活动中,由维修责任造成的严重威胁飞行安全的事件或具有重大直接经济损失的航空器、航空器部件、车辆、设备、设施损坏和人员伤残,但其程度未构成维修事故的事件。

During maintenance activities, the incident leading to potential flight safety hazard or economic losses caused by the maintenance responsibility such as aircraft, aircraft parts, vehicles, equipment, facilities damage and person injury, but not reach maintenance accidents.

### 维修性;Maintainability

根据设计要求,航空器通过维修所能保持和恢复其在使用中的可靠性程度。

Maintaining and restoring the reliability of aircraft operation through the repair and maintenance according to the design requirements.

### 维修站审定局方;Shop Certification Authority; SCA

依据 CAR 573 颁发 AMO 的局方。

AMO-issued authority under CAR 573.

### 维修证明文件;Maintenance Proving Documents

对航空器或其部件完成了规定维修工作的证明性材料。

Documents to prove the performing of the required maintenance work for aircraft or its component.

### 维修执行程序;Maintenance Implementation Procedures; MIP

被双边航空安全协议所认可的,针对航空产品所做的维修、替换和修改的程序文档。

Procedural document authorized by the BASA Executive Agreement related to the performance of maintenance, alterations and modifications on civil aeronautical products.

## 维修指导小组—第二特别工作小组; Maintenance Steering Group-2nd Task Force; MSG‑2

由美国航空运输协会创建的维修分析逻辑方法,又称"航空公司/制造人维修大纲计划文件",主要用于 1980 年以前设计、生产的航空器。MSG‑2 是针对维修方式的分析逻辑,其分析结果是指定各单体的定时、视情、监控等具体维修方式。

A maintenance analysis logic method created by the Air Transport Association. It's also known as "airline/manufacturer maintenance program planning." It is mainly used for those aircraft designed and manufactured before 1980. MSG‑2 is the analysis logic for maintenance mode, the analysis result is the specific maintenance mode of the specified individual part, such as hard-time, on-condition, condition monitoring maintenance.

## 维修指导小组—第三特别工作小组; Maintenance Steering Group-3rd Task Force; MSG‑3

针对维修工作的分析逻辑,其分析结果是为系统/分系统指定具体的维修工作小组。

MSG‑3 is the analysis logic for maintenance tasks, the analysis results is specific maintenance tasks for the systems/subsystems.

## 尾流湍流; Wake Turbulence

航空器穿越大气造成的一种现象。当空中交通管制使用此术语时,包括涡流、发动机排气湍流、喷气洗流、螺旋桨洗流和旋翼洗流。

Phenomena resulting from the passage of an aircraft through the atmosphere. The term when used by ATC includes vortices, thrust stream turbulence, jet wash, propeller wash and rotor wash.

## 尾橇; Tail Skid

某些飞机上固定在后机身下部的滑橇,起尾轮的作用。

On certain airplanes, a skid attached to the underside at the rear of the fuselage serving the functions of a tail wheel.

### 委任代表；Designated Representatives

民航行政机关委派的民航行政机关以外、在授权范围内从事适航管理中有关审定、检验工作的个人。委任代表为民航行政机关颁发适航证件进行技术检查并出具技术检查结果，作为民航行政机关颁发适航证件的依据。

Designated representatives refer to the individuals who are engaged in certification and inspection work related to airworthiness management within the scope of authorization other than the civil aviation administrative organs. The technical inspection results issued by the designated representative serve as the basis for the issuance of airworthiness certificates by civil aviation administrative organ.

### 委任单位代表；Authorizing Representative

民航行政管理部门委任的，民航适航部门以外的，在委任范围内从事有关适航工作的组织或机构。

Units or organizations not affiliated to the airworthiness authority, which are designated by the CAAC and represent the CAAC in performing airworthiness related functions within the limit of their authorities.

### 委任改装站；Designated Alteration Station；DAS

合格的、按 145 部运行的维修单位，或者按 121 部运行的商业运营人。

An organization that is an eligible domestic repair station under part 145, or commercial operator under part 121.

### 委任工程代表；Designated Engineering Representatives；DER

获责任审查部门授权委任的申请人的工程技术人员，可以在局方监管下代表其行使授权范围内的型号合格审定工作。

A qualified technical people under the authority of the responsible reviewing department to perform type certification within the scope of authorization on behalf of the Authority under the supervision of the authority.

## 委任工程代表型号资料审查表；DER Statement of Compliance

授权的委任工程代表填写的，用于证实型号资料已经过审查、符合要求并按授权予以批准或提出批准建议的表格。

A from filled by the authorized designated engineering representatives to certify that the type data has been reviewed and complies with the requirements, used to approve or propose approval suggestion.

## 委任适航代表；Designated Airworthiness Representative

民航局委任适航部门之外的，在授权范围内代表民航局从事有关适航工作的某些单位或机构。

The professionals not belonging to Flight Stand Department, but Representative Flight Stand Department for work in its authorization.

## 委任适航代表—维修；Designated Airworthiness Representative-Maintenance

按照 183 部委任的个人，拥有按 65 部颁发的机体和动力装置类别的机械员执照，除飞行机组人员外的空勤人员，或具有修理人员执照，服务于持有 145 部执照的维修单位的个人，并满足相关资格要求。

An individual appointed in accordance with part 183 who holds a mechanic's certificate with an airframe and powerplant (A&P) rating under part 65 certification: Airmen Other Than Flight Crewmembers, or a person who holds a repairman certificate and is employed at a repair station certificated under part 145, repair stations, and who meets the qualification requirements.

## 委任适航代表—制造；Designated Airworthiness Representative-Manufacturing

依照 183 部委任的个人，该人具备航空知识和经验，并满足相关资格要求。

An individual appointed in accordance with part 183 who possesses aeronautical knowledge and experience, meeting the qualification requirements.

## 委任证书；Certificate of Designation；COD

获得资格的委任人员关于委任种类的可展示证书，该证书用于展示。

A suitable for framing certificate that specifies the kind of designation for which the designee is qualified; it is used for display purposes.

### 委任职责;Delegated Function

对于 DER,委任职责指技术领域,包括判断与相关适航规章的符合性。

For DER, a delegated function applies to the technical areas involved in determining compliance with applicable airworthiness regulations.

### 未经批准的可疑航材;Suspected Unapproved Materials; SUP

未经民航局批准或认可的航空器部件或原材料,这些航材可能在表面上与民航局批准或认可的航材一样,但没有证据证明其制造或维修过程满足民航局批准或认可的数据,而购买者不能轻易地发现(如热处理、电镀、各种测试和检查的标准等)。

The aircraft components or raw materials not approved or accepted by the Civil Aviation Administration, which may be consistent apparently with those approved or accepted by the civil aviation administration, but there is no evidence to show that its manufacture or repair process meet civil aviation administration approved or accepted data, and can't be easily distinguished by the buyers (such as heat treatment, electroplating, all kinds of testing and inspection standards, etc. ).

### 未经批准的零件;Unapproval Part

涉及下述任一方式的零件、部件或材料:① 不符合 CCAR - 21 第八章规定的批准程序;② 不符合 CCAR - 145 规定的维修工作准则;③ 不符合批准或认可的型号设计;④ 不符合适航部门认可的行业技术标准或国家技术标准。未经批准的零件包括但不限于下述范围:① "假冒"或伪造标记的零件、部件或材料;② 由未获得该零件生产批准的制造人供给用户的零件,或未得到零件生产批准书持有人授权供货的供应商或批发商供给用户的零件(例如超出许可生产范围的产品);③ 由未按 CCAR - 145 规定批准的人员或设施进行过修理或维护并返回使用的零件。

Part、component、material in relation to the following way: ① do not meet the approved procedure of CCAR - 21 Chapter 8; ② do not meet the maintenance job standards of CCAR - 145; ③ do not meet the approval or recognized type design; ④ do not meet the CTSO or national technology standards. Unapproval part include, but are not limited to, the following: ① "counterfeit" or forged marked parts, components, or materials; ② parts supplied to users by manufacturers who have not obtained production approval for the part, or by suppliers or wholesalers who have not been authorized by the holder of the production approval for the part (such as products beyond the permitted production range); ③ parts repaired or maintained by personnel or facilities not approved in accordance with CCAR – 145 and returned for use.

### 未列入持续适航文件的可修件;Undefined Repairable Part

可通过大修或修理恢复适航性的零件,但持续适航性说明中的可修理零件清单除外。

Parts that could be retrieve to airworthiness by overhaul or repair, except for repairable parts list in the instructions for continued airworthiness.

### 未修正的风险;Uncorrected Risk

如果没有改正措施,受影响的机队会积累的风险。

The risk that accumulates over time in the affected fleet if no corrective action is taken for a certain safety issue.

### 位置源综合水平;Source Integrity Level; SIL

一般指静态(非变化)的值,且如果一个单一型位置源与 ADS – B 系统综合,则应在安装时设定。

A typically static (unchanging) value that may be set at the time of installation if a single type of position source is integrated with the ADS – B system.

### 稳定飞行;Stabilized Flight

依靠惯性稳定装置控制航空器姿态的航空器飞行,例如用自动驾驶仪控制的飞行。

Aircraft flight which depends upon inertia-stabilized devices to control the attitude of the craft, for example, the flight controlled by an autopilot.

## 问题纪要；Issue Paper；IP

用来确认和解决型号合格审定过程中发生的有关技术、规章和管理的重要或有争议的问题的一种手段,也是用来确定问题处理进展情况的手段,并且是证后对问题处理情况进行总结的基础。

Issue paper is used by the certification team to explain and record some significant and controversial issues which need to be further discussed and reviewed to determine the means of compliance or the verification procedures. Issue paper is also a foundation for Type Certification Board to make final decision and used as a method to supervise and control the technical review activities of certification team.

## 问题纪要汇编；Issue Book

将所有的问题纪要汇编成册并进行动态管理的汇总性文件。在型号合格审定过程中,型号合格审定审查组组长收集当下所有的问题纪要并汇编成册,供型号合格审定委员会(TCB)、型号合格审定审查组和申请人使用。

A summary document that assembles all issue papers and dynamically manages them. The project manager assembles issue papers and publishes them in the form of an issues book for distribution to the TCB members, project team members, and applicant during type certificate.

## 涡流除冰系统；Eddy Current Deicing System；ECDS

对通过嵌入并沿保护面前缘展向分布的平面线圈脉冲电流的反作用,为保护面提供的一种脉冲运动。

ECDS provides an impulsive movement of the protection surface by a reaction to electrical current pulsed through planar coils embedded within and spanwise along the leading edge of the protected surface.

## 涡流发生器；Vortex Generator

气动物体上的可动面,可用作分离气流用的扰流片。

A movable surface on an aerodynamic body that may be used as a spoiler to break down the airflow.

### 涡轮动力飞机；Turbine-powered Airplane

涡扇、涡喷、桨扇、超高涵道比的风扇动力飞机，不包括涡桨动力飞机。

Turbofan, turbojet, propfan, and ultra-high bypass fan-powered airplanes. The definition specifically excludes turbopropeller-powered airplanes.

### 涡轮发动机关键零件；Turbine Engine Critical Components

被涡轮发动机制造商指定为"关键件"的任何零件。

Any critical parts identified by turbin-engine manufacturing.

### 涡轮发动机起飞推力；Takeoff Thrust, with Respect to Turbine Engines

在具体高度和大气温度的静态条件下，处于经批准的正常起飞最大转子轴转速和最高燃气温度时所产生的喷气推力，限于在经批准的发动机规范所示的时间段内连续使用。

The jet thrust produced at the approved maximum take-off rotor speed and highest temperature during normal take-off, under the specific static condition of height and atmospheric temperature, limited to be used continuously in the time period shown in the approved engine specification.

### 涡轮噪声；Turbine Noise

涡轮风扇发动机火焰筒与主排气喷管之间产生的噪声，包括离散频率噪声和随机噪声。

The noise generated between the burner cans and the primary exhaust nozzle in a turbofan engine, containing both discrete frequencies and random noise.

### 涡喷发动机型号合格审定的额定起飞加力推力；Rated Takeoff Augmented Thrust, With respect to Turbojet Engine Type Certification

在依据 33 部为该发动机所制定的使用限制以内的标准海平面条件下，按静态确定的带液体喷射或带外加燃烧室内燃油燃烧的经批准喷气推力，使用时间限制在起飞运行时不超过 5 分钟。

According to Part 33 for the engine use limits, standard sea level conditions, according to the static determined with liquid injection or with additional combustion chamber combustion of fuel by the approval of the thrust of the jet, using a time limit no more than 5 minutes during the takeoff run.

### 污染跑道；A Contaminated Runway

在大于 25%的要求场长上，在使用的宽度范围内有一层深度大于 0.125 英寸（约 3.2 毫米）的积水或泥浆跑道，或者是有积雪或积冰的跑道。其他情况也可视为脏跑道。例如，跑道表面有一层积水或泥浆的那一段在飞机将要抬前轮和离地处，或在起飞滑跑的高速部分处，它对飞机的减速作用比在起飞初期低速时遇到上述情况要严重得多。在这种情况下，认为跑道是"脏"的而不是"湿"的可能要好些。

A runway where more than 25 percent of the required field length, within the width being used, is covered by standing water or slush more than 0.125 in (3.2 mm) deep, or that has an accumulation of snow or ice. However, in certain other situations it may be appropriate to consider the runway contaminated. For example, if the section of the runway surface that is covered with standing water of slush is located where rotation and liftoff will occur, or during the high speed part of the takeoff roll, the retardation effect will be far more significant than if it were encountered early in the takeoff while at low speed. In this situation, the runway might better be considered "contaminated" rather than "wet."

### 无内胎轮胎；Tubeless Tires

没有内胎的轮胎。

Tires not requiring tubes.

### 无损检测；Nondestructive Inspection/Nondestructive Test；NDI/NDT

一种不减弱材料或零件的强度、许用值、质量或适用性而检查该部件完整性的方法。

A method used to check the soundness of a material or a part without diminishing the strength, value, quality, or serviceability of the part.

### 无线电波束耦合器；Radio Beam Coupler

将航空器自动驾驶仪与 VOR 或 ILS 设施发射的信号相耦合的一种装置；

用于航线中和对进近的控制;能为自动感受信号和排定程序的顺序而预先编程。

A device coupling an aircraft autopilot to signals from a VOR or ILS facility; used for both en route and approach control, which can be preprogrammed for automatic sensing of signals and sequencing of program.

### 无线电测向;Radio Direction Finding

以接收无线电信号的方法来确定位置的过程:带有测向天线(通常为环形天线)的无线电接收机测取两个或更多个已知位置的固定发射机的方位;任何两条方位线的交叉都可确立观察者的位置。

The process of determining a fix by reception of radio signals: a radio receiver with a direction finding antenna (commonly a loop) takes bearings on two or more fixed transmitters of known position; the intersection of any two bearing lines establishes the observer's position.

### 无线电测向器;Radio Direction Finder

飞行咨询和空中交通管制系统中的一种无线电设施,它接收航空器的连续载波无线电信号,处理这些信号并在其自身的指示器上显示航空器相对于该设施天线的方位(以方位角度数表示)。

A radio facility in the flight advisory and air traffic control systems, which receives aircraft continuous wave carrier radio signals, processes these signals and displays the bearing of the aircraft in degrees of azimuth on its own indicator, relative to the facility antenna.

### 无线电方位;Radio Bearing

在 ICAO 中,指一定的电磁(无线电)波发射源的视方向与无线电测向台测定的基准方向之间的夹角。真无线电方位是基准方向为真北的方位。磁方位是基准方向为磁北的方位。

In ICAO, the angle between the apparent direction of a definite source of emission of electromagnetic (radio) waves and a reference direction as determined at a radio direction- finding station. A true radio bearing is one for which the reference direction is that of true north. A magnetic bearing is one for which the reference direction is that of magnetic north.

### 无线电高度;Radio Altitude (Height); RA

由无线电高度表系统测定的飞机与地面的距离。

The distance of an aircraft above the ground, as determined by a radio altimeter system.

### 无线电罗盘;Radio Compass

航空器机上使用的指示方向的无线电接收装置,它利用环形天线的方向特性来测向和指示相对于接收机调谐到的无线电发射台的方向。

A direction indicating radio-receiving apparatus used aboard aircraft, which makes use of the directional characteristics of a loop antenna for finding and indicating direction in relation to a radio-transmitting station to which the receiver is tuned.

### 误导性信息;Misleading Information; MI

在警告状态下,与飞机有关的地形危险情况的不正确描述。

An incorrect depiction of the terrain threat relative to the aeroplane during an alert condition.

# X

**系列;Series**

表示某个产品的相似型别。

The similar type with a product.

**系留;Tie Down**

当飞机停留在地面上时,进行固定、防止其移动或翻滚的一种方法。

A method which fixes the aircraft parking on the ground, preventing its moving or upset.

**系统;System**

相互连接,执行一个或多个功能的零部件及元素的组合。

The combination of components, parts, and elements that are interconnected to perform one or more functions.

**系统安全性;System Safety**

使用工程和管理的原理、标准和技术,在系统生命周期的所有阶段,在运行效果、时间和成本的限制范围内优化安全的各个方面。

The application of engineering and management principles, criteria, and techniques to optimize all aspects of safety within the constraints of operational effectiveness, time, and cost throughout all phases of the system lifecycle.

**系统安全性评估;System Safety Assessment**

为表明满足相关安全要求而对系统进行的系统性、综合性评估。

A systematic, comprehensive evaluation of the implemented system to show that the relevant safety requirements are met.

### 系统运行；System Operations

包含所有与局方管辖范围内的 NAS 的运行、维护和监管相关的功能。

System operations encompass all functions related to the operation, maintenance, and regulation of the NAS under the purview of Authority.

### 下滑道；Glide Slope; GS, G/S

由进近或着陆期间，飞机或驾驶员使用的地面电子信号确立的一条垂直下滑航道。

A vertical path defined by a ground based electronic signal used by an airplane or pilot during approach or landing.

### 下滑道截获高度；Glide Slope Intercept Altitude

在经批准的下降程序中建议或公布的真实高度（平均海平面高度），航空器在此高度上截获下滑道并开始下降。

The true altitude (MSL) proposed or published in approved let-down procedures at which the aircraft intercepts the glide slope and begins descent.

### 下滑道可用距离；Glide Slope Usable Distance

在规定高度上距下滑道信标发射机的最大距离，在此距离上飞机能继续朝向发射机飞行而不出现警告旗。

The maximum distance from the glide slope transmitter at a specified altitude, at which full scale fly-up is continuously obtained with no flag alarm activity when flying inbound to the transmitter.

### 下滑道宽度；Glide Slope Width

下滑道正常进近包线的宽度（以度计）。

The width in degrees of the glide slope normal approach envelopes.

### 下滑航迹；Glide Path

预定航线上的下滑道段，它与重合于跑道中心线的横向导引段相交，是使用下滑道设施的航空器应遵循的航迹。

The on-course portion of the glide slope that intersects the lateral guidance coincidence with runway centerline. It is the path that an aircraft using the glide slope facility will follow.

### 下滑航迹角;Glidepath Angle; GPA

最后进近降落路径相对于进近表面基准线的角度。

The angle of the final approach descent path relative to the ASBL.

### 下滑航迹结构;Glide Path Structure

下滑航迹的特征,包括弯曲、偏离、误差和宽度。

Characteristics of a glide path including bends, scalloping, roughness and width.

### 下滑角;Glide Angle; GA

正在下滑的航空器的飞行航迹与水平面的垂直夹角。

The vertical angle between the flight path of a gliding aircraft and the horizontal.

### 下滑信标台;Glide Beacon Station

航站区域电子导航系统中的一种仪表着陆系统导航设施,它在进近着陆期间为航空器提供垂直导引,其方法是发射用两个信号调制的甚高频无线电波方向图,这两个信号在以等强度接收时,依靠兼容的机载设备将其显示为"在航迹上"的指示。

An ILS navigation facility in the terminal area electronic navigation system, providing vertical guidance for aircraft during approach and landing by radiating a directional pattern of VHF radio waves modulated by two signals which, when received with equal intensity, are displayed by compatible airborne equipment as an "on-path" indication.

### 下降角;Angle of Descent

正在下降的航空器的飞行航迹与水平面的夹角。

The angle between a descending aircraft's flight path and the horizontal.

## 现场批准；Field Approval

局方对非型号设计分类的重要改变，且用于单架飞机的重要修理或重要改装技术资料的一种批准方法。

A method by which Authority approves technical data used to accomplish a major repair or a major alteration on a single aircraft, the repair or alteration provided is not classified as a major change in type design.

## (维修培训)现场审查；On Site Audit

对维修培训机构的培训及管理现场的规章符合性进行监督检查。

Supervision and inspection of regulation compliance on training and management site of the maintenance training organization.

## 现行通知；Actual Notice

一个法律术语，用于描述向另一个人提交法律文件的个人行为。在规章中规定，局方提供"现行通知"，可以以个人名义通过邮政服务或传真发送法规文件或复印件。

A legal term to describe the act of an individual personally delivering a legal document to another person. In rulemaking, the Authority can provide "actual notice" by personal delivery of the rulemaking document or sending a copy by the postal service or by facsimile.

## 现役航空器；Active Based Aircraft

持有有效适航证并以某机场为基地的航空器。

Aircrafts that have a current airworthiness certificate and are based at an airport.

## 限用类航空器；Restricted Aircraft

仅供专门作业用的某种类别的航空器。本条中的"专门作业"指：① 农业(喷洒药剂和播种等)；② 森林和野生动植物保护；③ 航测(摄影、测绘、石油及矿藏勘探等)；④ 巡查(管道、电力线和水渠的巡查等)；⑤ 天气控制(人工降雨等)；

⑥ 空中广告;⑦ 局方规定的任何其他用途。

Specialized job with only a certain type of aircraft. "Specialized job" means: ① agriculture (spraying of pesticides and planting, etc.); ② forest and wildlife protection; ③ aerial plotting (photography, mapping, oil and mineral exploration, etc.); ④ aerial inspections (pipes, power lines and drains inspections, etc.); ⑤ weather control (artificial rainfall, etc.); ⑥ aerial advertising; ⑦ other purpose defined by CAAC.

### 限制飞行包线;Limit Flight Envelope; LFE

最外围的飞行包线,通常与飞机设计限制或 EFCS 保护限制有关。

The most outside flight envelope, generally associated with airplane design limits or EFCS protection limits.

### 限制载荷;Limit Loads

预期运行条件下假设会产生的最大载荷。

The maximum loads assumed to occur in the anticipated operating conditions.

### 相对湿度;Relative Humidity

表示空气中的绝对湿度与同温度下的饱和绝对湿度的比值,是一个百分比。

The ratio of the amount of water vapor in the air at a specific temperature to the maximum amount that the air could hold at that temperature, expressed as a percentage.

### 相似设计方法;Similar Design Philosophy

使用相同的制造和加工过程方法的设计(如机械加工零件与组合零件):① 详细零件材料(如合金或热处理);② 载荷路径;③ 几何特性,包括截面特性,但不包括由座椅或飞机界面的空间限制造成的轻微差异;④ 连接方法,不包括由座椅空间限制造成的轻微差异。

A design which uses the same method of construction and manufacturing process (for example, machined part versus built-up part): ① detail part materials (for example, alloys or heat treat); ② load path; ③ geometry, including section properties, except for minor differences resulting from space limitations within the seat or aircraft interface; ④ attachment method, except for minor differences resulting from space limitations within the seat.

### 相似性; Similarity

适用于在特性和用途上与以前取证的飞机所用系统相似的系统。

Similarity is applicable to systems similar in characteristics and usage to systems used on previously certificated aircraft.

### 相应适航要求; Appropriate Airworthiness Requirements

缔约国为航空器、发动机或螺旋桨的等级所确定、通过或接受的全面而详细的适航规范。

The comprehensive and detailed airworthiness codes established, adopted or accepted by a Contracting State for the class of aircraft, engine or propeller under consideration.

### 详细检查; Detailed Inspection; DI

为了检测损伤、故障或不正常的现象,对一个特定的结构区域、系统、安装件或组件进行的仔细检查。

Careful check of a particular structural area, system, installation and component to detect damage, failure and abnormal indications.

### 详细目视检查; Detailed Visual Inspection; DVI

对特定结构区域、系统、安装或组装组件进行加强的目视检查,以发现损伤或失效。通常可使用检查员认为强度适当的良好光源,作为单一光束的可用照明。可使用检查辅助设备如镜子、放大镜等。可要求表面清洗和详细的接近方法。

An intensive visual examination of a specific structural area, system, installation or assembly to detect damage, failure or irregularity. Available lighting is normally supplemented with a direct source of good lighting at intensity deemed appropriate by the inspector. Inspection aids such as mirror, magnifying lenses may be used. Surface cleaning and elaborate access procedures may be required.

### (证后)项目工程师;Project Engineer; PE

由证后管理部门指定的,对获得型号合格证后的民用航空产品设计状态变更和制造过程中出现的设计构型偏离进行日常管理和监控的人员。

Project engineers are persons who are appointed by competent certification organization, and responsible for routine management and monitoring during post TC/TDA for aircraft design modifications, and design configuration deviation in manufacturing process.

### 项目研制保证;Project Development Assurance

用于以足够的置信度证明研制差错已被发现和纠正,且该项目满足所确定的一系列要求的所有经规划的系统性任务。

All of those planned and systematic tasks used to substantiate, to an adequate level of confidence, that development errors have been identified and corrected that the items satisfy a defined set of requirements.

### 消耗件手册;Consumable Products Manual; CPM

为适应主制造商设备和部件及其供应商部件的翻修和修理,消耗件手册应包括主制造商及其供应商手册和程序中要求的所有消耗器材。

The consumable products manual shall include all the consumable materials that may be called for in the manufacturers and their vendor's manuals and processes, for the overhaul or repair of the prime manufacturer's equipment and components and their vendor's components.

### 小概率;Remote

在单架飞机的整个运行寿命期间不太可能发生,但是考虑所有该型号飞机包括其安装的发动机的总运行寿命期间的话,可能发生几次。在使用这个词时,其对应数值通常被认为概率在 $10^{-7} \sim 10^{-5}$(按飞行小时计)。

Remote means that unlikely to occur during each aircraft's total operational life, but may occur several times when considering the total operational life of a number of aircraft of the type in which the engine is installed. When this term is used，numerical values may normally be interpreted as a probability in the range of $10^{-7} \sim 10^{-5}$ per hour of flight.

### 小修；Minor Repair

大修以外的修理。

A repair other than a major repair.

### 协议维修单位；Contracted Maintenance Organization

通过与运营人正式签订协议接受委托和授权,根据运营人的维修方案、维修技术要求和改装方案,选择或者安排实施维修的工作,并至少在运营人基地提供航线维修的维修单位。

Maintenance organizations those sign an agreement to accept the entrustment and authorization，select or carry out maintenance according to the operator's maintenance programme，maintenance technical requirements and alteration programme，and provide line maintenance in the operator's base at least.

### 斜交轮胎；Bias Tire

帘布层延伸至轮胎卷边且与胎面中心线呈远小于 90 度角交替排列的充气轮胎。

A pneumatic tire in which the ply cords extend to the beads and are laid at alternate angles substantially less than 90 degrees to the centerline of the tread.

### 斜掠翼；Swept Wing

后掠或前掠的机翼。

A wing with either sweepback or sweepforward.

### 新件；New Part

没有使用时间或循环经历的航空器部件(制造厂型号审定过程中的审定要求经历或台架试验除外)。

Aviation part without using time or cycle(except of the time required by certification requirements or bench test).

## 新生产的飞机;New Produced Aircraft

仅在生产认证或型号认证下生产的一架新飞机。

A new aircraft manufactured under a production certificate or a type certificate only.

## 新运营人;New Operator

首次使用某种型号航空器的运营人。需要特别说明的是:不论某运营人首次使用的某型号航空器是全新的还是使用过的,都将该运营人视为新运营人。

The first use of a certain type of aircraft. Special note: the operator will been considered as a new operator whether the aircraft is used for the first time or used.

## 信息通告;Information Bulletin; IB

民航局各职能部门下发的反映民用航空活动中出现的新情况以及对国内外有关民航技术存在的问题进行通报的文件。

The briefing issued by the various responsible departments in CAAC to reflect the new situation in the civil aviation activities and the existing technical problems of civil aviation at home and abroad.

## 信用;(DO - 178B) Credit

RTCA/DO - 178B 软件生命周期数据支持的对一个或多个 RTCA/DO - 178B 的目标符合性。该符合性表明设备符合合格审定基础,可能会获得证书。有三种类型的信用。① 全部信用: 完全符合 RTCA/DO - 178B 目标,并不要求申请人或集成商进一步活动;② 部分信用: 部分满足 RTCA/DO - 178B 目标,需要申请人或集成商通过更多的活动表明符合性;③ 无信用: 不符合 RTCA/DO - 178B 目标,申请人或集成商必须完成对该目标的符合性。

The compliance to one or more RTCA/DO - 178B objectives supported by RTCA/DO - 178B software life cycle data. This compliance shows the equipment meeting the certification basis and that may receive a certificate. There are three types of credit. ① Full credit: fully meets the RTCA/DO -

178B objective and requires no further activity by the applicant or integrator; ② partial credit: partially meets the RTCA/DO - 178B objective and requires more activity by the applicant or integrator to complete compliance; ③ no credit: does not meet the RTCA/DO - 178B objective, and the applicant or integrator must complete the objectives for compliance.

### 形态(用于飞机时);Configuration (as Applied to the Aeroplane)

影响飞机空气动力特性的活动部件位置的特定组合,如襟翼和起落架。

A particular combination of the positions of the moveable elements, such as wing flaps and landing gear, that affect the aerodynamic characteristics of the aeroplane.

### 型号;Type

① 用于航空人员的合格审定、定级、特权和限制时,指航空器的特定构造和基本型式,包括不改变其操纵特性或飞行特性的改型,例如 DC - 7、1049 和 F - 27。② 用于航空器的合格审定时,指设计相似的航空器,例如 DC - 7 和 DC - 7C,1049G 和 1049H,及 F - 27 和 F - 27F。③ 用于航空器发动机的合格审定时,指设计相似的发动机,例如 JTBD 和 JTBD - 7 是同一型号的发动机,JT9D - 3A 和 JT9D - 7 是同一型号的发动机。

① As used with respect to certification, ratings, privileges, and limitations or airmen, means a specific structure and basic model of aircraft, including modifications that do not change its handling or flight characteristics, such as DC - 7, 1049, and F - 27. ② As used with respect to the certification of aircraft, means aircrafts which are similar in design, such as DC - 7 and DC - 7C, 1049G and 1049H, F - 27 and F - 27F. ③ As used with respect to certification of aircraft engines, means engines which are similar in design, for example, JTBD and JTBD - 7 are the same type, JT9D - 3A and JT9D - 7 are the same type.

### 型号合格审查组;Type Certificate Team; TCT

由责任审查部门负责组建,经型号合格审定委员会审议批准的审查团队,一般包括工程审查代表、制造符合性检查代表和设计保证系统审查代表。

A review team established by the responsible review department and approved by the Type Certification Board, generally including engineering representatives, manufacturing inspection representatives and design assurance system representatives.

### 型号合格审定；Type Certification

中国民用航空局对民用航空产品(指民用航空器、航空发动机或者螺旋桨)进行设计批准的过程(包括颁发型号合格证及对型号设计更改的批准)。

Type certification is the process of design approval of civil aviation products (referring to civil aircraft, engine or propeller) by the Civil Aviation Administration of China (including the issuance of type certificate and the approval of type design changes).

### 型号合格审定基础；Type Certification Basis

经型号合格审定委员会(TCB)确定的，对某一民用航空产品进行型号合格审定所依据的标准。型号合格审定基础包括适用的适航规章、环境保护要求及专用条件、豁免和等效安全结论。

The basic standard which is determined by TCB, and used in the type certification to the civil aviation product, including appropriate regulations, environmental protect requirements, specific condition, exemption and equivalent safety conclusion.

### 型号合格审定委员会；Type Certification Board；TCB

型号合格审定项目的管理团队，负责监控型号合格审定项目的审查工作，解决审查中出现的重大问题，并确定完成型号审查的项目计划。

An administration management team responsible for acquainting the applicant with the certification process, resolving significant problems, and establishing a schedule for the overall accomplishment of the type certification project.

### 型号合格审定委员会会议；Type Certification Board Meeting；TCBM

型号合格审定委员会(TCB)和申请人协商项目计划或解决问题的正式会

议,包括初步会议、中间会议、航前会议及最终会议。

Any formal meeting between the TCB and the applicant to coordinate the move to the next project phase or to resolve issues preventing progress to the next phase. Examples include preliminary, interim, pre-flight, and final TCBM.

### 型号合格证;Type Certificate; TC

中国民用航空局根据《民用航空产品和零部件合格审定规定》(CCAR‑21)颁发的、用以证明民用航空产品符合相应适航规章和环境保护要求的证件。型号合格证包括以下内容:型号设计、使用限制、数据单、有关适航要求和环境保护要求,以及对民用航空产品规定的其他条件或限制。

Type certificate (abbreviated to TC) is a certificate issued by CAAC according to the "Certification Procedures for Civil Aviation Products and Parts" (CCAR‑21) to approve the type design of civil aviation products which have complied with applicable airworthiness regulations and environmental protection requirements. A TC includes: type design, operation limitation, data sheet, the relevant airworthiness requirements and environmental requirements, and any other prescribed conditions or limitations for civil aviation product.

### 型号合格证持有人;Type Certificate Holder; TCH

可以使用型号合格证来制造新飞机、发动机、螺旋桨的个体,此人需要具有局方接受的书面许可协议。

A type certificate holder is a person allowed to use the type certificate to manufacture a new aircraft, aircraft engine, or propeller, who must be provided with a written licensing agreement acceptable to the Authority.

### 型号合格证更改;Amended TC(Type Certificate)

由型号合格证持有人提出并获得批准的 TC 更改。只有型号合格证持有人可以申请型号合格证更改。

Amended TC refers to approved TC change proposed by the TC holder. Only TC holders can apply for Amended TC.

### 型号合格证数据单;Type Certificate Data Sheet; TCDS

与型号合格证或型号设计批准书同时颁发,并构成型号合格证或型号设计批准书组成部分的文件。它记载了经批准的型号设计的基本数据和使用限制。

A part of the TC that documents the conditions and limitations necessary to meet the airworthiness requirements. It provides a concise definition of the configuration of a TC'd product.

### 型号检查报告;Type Inspection Report; TIR

按 TIA 授权进行的检查及地面和飞行试验的记录,用以确认对相关规章要求(CCAR-21 第 21.33 条和第 21.35 条)的符合性。TIR 由制造检查和试飞人员完成。对颁发了 TIA 的审定项目,TIR 还记录试验件标识信息和局方审定活动的其他适用信息。

A record of inspections and ground and flight tests conducted under the Authority of the TIA to verify compliance with the applicable regulations (CCAR-21.33 and 21.35). TIR is completed by manufacturing inspection and flight test personnel. For approval projects that have been issued by TIA, TIR also records the identification information of test articles and other applicable information of certification activities.

### 型号检查核准书;Type Inspection Authorization; TIA

由型号合格审定审查组组长签发的,批准审查代表(含委任代表)对航空器原型机进行审定飞行试验前检查、现场目击或进行飞行试验的文件。型号检查核准书(TIA)中明确了检查和审定飞行试验审查的具体要求。对结构试验和工艺试验的检查不使用型号检查核准书(TIA),用制造符合性检查请求单。

The TIA is a document issued by the leader of the type certification team, approving the review representative (including the designated representative) to conduct preflight test inspection, on-site witnessing, or flight test on the prototype aircraft. The TIA specifies the specific requirements for inspection and certification flight test. For the inspection of structural and process tests, the TIA is not used, but the manufacturing compliance inspection request form is used.

### 型号认可;Type Validation

进口产品符合进口国的需求或适航标准的型号审定。

Type validation is type certification of an imported product to the importing country's applicable requirements or airworthiness standards.

### 型号认可证后;Post-type Validation

型号获认可后的设计更改,维修使用的数据和包括服务信息在内的对适航资料的批准程序。持续适航活动也是型号认可证后活动。

Process leading to approval of post-type certification design changes, data used in repairs, and airworthiness data including service information. Continued airworthiness activities are also post-type validation activities.

### 型号设计;Type Design

根据《民用航空产品和零部件合格审定规定》(CCAR-21)第21.31条规定,型号设计包括: ① 定义航空器构型和设计特征符合有关适航规章和环境保护要求所需的图纸、技术规范及其清单; ② 确定民用航空器结构强度所需要的尺寸、材料和工艺资料; ③ 适航规章中规定的持续适航文件中的适航性限制部分; ④ 通过对比法来确定同一型号后续民用航空器的适航性和适用的环境保护要求的其他资料。

According to the civil aviation products and parts certification requirements of (CCAR-21) 21.31, type design including: ① the drawings, technical specifications and its list required to definite the aircraft configuration and design features complying with the relevant airworthiness regulations and environmental protection requirements; ② structure size, material and technology information used to determine the strength of civil aircraft; ③ the airworthiness limitations section within the continuing airworthiness regulations; ④ by comparing the same type civil aircraft's subsequent airworthiness and environmental protection requirements applicable to other information.

### 型号设计批准书数据单;TDA Data Sheet

与型号合格证或型号设计批准书同时颁发并构成型号合格证或型号设计

批准书组成部分的文件。它记载了经批准的型号设计的基本数据和使用限制。

Documents issued with the Type Certificate or Type Design Approval simultaneously as part of Type Certificate or Type Design Approval, which records the basic data and operation limitations of the approved type design.

### 型号设计资料;Type Design Data

型号设计包括的资料称为型号设计资料。

All information including in type design is called type design information.

### 型号审定委任单位代表;Designated Type Certification Organization Representative, DTCOR

指根据 AP‐183‐AA‐2018‐11《适航委任单位代表管理程序》获得民航行政机关的授权,代表民航行政机关执行部分或全部型号审查工作的单位或机构。型号审定委任单位代表是委任单位代表的一种。

The organization or institution authorized by the civil aviation administrative organ to carry out part or all of the type certification on behalf of the civil aviation administrative organ in accordance with AP‐183‐AA‐2018‐11 Airworthiness Designated Organization Representative Management Procedure. The designated type certification organization representative is one of the authorized organization representatives.

### 型号资料;Type Data

型号设计资料与符合性验证资料的统称。

The data including type design data and compliance verification data.

### 型号资料批准表;Type Data Approval Form

由审查代表填写,用于证实型号资料已经过审查、符合要求并予以批准的表格。

The sheet to be filled in, and to validate type data have been reviewed, conformed, approved.

### 型号资料评审表；Type Data Review Form

由审查代表或授权的委任工程代表填写，用于记录型号资料审查过程以及向申请人反馈型号资料审查意见的表格。

The sheet to be filled in, and to record the process of type data review, giving feedback to applicant with review questions.

### 修订的补充型号合格证；Amended STC；ASTC

由补充型号审定合格证持有人提出并获得批准的 STC 修订。只有补充型号审定合格证持有人能申请修订补充型号审定合格证。

An approval for a change to a STC, made by the STC holder. Only the holder of the STC may apply for an amended STC.

### 修订的补充型号合格证项目；Amended STC(SA) Project

导致已有的补充型号合格证修改的审定项目，该项目是用于型号设计的重大改变，不论其补充型号是否发生了物理改变。只有补充型号审定合格证持有人可以申请补充型号审定合格证修订。

A certification project that will lead to the amendment of an existing supplemental type certificate. This project designator is used for major changes to type design whether or not the supplemental type certificate is physically altered. Only the holder of the STC may apply for an amended STC.

### 修订的型号合格证；Amended TC；ATC

由型号合格证持有人提出并获得批准的 TC 修订。只有型号审定合格证持有人可以申请修订型号审定合格证。

An approval for a change to a TC, made by the TC holder. Only the holder of the TC may apply for an amended TC.

### 修订的型号合格证项目；Amended TC (AT) Project

型号审定项目通过对已存在的型号合格证增加新的产品，或者修订已存在的型号合格证，实现对一个新的飞机、发动机或者螺旋桨型号的批准。该项目是用于型号设计的重大更改，不论其型号合格证是否发生了物理改变。只有型号审定合格证持有人可以申请型号审定合格证修订。

A certification project that will lead to an approval of a new aircraft，engine or propeller model by adding the new product to an existing type certificate，or amending the existing type certificate. This project designator is used for major changes to type design whether or not the type certificate is physically altered. Only the holder of the TC may apply for an amended TC.

### 修理；Repair

在某一航空产品损坏或被磨损之后，将其恢复到适航状态，以确保该航空器继续符合为颁发该型航空器型号合格证适用的相应的适航性要求。

The restoration of an aeronautical product to an airworthy condition to ensure that the aircraft continues to comply with the design aspects of the appropriate airworthiness requirements used for the issuance of the type certificate for the respective aircraft type，after it has been damaged or worn.

### 修理规范委任工程代表；Repair Specification DER

被授权在修理规范中管理和批准技术数据的委任工程代表。

The specific authority to manage and approve technical data in repair specifications granted by DER.

### 修理级别；Repair Classifications

① A 级：一种永久修理方式，可以获得等同于周边未修理区域的基准线区域检查，足以确保持续适航（可检查性）。运营人需要向局方展示维护或检查方案至少与基准线区域检查一样严格；② B 级：一种通过补充检查来确保持续适航的永久修理方式；③ C 级：一种在某一指定时间前需要重做或更换的临时修理方式，可能需要补充检查以确保在这个时间限制前的持续适航。

① Category A：a permanent repair for which the BZI is adequate to ensure continued airworthiness（inspectability）equal to the unrepaired surrounding structure. The operator should demonstrate to the Authority that its maintenance or inspection program is at least as rigorous as the BZI. ② Category B：a permanent repair that requires supplemental inspections to ensure continued airworthiness. ③ Category C：a temporary repair that need to be reworked or replaced before an established time limit. Supplemental inspections may be necessary to ensure continued airworthiness before this limit.

### 修理评估方案；Repair Assessment Program

基于制造厂发布的修理评估指南或其他局方批准的修理评估指南开发的特定方案。

The specific program，as developed using the manufacturer's repair assessment guidelines（RAG）or other Authority-approved RAG.

### 修理评估指南；Repair Assessment Guidelines；RAG

飞机制造厂发布的指导如何进行某种特定飞机型号修理评估的指南。

Airplane manufacturer's model-specific guidelines on how to perform the repair assessment.

### 修理设计批准书；Repair Design Certificate；RDC

RDC 记录恢复航空产品适航状态至初始设计特征的修理设计数据的批准书，同时记录发生过超出容限的损坏或缺陷的航空产品的序列号信息。

An approval letter recording the repair design data for restoring the airworthiness status of the aviation product to its initial design features，and recording the serial number information of the aviation product that has undergone damage or defects beyond tolerance.

### 修理站；Repair Station

获得 145 部许可的航空器维修供应商。

An aircraft maintenance provider certificated by the Authority in accordance with part 145.

### 修理站手册；Repair Station Manual；RSM

描述修理站运营中政策和程序要求的手册。

A manual that describes the procedures and policies of the operations of a repair station.

**修理站运营规范; Repair Station Operations Specification; Repair Station OPSPECS**

标志出修理站等级、授权、限制的局方文件。

An Authority document identifying the repair stations ratings, authorizations, and limitations.

**修理站质量控制手册; Repair Station Quality Control Manual; QCM**

描述修理站检查和质量控制程序的手册。

A manual that describes the inspection and quality control procedures used by the repair station.

**修理周期; Repair Time**

从厂家收到产品及采购订单到产品被修复至可装运状态的时间。

The time begins from repair station receiving product and relative purchase order, ends to product repaired and under available of shipment.

**修正案; Amendment**

立法机关对现有法律的修改,这样做可以避免重新立法,节省资源,同时不会改变现行法律的总条数,有利于法律的稳定性。

Amendment is changes to the existing laws made by legislature, which can avoid re-legislate and save resources, not changing the total number of existing laws, which is conducive to the stability of the law.

**修正案对应文件; Amendment Pair**

14 CFR 修正案的数量和对应的 CS/JAR 在特定的日期进行更改和修正生效。在这个指令中,修正案特指对飞机、旋翼机、气球、发动机、螺旋桨的适航性标准。当一个新的 14 CFR 修正案或一个新的 JAR 更改修正有效时,一个新的修正案就成立。

The number of the 14 CFR amendments and numbers of both comparable CS/JAR change and amendment in effect on a specific date. For this order, the term is specific to airworthiness standards for airplanes, rotorcraft, balloons, engines, and propellers. A new amendment pair is established when either a new 14 CFR amendment or a new JAR change or amendment to a change becomes effective.

### 虚假警报；False Advisory

由虚假航线或空中告警与防撞系统故障引起的警报。

An advisory caused by a false track or a TCAS malfunction.

### 需立即采取措施的警告；Time-critical Warning

警告的一种，用来保持飞机安全运行的最紧急警告级别。

A subset of warning, which is the most urgent warning level to maintain the immediate safe operation of the airplane.

### 许可生产项目单；Production Limitation Record

生产许可证的一部分。许可生产项目单列出准许生产许可证持有人生产的每一民用航空产品的型号合格证、补充型号合格证、改装设计批准书、型号认可证或者补充型号认可证的编号和型别。

The production limitation record is part of PC. TC, STC, MDA, VTC or VSTC of each civil aviation product produced by the PC holder shall be listed separately for the production limitation record.

### 许用值；Allowable

在概率基础上（如分别具有 99％概率和 95％置信度，与 90％概率和 95％置信度的 $A$ 或 $B$ 基准值），由层压板或单层级的试验数据确定的材料值。确定这些值所需的数据量由统计显著性（或基准）决定。

Material values that are determined from test data at the laminate or lamina level on a probability basis (for example, $A$ or $B$ basis values, with 99％ probability and 95％ confidence, or 90％ probability and 95％ confidence, respectively). The amount of data required to derive these values is governed by the statistical significance (or basis) needed.

### 巡航；Cruise；CRZ

飞机的高度和速度基本没有变化的飞行阶段。

The phase of flight where altitude and speed are basically constant.

## 巡航高度;**Cruise Altitude; CA**

在水平巡航飞行期间,飞机所保持的高度或飞行高度层。

An altitude or flight level maintained during enroute level flight.

## 训练规范;**Training Specifications**

由中国民用航空局或民航地区管理局向经审定合格的飞行训练中心颁发的飞行训练中心合格证的补充文件,包括对训练中心的训练批准和限制。

Supplement document of flight training center certification issued by the civil aviation administration of China or the regional civil aviation administration to the flight training center qualified of certification,including approval and restrictions to training center training.

# Y

**压舱物;Ballast**

为保持飞机重心在许可范围内装载的配重物。

A weight installed in an aircraft to maintain the center of gravity within allowable limits.

**压力高度;Pressure Altitude; PA**

① 一般指标准基准平面之上的高度,通过把标准压力递减率应用到某一高度的大气压力来确定;② 指示压力高度或校准压力高度;③ 低压腔内的模拟压力高度。

① Generically，means altitude above the standard datum plane，as determined by applying the standard pressure lapse rate to the atmospheric pressure at altitude; ② either indicated pressure altitude or calibrated pressure altitude; ③ a simulated pressure altitude inside a low-pressure chamber.

**亚声速飞机;Subsonic Airplane**

最大使用限制速度 $M_{mo}$ 不超过马赫数 1 的飞机。

An airplane for which the maximum operating limit speed $M_{mo}$ does not exceeds a Mach number of 1.

**延程运行备降机场;ETOPS Alternate Airport**

列入合格证持有人运行规范并且在签派或放行时指定的,在延程运行改航时可使用的合适机场。这一定义适用于飞行计划,并不以任何方式限制飞行过程中飞行员的指令。

An adequate airport listed in the certificate holder's operations specifications that is designated in a dispatch or flight release for use in the event of a diversion during ETOPS. This definition applies to flight planning and does not in any way limit the authority of the piloting-command during flight.

### 延程运行关键系统;Extended Range Key System

包括动力系统在内的飞机系统,当其失效或发生故障时会危及延程运行安全,或危及飞机在延程运行改航时的持续安全飞行和着陆。

Aircraft system including power plant system when failure or fault occurs, it will endanger extended range operations, or will endanger continuous safe flight/landing(CSF/L) of aircraft during enroute diversion of extended range operation.

### 延程运行区域;ETOPS Area of Operation

下列区域之一:① 对双发涡轮发动机提供动力的飞机,指在标准条件下静止大气中,以一台发动机不工作的巡航速度飞行时间超过 60 分钟才能抵达一个合适机场的区域;② 对两台以上涡轮发动机提供动力的载客飞机,指在标准条件下静止大气中,以一台发动机不工作的巡航速度飞行超过 180 分钟才能抵达一个合适机场的区域。

One of the following areas:① for turbine-engine-powered airplanes with two engines, an area beyond 60 minutes from an adequate airport, computed using a one-engine-inoperative cruise speed under standard conditions in still air;② for turbine-engine-powered passenger-carrying airplanes with more than two engines, an area beyond 180 minutes from an adequate airport, computed using a one-engine-inoperative cruise speed under standard conditions in still air.

### 延程运行双重维护;ETOPS Dual Maintenance

在计划维修或非计划维修时,对单独的 ETOPS 重要系统相同元件的维护。对"实质上类似"ETOPS 重要系统的双重维护指在同一维护工作中,对 2 台发动机的发动机驱动组件的维护工作。

Maintenance actions performed on the same element of identical, but separate ETOPS significant systems during a scheduled or unscheduled maintenance visit. For "substantially similar" ETOPS significant systems, the term means maintenance actions performed on engine-driven components on both engines during the same maintenance visit.

### 延程运行重要系统;ETOPS Significant system

飞机的一种系统,包括推进系统,其失效或故障可能对 ETOPS 飞行的安全性或者对 ETOPS 改航期间的飞机继续安全飞行和着陆有不利影响。

An airplane system, including the propulsion system, the failure or malfunctioning of which could adversely affect the safety of an ETOPS flight, or the continued safe flight and landing of an airplane during an ETOPS diversion.

### 延期维护;Deferred Maintenance

推迟修理或更换设备或仪表的元器件。

The postponement of the repair or replacement of an item of equipment or an instrument.

### 延伸航程运行;Extended Range Operation; ETOP

具有两台涡轮动力装置的固定翼航空器从航路上任一点以一台动力装置不工作的巡航速度(在国际标准大气和无风条件下),到一可用备降机场的飞行时间超过经营人所在国批准的门限时间的任何飞行。

Any flight by an aeroplane with two turbine power-units where the flight time at the one power-unit inoperative cruise speed (in ISA and still air conditions), from a point on the route to an adequate alternate aerodrome, is longer than the threshold time approved by the state of the operator.

### 延伸跨水运行;Extended Over-water Operation

航空器距最近海岸线的水平距离超过 93 千米(约 50 海里)的跨水运行。

An aircraft operation over water at a horizontal distance of more than 93 km (about 50 n mile) from the nearest shoreline.

## 延误；Delay

延误是指航班降落时间（航班实际到港挡轮挡时间）比计划降落时间（航班在时刻表上的时间）延迟 15 分钟以上或航班取消的情况。

Delay refers to the situation where the landing time (actual arrival time of the flight) is delayed by more than 15 minutes compared to the planned landing time (time on the flight schedule) or the flight is cancelled.

## 严重不符合项；Major Noncompliance

不符合项可能导致质量系统的整体失效，或显著降低质量系统保证受控活动或产品符合要求的能力。可以包括下述一种或多种情形：① 可能导致质量系统对满足适用规章的要求出现显著缺失或整体失效；② 可能妨碍航空器继续安全飞行或着陆，减少安全裕度、降低飞行性能或导致丧失某些飞行操作能力；③ 可能降低飞机性能或削弱机组对不利操作条件或后续故障的处理能力。

Such noncompliance may result in an overall failure of the quality system or significantly reduce the ability of the quality system to ensure that controlled activities or products meet requirements, including one or more of the following: ① may result in a significant deficiency or total failure of the quality system to meet applicable regulatory requirements; ② may prevent the aircraft from continuing fly or landing safely, reduce safety margins, reduce flight performance or cause the loss of certain flight operation capabilities; ③ may degrade aircraft performance or impair the crew's ability to handle adverse operating conditions or subsequent failures.

## 严重结冰；Severe Icing

通常作为飞行机组向空中管制报告遇到的结冰强度的一种描述。

A description used operationally by flight crews reporting encountered icing intensity to traffic control.

## 严重伤害；Serious Injuries

具有下列情况的损害称为严重伤害：① 从受到伤害开始之日起七日内，需要住院治疗超过 48 小时的；② 导致骨折（除简单的手指、脚趾或鼻子骨折外）；

③ 涉及撕裂伤,导致严重出血,神经、肌肉和肌腱损伤;④ 涉及任何内部器官损伤;⑤ 涉及二度或三度烧伤或灼伤,影响超过身体表面百分之五的;⑥ "致命伤害",指会在事故发生后 30 天内导致死亡的伤害。

Any injury that: ① requires hospitalization for more than 48 hours, commencing within seven days from the date the injury was received; ② results in the fracture of any bone (except simple fractures of fingers, toes or nose); ③ involves lacerations that cause severe hemorrhages, nerve, muscle or tendon damage; ④ involves injury to any internal organ; ⑤ involves second- or third-degree burns or any burns affecting more than five percent of the body surface; ⑥ "fatal injury" is defined as an injury that results in death within 30 days of the accident.

### 严重性;Severity

发生危险情况的后果,用人员伤亡、飞机损伤、安全裕度降低或机组工作量增加等表示。

The consequence if the hazard occurs, expressed in terms of injury to personnel, damage to aircraft, reduction in safety margins, or increase in crew workload.

### 研发(工程)模型;Development (Engineering) Model

为满足规范的性能要求和/或为生产设备制定技术要求而设计的模型。

A model designed to meet performance requirements of the specification and/or establish technical requirements for production equipment.

### 研制保证;Development Assurance

所有具有足够的置信度水平,用来证明研制错误已被识别和纠正,使系统满足适用审定基础的计划的和系统性的行为。

All planned and systematic actions used to substantiate, at an adequate level of confidence, that errors in requirements, design and implementation have been identified and corrected such that the system satisfies the applicable certification basis.

**研制错误；Development Error**

在决策、设计或应用等需求方面的错误。

A mistake in requirements of determination, design or implementation.

**衍生风险；Substitute Risk**

风险控制无意中带来的新风险。

A risk unintentionally created as a consequence of safety risk control(s).

**衍生型；Variant**

一种特别构型的航空器，局方已明确其训练和资格认定显著不同于同一制造厂家、型号和系列的其他航空器。

A specifically configured aircraft for which the Authority has identified training and qualifications that are significantly different from those applicable to other aircraft of the same manufacturer, model, and series.

**衍生型号航空器；Derivative Aircraft**

继首次获得局方型号合格证（或认可证）之后，通过设计更改改变构型并在同一型号合格证（或认可证）数据单中单独列出的航空器型号。若有些航空器型号系列在首次申请局方型号合格证（或认可证）时就同时申请了多个构型的航空器型号，则在首个构型航空器型号之后单独列出的型号均视为衍生型号航空器。

The aircraft type which is separately listed in the data sheet of the same type certificate (or validation of type certificate) after obtaining the type certificate (or validation of type certificate) of the Authority for the first time and changing the configuration through design modification. Some aircraft model series may apply for multiple aircraft models at the same time when they first apply for the type certificate (or accreditation certificate) of the Authority, and the models listed separately after the first aircraft model are regarded as derivative aircraft.

**衍生需求；Derived Requirements**

研制过程中，由设计或实施的决策所引起的、不能直接追溯到更高层次需求的附加需求。

Additional requirements resulting from design or implementation decisions during the development process which are not directly traceable to higher-level requirements.

### 眼位参考点；Eye Reference Point；ERP

可提供有限参考的驾驶舱中的一点,通过该点可精确判断定义驾驶舱和显示器布局的几何体。

The point in the cockpit that allows for a finite reference enabling the precise determination of geometric entities that define the layout of the cockpit and displays.

### 要求的特殊飞机和机组授权；Special Aircraft and Aircrew Authorization Required；SAAAR

飞机可配备超出公共 RNP 标准的最低标准要求和经训练的机组,以完成更高级别的仪表进近。

Aircraft may be equipped beyond the minimum standard for public RNP criteria and aircrews trained to achieve a higher level of instrument approach performance.

### 要求的障碍净空；Required Obstruction Clearance；ROC

在仪表程序航段的 OEA 空域内,航空器和地面最高的障碍物之间的最小垂直距离(以英尺计)。

The minimum vertical clearance (in feet) that exists between aircraft and the highest ground obstruction within the OEA of instrument procedure segments.

### 叶片；Blade

风扇、压缩机和涡轮的翼型部分(不包括叶台和叶根)。

The airfoil sections (excluding platform and root) of the fan, compressor and turbine.

### 叶片冲击损伤；Blade Impact Damage

在飞行中或在地面上,由雪堆、跑道灯、维护平台、鸟类等物体撞击叶片所导致的可见或不可见的损伤。

Damage (visible or not) resulting from a blade striking or being struck, while in flight or on the ground, by an object such as a snow bank, runway light, maintenance stand, birds.

## 夜间；Night

黄昏结束至晨曦出现之间的时间，或由适航当局规定的由日落到日出之间的其他时间。暮光在晚上日轮中心低于地平线 6 度时结束，曙光在早上日轮中心高于地平线 6 度时开始。

The hours between the end of evening civil twilight and the beginning of morning civil twilight or such other period between sunset and sunrise, as may be prescribed by the appropriate authority. Civil twilight ends in the evening when the centre of the sun's disc is 6 degrees below the horizon and begins in the morning when the centre of the sun's disc is 6 degrees below the horizon.

## 一般飞行事故；General Flight Accident

下列情况之一者：① 人员重伤，重伤人员在 10 人及以上；② 最大起飞重量为 5.7 吨（含）以下的航空器严重损坏，或迫降在无法运出的地方；③ 最大起飞重量为 5.7～50 吨（含）的航空器一般损坏，其修复费用超过事故当时同型或同类可比新航空器价格的 10%（含）；④ 最大起飞重量为 50 吨以上的航空器一般损坏，其修复费用超过事故当时同型或同类可比新航空器价格的 5%（含）。

One of following cases: ① personnel serious injury, the number is ten or above; ② serious damage of the aircraft which is 5.7 t or below in maximum take-off weight, or where the aircraft can not tow out after landing; ③ general damage of the aircraft which is 5.7～50 t(including) in maximum take-off weight, and the repair cost is 10%(including) more than the price of a new aircraft with the same or similar model at that moment; ④ general damage of the aircraft which is above 50 t in maximum take-off weight, and the repair cost is 5%(including) more than the price of a new aircraft with the same or similar model at that moment.

## 一般改装；General Alteration

除重大改装以外的改装。

The alteration except major alteration

## 一般高原机场；General Plateau Airport

海拔高度在 1 500 米(约 4 922 英尺)及以上,但低于 2 438 米(约 8 000 英尺)的机场。

The altitude of a airport is between 1 500 m(4 922 ft) to 2 438 m(8 000 ft).

## 一般目视检查；General Visual Inspection；GVI

为查找明显的损伤、故障或缺陷而对内部、外部区域、安装件或组件进行的目视检查。

Visual check performed on internal, external area, installation parts or components to look for obvious damage, failure or defect.

## 一般事件；General Event

在民用航空器运行阶段或者在机场活动区内发生的与航空器有关的航空器损伤、人员受伤或者其他影响安全,但其严重程度未构成事故征候的事件。

Aircraft injuries, personnel injuries or other situations affecting safety that occur during the operation of civil aircraft or in the airport activity area, whose severity does not constitute incidents.

## 一般维修事故；General Maintenance Accident

维修造成下列情况之一者为一般维修事故：① 造成航空器及部件在地面损坏,直接经济损失超过事故当时同型或同类可比新航空器(最大起飞重量小于或等于 5.7 吨的航空器除外)整机价格的 0.5％或直接经济损失达 50 万元(含)～100 万元,以低限为准；② 地面设备、厂房设施损坏,直接经济损失达 50 万元(含)～100 万元；③ 人员重伤。

The general maintenance accident is considered by maintenance reasons to cause one of the below conditions： ① leading to aircraft and components damage on the ground, the direct economic loss of which is 0.5％ more than the price of a new aircraft with the same or similar model (except for the aircraft with maximum take-off weight less than or equal to 5.7 t) at that moment, or between 500 000 RMB(including) and 1 000 000 RMB, whichever comes lower; ② the damage for the ground equipment and factory facilities, the direct economic loss of which is between 500 000 RMB(including) and 1 000 000 RMB; ③ personnel serious injury.

## 一般修理；General Repair

除重大修理以外的修理。

To repair except major repair.

## 一次雷达；Primary Radar

一种雷达系统，从雷达配置点发射出的无线电脉冲的小部分被物体反射，再在此配置点被回收。

A radar system in which a minute portion of a radio pulse transmitted from a site is reflected off an object and then received back at that site.

## 一级腐蚀；Level 1 Corrosion

通过连续检查发现的局部腐蚀现象，可以采用返工等制造商结构维修手册、服务通告等所认可的维修手段。

Localized corrosion occurring between successive inspections and can be repaired by means of rework, such as approved in the manufacture's structural maintenance manual, service bulletin, etc.

## 一台发动机不工作的巡航速度；One-engine-inoperative-cruise Speed

由合格证持有人选定且经局方批准的在飞机额定限制范围内的速度，用于：① 计算一台发动机不工作时所需燃油储备；② 确定在延程运行中飞机能否在批准的最长改航时间内飞抵延程运行备降机场。

A speed within the certified operating limits of the airplane that is specified by the certificate holder and approved by the authority for: ① calculating required fuel reserves needed to account for an inoperative engine; ② determining whether an ETOPS alternate is within the maximum diversion time authorized for an ETOPS flight.

### 仪表；Instrument

靠内部机构，以视觉或音响方式显示航空器的姿态、高度或航空器及其部件的工作状态的一种装置。它包括航空器飞行自动控制的电子装置。

A device using an internal mechanism to show the attitude，altitude，or operation of an aircraft or aircraft part visually or aurally. It includes electronic devices for automatically controlling an aircraft in flight.

### 仪表板；Instrument Panel(Board)

装有仪表或仪表刻度盘的板或盘，一般指装在航空器上的。这种板安装在驾驶员前面，容易被驾驶员看清，亦称"仪表盘"。

A panel or board containing instruments or instrument dials，typically on an aircraft. Such a panel is mounted ahead of，and easily visible to the pilot，also called an "instrument board".

### 仪表飞行规则；Instrument Flight Rules；IFR

规定进行仪表飞行程序的规则。

Rules that govern the procedures for conducting instrument flight.

### 仪表飞行规则/目视飞行规则计划图；IFR/VFR Planning Chart

以灵活的形式提供飞行计划所需的航空信息，可用于仪表飞行规则或目视飞行规则的计划制订。

Provides aeronautical information required for flight planning in a flexible format which can be used for IFR or VFR planning.

### 仪表飞行规则条件；IFR Conditions

低于目视飞行规则最低飞行条件的气象条件。

Weather conditions below the minimum for flight under visual flight rules.

### 仪表飞行气象条件；Instrument Meteorological Conditions；IMC

用能见度、离云距离和云幕高表示的、低于目视气象条件规定的最低标准的气象条件。

Meteorological conditions expressed in terms of visibility，distance from cloud, and ceiling, less than the minima specified for visual meteorological conditions.

### 仪表进近；Instrument Approach

当能见度低于 3 英里(约 4.8 千米)和(或)云底高度处于或低于最小初始高度时，按照仪表飞行规则飞行计划飞行的航空器向机场意图着陆的进近。

An approach to an airport, with intent to land, by an aircraft flying in accordance with an IFR flight plan, when the visibility is less than 3 mi (4.8 km) and/or when the ceiling is at or below the minimum initial altitude.

### 仪表进近程序；Instrument Approach Procedure；IAP

参考飞行仪表进行的一系列预定机动，机动时能够越过规定的障碍物，并能保证接收到导航信号。仪表进近程序从初始进近定位点开始，或当适用时，从定义的到达航路开始点：① 可以到开始完成着陆的一个点；② 如果没有完成着陆，则到达一个适用待机或航路越障准则的位置。

A series of predetermined maneuvers by reference to flight instruments with specified protection from obstacles and assurance of navigation signal reception capability. It begins from the initial approach fix, or where applicable, from the beginning of a defined arrival route to a point: ① to which a landing can be completed; ② if a landing is not completed, to a position at which holding or en route obstacle clearance criteria applies.

### 仪表进近跑道；Instrument Approach Runway

配备有至少可提供直接进近着陆所需方向导引的电子辅助设施的跑道。

A runway served by an electronic aid providing at least directional guidance adequate for a straight-in approach.

### 仪表进近系统；Instrument Approach System

用以引导航空器安全着陆的一种空中导航系统，起始于初始进近点，终止于足以接近地面，能做目视着陆的一点。

An air navigation system used to guide an aircraft to a safe landing beginning at an initial-approach point and ending at a point near enough to the ground to permit a visual landing.

### 仪表系统；Instrument System

由阅读装置及以发动机接口点至各自阅读装置的相关部件（例如传感器、连接电缆、光纤）组成的系统。

A system comprising a readout device and its associated components （such as sensors，interconnecting cables，tubing，fiber optics） from the engine interface point to the respective readout device.

### 仪表运行；Instrument Operation

航空器按照仪表飞行规则飞行计划进行的运行，或者按照由航站管制设施或空中航路交通管制中心提供的仪表飞行规则间隔距离的运行。

An aircraft operation in accordance with an IFR flight plan or an operation where IFR separation between aircraft is provided by a terminal control facility or air route traffic control center.

### 仪表着陆系统；Instrument Landing System；ILS

航空器上提供着陆所需横向、纵向和垂直方向引导的系统。

A system in the aircraft，which provides the lateral，longitudinal，and vertical guidance necessary for a landing.

### 仪表着陆系统差分扫描线；ILS Differential Trace

记录上的扫描线，它是无线电遥测经纬仪（RTT）交叉指针式指示器（DDM）和航空器接收机交叉指针式指示器的代数和，由机载经纬仪记录系统（TRS）内的差分放大器给出。

The trace on the recording which is the algebraic sum of the radio telemetering theodolite （RTT） crosspointer （DDM） and the aircraft receiver crosspointer （DDM），produced by the differential emplifier within the airborne theodolite recording system （TRS）.

### 仪表着陆系统基准点；ILS Reference Datum

指定高度的一点，位于跑道中心线和 ILS 下滑道向下延伸的直线部分通过的入口的交叉点上方。

A point at specified height located vertically above the intersection of the runway centerline and the threshold through which the downward extended straight portion of the ILS glide path passes.

### 仪表着陆系统下滑道扇形区宽度(正常进近包线);ILS Glide Path Sector Width (Normal Approach Envelope)

含有下滑道的垂直平面内扇形区的宽度,由下滑道之上和之下若干点的空间位置予以限制,此处获得的读数为 75 微安。

The width of a sector in the vertical plane containing the glide path and limited by the loci of points above and below the path at which a reading of $75\ \mu A$ is obtained.

### 仪表着陆系统性能效率报告; ILS System Performance Activity Report; ILS SPAR

在规定期间对设施性能的统计分析,它决定了设施保持在规定允差内的概率。设施部件的概率是决定例行飞行检查频度的基础。

A statistical analysis of facility performance over a specified period of time which determines a probability of the facility remaining within established tolerances. The facility component probability is the basis for determining the frequency of routine flight inspections.

### 仪表着陆系统远距指点标;ILS Outer Marker; OM

位于或接近仪表着陆系统(ILS)进近的下滑道拦截高度处的指点信标。在 400 赫兹的音调上,以每秒钟 2 个破折号来键入发射,并由兼容的机载设备在听觉和视觉上接收。

A marker beacon at or near the glideslope intercept altitude of an ILS approach. It is keyed to transmit two dashes per second on a 400 Hz tone, which is received aurally and visually by compatible airborne equipment.

### 仪表着陆系统中距指点标;ILS Middle Marker; MM

一个指点信标,规定为沿着通常位于或接近决断高度点的仪表着陆系统

（ILSⅠ类）下滑道处的一个点。键入发射交替点和破折号，在1 300 赫兹的音调上，以每分钟 95 个点/破折号的组合来交替键入，并由兼容的机载设备在听觉和视觉上接收。

A marker beacon that defines a point along the glideslope of an ILS normally located at or near the point of decision height (ILS Category Ⅰ). It is keyed to transmit alternate dots and dashes, with the alternate dots and dashes keyed at the rate of 95 dot/dash combinations per minute on a 1 300 Hz tone, which is received aurally and visually by compatible airborne equipment.

### 仪器校验；Instrument Calibration

特定仪器与一级标准或二级标准（传递标准）进行的比较。该标准应能追溯至美国国家标准与技术研究院（NIST）或其他国家标准化组织保持的标准。

The comparison of a particular instrument with either a primary or secondary (transfer) standard. The standard should be traceable to a standard maintained by NIST or another national standards organization.

### 移动式交通灯；Portable Traffic Light

一种由机场交通管制员选择，发射白、绿或红色耀眼窄细光束的指挥灯。所发射的灯光颜色和类型可用来批准或不批准预期的驾驶员行动。

A directional light which emits a brilliant narrow beam of white, green, or red light as selected by the airport traffic controller. The color and type of light is used to approve or disapprove anticipated pilot actions.

### 已建立的工业方法；Established Industry Practice

一种被广泛执行的操作方法，能始终如一地履行特定的功能。已建立的工业方法的例子包括重复校验系统和内部审核系统。

A widely followed method of operating that achieves consistent performance of specific functions, including a calibration recall system and an internal audit system.

### 异地培训；Off-site Training

在证件批准地点的范围外进行的培训。

Training performed in a place which is outside of place approved by its certificate.

## 异地取证改装;Off-site Certificated Modification

在 DMDOR 持有人管理手册列出的取证改装地址以外的地点,或不是由 DMDOR 持有人实施的取证改装。

The certificated modification is performed at the location where is not listed in the management manual of the DMMOR holder, or is not performed by the DMMOR holder.

## 异地运行;Operation Out of Main Base

飞机在航空运营人主基地和各分基地以外跨区驻扎运行。

Aircraft stationed and operated across regions outside the main base and satellite bases of the aviation operator.

## 翼尖失速;Tip Stall

机翼或旋翼桨叶尖部的失速,这种失速在飞机上常造成滚转、偏航、俯仰和丧失有效的副翼操纵。

A stall at the tip of a wing or rotor blade in an airplane, which often results in rolling, yawing, pitching, and loss of effective aileron control.

## 翼梢小翼或端板式垂尾;Winglet or Tip Fin

从某个升力面延伸出来的竖立表面,可以有或无操纵面。

A vertical board extending out of the lift surface, with or without control surfaces.

## 翼展;Wingspan

航空器一侧翼尖至另一侧翼尖的距离。

The distance from one wingtip of an aircraft to the opposite wingtip.

## 音频警告;Aural Alert

一个离散的声音、音调或口头声明,用于阐明一个条件、情况或事件。

A discrete sound, tone, or verbal statement used to enunciate a condition, situation, or event.

## 音响式无线电指向标；Aural Radio Range

提供导航所需声音信号的四航道指向标。

A four-course range station which provides audible signals for navigational purposes.

## 引导指点标；Aiming Marker

位于跑道边上的明显标志，作为判断和建立着陆航空器下滑角的定位点，它通常距跑道入口 1 000 英尺。

A clear sign located on the runway edge, usually 1 000 ft from the runway threshold, serves as a positioning point for determining and establishing the landing aircraft's glide angle.

## 引进；Introduction

航空运营人通过签订合法的协议或合同等方式，获得航空器的所有权或使用权。需要特别指出的是，"引进"并不单指从国外进口航空器，也包括航空器在国内所有权或使用权的转移，不包括湿租。

Air operators through a legal agreement or contract to obtain the aircraft ownership of or right to use. Needs to be noted that, "introduction" does not only refer to imports from abroad, but includes the transfer of aircraft or use rights in domestic ownership, not including a wet lease.

## 隐蔽失效；Latent Failure

在被机组或维修人员发现前是隐蔽的。重大隐蔽失效是指在与一个或多个失效或事故同时发生时，会导致危险或灾难性失效的情况。

A failure is latent until it is made known to the flightcrew or maintenance personnel. A significant latent failure is one which would, in combination with one or more specific failures or events, result in a hazardous or catastrophic failure condition.

## 迎角；Angle of Attack；A/A

翼弦与来流矢量在飞机对称面内投影的夹角。

The angle between the projection of the wing chord to flow vector aircraft symmetry plane.

## 影响限制线；Impact Limit Line

预先在航空器上建立的一条分界线，在其中的任何部分或其货物不允许被影响。影响限制线主要目的是为提供公共安全。

A line predetermined of the launch establishing a boundary beyond which a launch vehicle, or any portion thereof, or its cargo will not be allowed to impact. A major function of the impact limit line is to provide safety for public.

## 应答机；Transponder

一种接收机和发射机的组合，用以接收信号和自动发射回答信号，通常使用不同的频率。应答机可以是机载的（如在空中交通管制雷达信标系统中）、地面的（如在塔康系统中）、舰载的或星载的。

A receiver-transmitter combination designed to receive a signal and automatically transmit a reply signal, usually on a different frequency. A transponder may be airborne (as in the air traffic control radar beacon system), ground based (as in the TACAN System), ship-borne, or satellite-borne.

## 应急定位发射器；Emergency Locator Transmitter；ELT

一种安装在航空器结构上的无线电发射机，靠自身的电源以121.5兆赫和243兆赫工作，发射清晰的向下扫描的音调以供寻的导引，并设计成在事故后能自行工作，不需人工操作。

A radio transmitter, attached to the aircraft structure, which operates from its own power source on 121.5 MHz and 243 MHz, transmitting a distinctive downward swept audio tone for homing purposes, and is designed to function without human action after an accident.

### 应急放油程序;Jettisoning Fuel Procedures

为将航空器应急放油的潜在危险降至最低而建立的一种空中交通管制程序。最大的危险存在于应急放油的航空器两侧各 0.5 海里、水平后方 5 海里、上方至少 1 000 英尺、下方至少 2 000 英尺的空域之内。

Air traffic control procedures established to minimize potential hazards due to aircraft jettisoning fuel. The greatest hazard exists within the airspace at least 1 000 ft above, 2 000 ft below, five n mile horizontally behind, and 0.5 n mile on either side of the aircraft jettisoning fuel.

### 应急设备;Emergency Equipment

在发生意外事故或其他紧急情况时,保护机上乘客并帮助其安全逃生、生还的设备。

Equipment that protects occupants aboard an aircraft and assists with safe escape, survival during an accident or other emergency.

### 硬翼飞机;Hard Wing Airplanes

没有大翼前缘高升力装置的飞机。

Airplanes without wing leading-edge high-lift devices.

### 永久动力损失;Sustained Power Loss

永久的发动机功率或推力在主要参数(例如风扇转子速度、发动机压力比)上的下降。

A permanent reduction in power or thrust at the engine's primary power set parameter (for example, fan rotor speed, engine pressure ratio).

### 用于审定目的的试验;Test for Certification

按照局方批准的试验大纲进行的,用来表明对规章符合性的系统审定试验。

System certification test conducted under an authority approved test plan for the purpose of showing compliance to the regulations.

### 优先处理;Priority Handling

局方事故调查办公室必须处理这个事件优先于其他所有的不紧急事件,意味着这个事件必须再指定办公室成员,因为一个成员培训或者离开,如果不尽量去处理可能导致事件被延迟。

Priority handling means the Authority investigative office must process this case before all other non-emergency matters. This may mean that cases must be reassigned within an office, because, generally, it is not diligent handling if case processing is delayed because an employee is in training or on leave.

### 油门杆角度;Power Lever Angle/Throttle Level Angle; TLA

驾驶舱内发动机主控手柄的角度。

The angle of the pilot's primary engine control lever(s) on the flight deck.

### 有限翼;Finite Wing

具备有限展长的机翼。

A wing having finite span.

### 有效感觉噪声级;Effective Perceived Noise Level; EPNL

对声谱的不规则性(如由离散频率分量引起的不规则性)进行调整后计算出的时间积分感觉噪声级。有效感觉噪声级的单位为分贝,为清晰起见,加上识别前缀,成为 EPNdB。

Time-integrated perceived noise level calculated with adjustments for irregularities in the sound spectrum, such as that caused by discrete-frequency components (tone-correction). The unit of effective perceived noise level is the decibel, with identifying prefix for clarification, EPNdB.

### 与指定高度的偏差;Assigned Altitude Deviation; AAD

C 模式应答机的高度与指定高度/高度层的差值。

The difference between the transponder mode C altitude and the assigned altitude/flight level.

### 预防性维修;Preventive Maintenance

简单的或小型的保养工作,以及不涉及复杂装配工作的小型标准件更换。

Simple or minor preservation operations and the replacement of small standard parts not involving complex assembly operations.

### 预期的最大寿命；Expected Maximum Lifetime

预期的最长时间，在此期间电子系统、子系统、模块或组件的工作令人满意。

The expected maximum period of time over which an avionics system，subsystem，module，or component performs satisfactorily.

### 预期功能；Intended Function

为满足特定要求，设备必须展示并达到的规定特性。

The defined characteristic（s）that equipment must demonstrate and achieve to meet the specific requirements.

### 预期运行条件；Anticipated Operating Conditions

可从经验获得的运行条件，在飞机合格运行的条件下合理设想飞机运行过程中的运行环境，这些条件与大气的气象状况、地形、航空器工作情况、人员效率和影响飞行安全的所有因素相关。预期运行条件不包括：① 可以由操作程序有效避免的极端条件；② 很少发生的极端条件，要求在这种极端条件下达到标准，此经验表明必要且切合实际的适航性要求更高。

Conditions which are known from experience or which can be reasonably envisaged to occur during the operational life of the aircraft，taking into account the operations for which the aircraft is made eligible，the conditions considered being relative to the meteorological state of the atmosphere，the configuration of terrain，the functioning of the aircraft，the efficiency of personnel and all the factors affecting safety in flight. Anticipated operating conditions do not include：① those extremes which can be effectively avoided by means of operating procedures；② those extremes which occur so infrequently that to require the standards to be met in such extremes would give a higher level of airworthiness than experience has shown to be necessary and practical.

### 预先的立法建议通告（美国）；Advance Notice of Proposed Rulemaking；ANPRM

告知公众，FAA 正在考虑一个地区规章制定和要求书面评论的适当范围制定或特定主题。一个预先的立法建议通告也可以包含或者不包含规则的潜在变化的文本。

Tells the public that FAA is considering an area for rulemaking and requests written comments on the appropriate scope of the rulemaking or on specific topics. An advance notice of proposed rulemaking may or may not include the text of potential changes to a regulation.

### 裕度;Margin

设备敏感性或鉴定水平与飞机内部环境之间的差异。可以规定裕度要求，以说明设计、分析或测试中的不确定性。

Difference between equipment susceptibility or qualification levels and the aircraft internal environment. Margin requirements may be specified to account for uncertainties in design，analysis，or test.

### 原材料;Material

符合确定的工业或国家标准或规范，用于按照航空器或其部件制造厂家提供的规范进行维修过程中的加工或辅助加工的材料。这些标准或规范必须是公开发布并在航空器或其部件制造厂家的持续适航文件中明确的。

Materials that comply with established industrial or national standards or specifications and are used for processing or auxiliary processing during the maintenance process in accordance with the specifications provided by the manufacturer of aircraft or its components. These standards or specifications must be publicly released and clearly defined in the continuous airworthiness documents of the aircraft or its component manufacturers.

### 原型发动机;Prototype Engine

某一型号的第一台未经批准的用于型号批准测试的发动机。

The first engine，of a type and arrangement not previously approved，to be submitted for type-approval test.

### 原型模型(预生产模型);Prototype (Preproduction) Model

适合于全面评价机械和电气结构、设计与性能的一种模型。它应当是最终的机械和电气结构,包括最终的构型,采用经批准的最终零件,并完全代表最终的设备。

A model suitable for complete evaluation of mechanical and electrical form, design and performance. It shall be of final mechanical and electrical form, including the final configuration, employ approved final parts and be completely representative of final equipment.

### 原有的审定基础;Original Certification Basic

局方颁发产品型号合格证时所确定的审定基础。

Recognized certification basic when the Authority issues type certification.

### 远程导航系统;Long-range Navigation System; LRNS

主导航装置批准用于仪表飞行规则的电子导航装置,至少应具有一个导航输入源,例如惯性导航系统、全球定位系统、欧米加/甚低频,或罗兰 C。

An electronic navigation unit that is approved for use under instrument flight rules as a primary means of navigation, and has at least one source of navigational input, such as inertial navigation system, global positioning system, Omega/very low frequency, or Loran C.

### 远程通信系统;Long-range Communication System; LRCS

使用中转卫星、数据链、高频的通信系统或超越视线延伸的其他经批准的通信系统。

A system that uses satellite relay, data link, high frequency, or another approved communication system which extends beyond line of sight.

### 远程无线电导航系统;Long-range Radio Navigation System

一种远程连续波低频导航系统,同时提供方位和距离信息。

A long distance continuous wave LF navigation system providing simultaneous bearing and distance information.

### 远距发射/接收设施;Remote Transmitter/Receiver Facility; RTR

一种远距的空地通信设施,它拥有发射机和(或)接收机及辅助设备,为机场空中交通管制塔台(ATCT)、联合台站/塔台(CST)、雷达进近管制(RAPCO)、RATCC、飞行服务站(FSS)、国际飞行服务站(IFSS)或航空器调度中心(CAD)服务,并使用直通电缆、金属导线或 VHF/UHF/MW 无线电通信线路控制电路。RTR 不起中心设施的作用。

A remote air/ground communications facility having transmitters and/or receivers and ancillary equipment serving an airport traffic control tower (ATCT), combined station/tower (CST), radar approach control (RAPCO), RATCC, Flight service Station (FSS), international flight service station (IFSS), or Central Aircraft Dispatch (CAD); and employing direct cables, metallic lines, or VHF/UHF/MW radio link control circuits. RTR facilities do not serve center facilities.

**远距离空地通信台;Extended Range Air/Ground Communication Station**

国际飞行咨询系统中的一种无线电通信设施,它通常利用中等增益部件为距该设施 350 海里以内的航空器提供甚高频空地通信。

A radio communication facility in the international flight advisory system using moderate gain components to provide, normally, VHF air/ground communications to an aircraft up to 350 n mile from the facility.

**远距指点标;Outer Marker; OM**

航站区域导航系统中的一种仪表着陆系统导航设施,位于跑道中心线延长线上,距跑道边界 4~7 英里(约 6.4~11.3 千米),发射调制在 400 赫的 75 兆赫扇形辐射图,以每秒两划键控,由兼容的机载设备接收,用音响和目视两种方式向驾驶员指出,他正在越过该设施,可以开始最终进近。

An ILS navigation facility in the terminal area navigation system, located 4~7 mi (6.4~11.3 km) from the runway edge on the extended centerline transmitting a 75 MHz fan-shaped radiation pattern, modulated at 400 Hz, keyed at two dashes per second, and received by compatible airborne equipment indicating to the pilot, both aurally and visually, that he is passing over the facility and can begin his final approach.

### 云上;Over-the-top

在形成云幕的云层或其他遮暗层之上。

Over the clouds forming a cloud ceiling or other obscuring layer.

### 云上目视飞行规则飞行;VFR Over-the-top

当航空器不按仪表飞行规则的飞行计划飞行时,按目视飞行规则实施的云上飞行。

The flight of aircraft, referring to when aircraft is not on the basis of the flight plan of instrument flight rules, but the visual flight rules of the implementation of the flying above the clouds.

### 云上仪表飞行规则;IFR Over-the-top

对航空器飞行而言,指航空器经空中交通管制部门许可,按仪表飞行规则飞行计划保持"目视飞行规则条件"或"云上目视飞行规则条件"的云上飞行。

With respect to the operation of aircraft, it means the operation of an aircraft over-the-top on an IFR flight plan when cleared by air traffic control to maintain "VFR conditions" or "VFR conditions on top".

### 运输安全管理局(美国);Transportation Security Administration; TSA

美国运输安全的管理机构。

Transportation Security Administration in USA.

### 运输类航空器;Transport Category Aircraft

按照运输类航空器(包括飞机和旋翼航空器)进行型号合格审定的航空器(无论大小)。

Aircrafts (large or small) that are demonstrated to meet and are certificated to the regulatory requirements for transport category airplanes or transport category rotorcraft.

### 运行;Operation

就飞机而言,是指出于航行目的使用、导致使用或授权使用飞机,包括驾驶飞机,无论是否有合法控制权(作为所有者、承租人或其他人)。

With respect to aircraft, operation means use, cause to use or authorize to use aircraft, for the purpose of an navigation including the piloting of aircraft, with or without the right of legal control (as owner, lessee, or otherwise).

## 运行飞行计划;Operational Flight Plan

经营人根据固定翼飞机的性能、其他运行限制及所飞航路与有关机场的预期条件,为安全飞行所制订的计划。

The operator's plan for the safe conduct of the flight based on considerations of aeroplane performance, other operating limitations and relevant expected conditions on the route to be followed and at the aerodromes concerned.

## 运行规范;Operation Specifications

依照局方规章,对运营人进行合格审定后颁发的运行批准、标准和限制。

The operational approval in accordance with the Authority Regulations issued after the certification of operators running, standards and limitations.

## 运行类型清单;Kinds of Operations List; KOL

用于明确特定运行类型[例如,目视飞行规则(VFR)、仪表飞行规则(IFR)、白天或晚上]。KOL 同样明确可能会影响飞行操作限制的已安装设备。

The KOL specifies the kinds of operations [for example, visual flight rules (VFR), instrument flight rules (IFR), day or night] in which the aircraft can be operated. The KOL also indicates the installed equipment that may affect any operating limitation.

## 运行批准;Operational Approval

飞行标准使用的五步程序,用来授权运营人在特定的运行环境下,使用特定的飞机和相关设备执行运行。

Operational approval is a 5-step process used by Flight Standards to authorize an operator to conduct operations using a specific aircraft and associated equipment in a specific operating environment.

## 运行手册；Operations Manual

运行人员在履行其职责时所用的，包含程序、指令和指南的手册。

A manual containing procedures, instructions and guidance used by operational personnel in the execution of their duties.

## 运行循环；Operating Cycle

一个由起飞、爬升、巡航、降落和着陆构成的完整飞行航段。

A complete flight segment consisting of a takeoff, climb, enroute portion, descent, and a landing.

## 运行延误；Operation Delay

由出现恶劣的气象条件、航空器设备故障、空中交通管制不畅等客观情况导致的延误。

The delay caused by such objective reasons as severe weather situations, aircraft equipment malfunctions, and ineffective air traffic control.

## 运营基地；Base of Operations

设立在不同于合格证持有人主运营基地的地点，具有飞行运行或者适航维修，或者两者兼有的运行资源和能力，且连续 6 个日历月内定期载客运行达到 10 班，非定期或者全货机运行达到 15 班的基地。

A location different from the principal base of operations of a certificate holder and has the operational resources and capabilities of flight operation or airworthiness maintenance, or the combination of them, and has no less than 10 scheduled passenger-carrying flights and no less than 15 unscheduled or cargo-only flights in any six consecutive calendar months.

## 运营人；Operator

按照 CCAR - 121 运行的公共航空运输承运人或者按照 CCAR - 135 运行的商业运输运营人。

Public air transport carriers operating under CCAR - 121 or commercial transport operators operating under CCAR - 135.

**运营人维修管理手册;Operator's Maintenance Control Manual**

描述运营人的程序的文件,这些程序对确保及时地、有序地和令人满意地完成对经营人的航空器的所有定期和不定期维修是必要的。

A document which describes the operator's procedures necessary to ensure that all scheduled and unscheduled maintenance is performed on the operator's aircraft on time and is in a controlled and satisfactory manner.

# Z

## 灾难性的;Catastrophic

导致多人死亡的失效状态,通常伴随机毁人亡。

Failure conditions that result in multiple fatalities, usually with the loss of the airplane.

## 灾难性失效状态;Catastrophic Failure Condition

此种失效状态可能妨碍继续安全飞行和着陆。

A failure condition that would prevent continuing safe flight and landing

## 载荷系数;Load Factor

规定载荷与航空器总重之比。规定载荷用任一下列术语表示:气动力、惯性力、地面反作用力或水面反作用力。

The ratio of the specified load to the total weight of the aircraft. The specified load is expressed in terms of any of the following: aerodynamic forces, inertia forces, ground or water reactions.

## 载旗运行(美国);Flag Type Operation

在下列情况下进行的载旗运行:① 设计载客量至少为 31 人的航空器,在美国境内所有州、哥伦比亚地区或美国境内所有领土或领地的可用机场运行;② 设计载客量大于 9 人但小于 31 人的航空器,在美国境内所有州(除阿拉斯加州)、哥伦比亚地区或美国境内所有领土或领地的可用机场运行。

Any flag operation conducted with: ① an airplane designed for at least 31 passenger seats at any land airport in any state of the United States, the district of Columbia, or any territory or possession of the United States; ② an airplane designed for more than 9 passenger seats but less than 31 passenger seats at any land airport in any state of the United States (except Alaska), the district of Columbia, or any territory or possession of the United States.

### 再次适航批准; Recurrent Airworthiness Approval

为之前是适航的,发现问题后仍然是适航的生产批准持有人的产品,颁发适航标签。

Issuance of approval tag for products or articles based on a prior finding by a PAH that the product or article was airworthy, and a current finding that the product or article remains airworthy.

### 早期延程运行; Early ETOPS

在预期要按 ETOPS 进行合格审定的飞机-发动机组合尚未得到非 ETOPS 使用服务经验的情况下获取 ETOPS 型号设计批准。

ETOPS type design approval obtained without gaining non-ETOPS service experience on the candidate airplane-engine combination certified for ETOPS.

### 噪声; Noise

任何令人讨厌的声音,因为它干扰了说话和影响了听力。

Any sound which is undesirable because it interferes with speech and hearing.

### 噪声等效功率; Noise Equivalent Power; NEP

对光辐射探测器或探测系统的灵敏度测量。

Measure of the sensitivity of an optical detector or detector system.

### 责任审查部门; Responsible Reviewing Department

主管单位指定的负责完成型号合格审定项目具体审查任务的机构。

Responsible reviewing department is the organization appointed by competent certification entity to execute type certification activities for specific certification project.

### 责任审定单位;Responsible Certification Organization

负责具体型号合格审定项目证件申请受理、颁发和管理的单位。责任审定单位对相应的型号审查活动进行指导和监控。

Responsible certification organization is responsible for the acceptance of application, issuance and management of certificate for a specific type certification project. Any activities of type certification should be guided and monitored by the competent certification entity.

### 增强型飞行能见度;Enhanced Flight Visibility; EFV

从飞行中的航空器驾驶舱,由一名驾驶员用增强型飞行视景系统,在白天或夜晚能够清楚区分和判别醒目地形目标的前向平均水平距离。

The average forward horizontal distance, from the cockpit of an aircraft in flight, at which prominent topographical objects may be clearly distinguished and identified during the day or at night by a pilot using an enhanced flight vision system.

### 增强型飞行视景系统;Enhanced Flight Vision System; EFVS

一种通过使用图像传感器,提供前向外部场景地形(一个地方或区域的自然或人工设置的特征,特别是在某种程度上能够表示它们的相对位置和相对高度的特征)显示的电子装置,例如前向红外观察器、毫米波射线探察仪、毫米波雷达、低照度图像增强装置。

An installed airborne system which uses an electronic method to provide a display of the forward external scene topography (natural or artificial features of a place or region, especially those that can represent their relative position and height to some extend) through the use of imaging sensors, such as forward looking infrared, millimeter wave radiometry, millimeter wave radar, and/or low light level image intensifying.

### 增强型目视系统;Enhanced Vision Systems; EVS

一种电子手段,通过使用成像传感器如红外前视系统、毫米波辐射计、毫米波雷达、微光图像增强技术,显示前部外侧地形场景。

An electronic method to provide a display of the forward external scene topography through the use of imaging sensors, such as a forward looking infrared, millimeter wave radiometry, millimeter wave radar, low light level image intensifying.

### 增强醒目性；Conspicuity Enhancement

航空器的喷漆和标记、昼夜的外部照明和其他增强航空器目视可探测性的措施。

Aircraft paint and markings, day/night exterior lighting, and other means of enhancing the visual delectability of aircraft.

### 增压舱；Pressurized Cabin

航空器的一个部分（通常为人员舱和货舱），其中高于外界大气压力的空气压力可用人工方法维持和控制。此种航空器称为"增压的"或具有"增压"的航空器。

The portion of an aircraft (usually the personnel and cargo compartments) in which an air pressure greater than the outside atmospheric pressure can be maintained and controlled by artificial means. Such an aircraft is described as being "pressurized" or having "pressurization".

### 增益；Gain

从一点传输至另一点时信号功率增加的通用术语。增益通常按输出电平与输入电平之比给出，单位用分贝表示。

A general term used to denote an increase in signal power in transmission from one point to another. Gain is usually given as the ratio of output level to input level and expressed in decibels.

### 展弦比；Aspect Ratio; AR

展弦比是翼展（翼尖到翼尖）和平均翼弦的比值，是影响由机翼产生的升力和阻力的因素。

Aspect ratio is determined by dividing the wingspan (from wingtip to wingtip), by the average wing chord. The aspect ratio is another factor that affects the lift and drag created by a wing.

### 真北；True North

地理北；沿真子午线朝向地理北极的方向。

Geographic north；the direction to the geographic north pole along true meridian lines.

### 真高度；True Altitude

平均海平面之上的高度。

The altitude above mean sea level.

### 真空速；True Airspeed

航空器相对未扰动空气的空速。真空速等于当量空速乘以海平面空气密度与该高度上的空气密度之比的平方根。

The airspeed of an aircraft relative to undisturbed air. True airspeed is equal to equivalent airspeed multiplied by the square root of the ratio of air density at sea level to air density at altitude.

### 整合系统；Federated System

飞机设备结构，包括主要执行具体功能且与特定接口相连的航线可更换件或飞机系统数据总线。

Aircraft equipment architecture consisting of primarily line replaceable units that perform a specific function，connected by dedicated interfaces or aircraft system data buses.

### 整体油箱；Integral Tank

壁板全部由航空器主结构件组成的燃油容器。

Fuel container with all the wall panels constituted by aircraft main structure.

### 正常飞行包线；Normal Flight Envelope

飞机制造商定义的与被设计的飞机执行飞行操作相一致的高度和运行速度的范围。

The range of altitude and operating speeds that are defined by the airplane manufacturer as consistent with conducting flight operations for which the airplane is designed.

### 正常进近范围(下滑道);Normal Approach Envelope (Glide Slope)

为产生机载导航设备从正半尺度到负半尺度的偏离指示,所需的航空器相对于地面天线基准点的角位移。

The angular aircraft displacement relative to the ground antenna reference point required to produce a positive half-scale to a negative half-scale deviation indication of the airborne navigation equipment.

### 正常类飞机;Airplane in Normal Category

座位设置(不包括驾驶员)为 9 座或以下,最大审定起飞重量为 5 700 千克(约 12 500 磅)或以下,用于非特技飞行的飞机。

Seat number (not including pilot) for 9 or less, the maximum certificated takeoff weight of 5 700 kg (12 500 lb) or less for non-aerobatic airplane.

### 正常起落次数;The Normal Times of Take-off and Landing

航空器完成起飞离地到着陆并使用刹车的完整飞行过程计为一次正常起落。

The complete flight process of the aircraft from take-off to landing, with the use of brakes, is considered as a normal take-off and landing.

### 正常运行包线;Normal Operating Envelope

海平面与经批准的最大运行高度之间的海拔高度,失速警告和最大操纵速度($V_{MO}$)/最大马赫操纵速度($M_{MO}$)之间的空速,以及适合该飞机类型的侧滑角。

Altitudes between sea level and the maximum approved operating altitude, airspeeds between stall warning and $V_{mo}/M_{mo}$, and sideslip angles appropriate for the type of airplane.

### 证件;Certificate

局方签发的文件(即合格证或批准书),用以认可申请人或生产批准持有人

(PAH)建立的质量系统,允许按照经局方批准的设计生产产品或零部件。

A document (that is, a certificate or approval) issued by the Authority that recognizes an applicant's or PAH's established quality system and allows for the production of products, articles, or parts in accordance with an Authority-approved design.

### 证件管理;Certificate Management

局方用来确保 PAH 其产品或零部件制造遵守相关法规的方法。

The methods by which the Authority ensures a PAH remains in compliance with those pertinent regulations that govern the manufacturing of its particular products, articles, or parts.

### 证明项;Item of Proof; IOP

每份证据的原始材料或经核准的副本,以证明存在违规的行为。

The items of proof consist of originals or certified copies of each piece of evidence gathered to prove the apparent violations.

### 证书管理;Certificate Management

局方保证生产批准书持有人保持与其有关的监管其特定产品的或物品生产的规章的一致性的方法。

The method by which the Authority ensures that a PAH remains in compliance with those pertinent regulations that govern the manufacturing of its particular products or articles.

### 证书管理办公室;Certificate Management Office; CMO

负责对指定运营人的运行证书进行管理的飞行标准办公室。

A Flight Standards office that is responsible for administration of operating certificate for a particular operator.

### 证书管理航空器审定办公室; Certificate Management Aircraft Certification Office; CMACO

管理产品型号证书的航空器审定办公室。该办公室也对已颁发许可并一直在使用中的所有产品进行持续适航管理。

The ACO managing the product's TC，also manages the continued airworthiness for all products they approve for as long as the products are in service.

### 证书管理信息系统；Certificate Management Information System；CMIS

集成证书管理多方面功能的电子数据系统，该系统的可用功能包括任务管理，以及评估工作的计划、安排和执行。

Electronic data system that incorporates several aspects of certificate management functions. Functions available within CMIS are certification management tasks，and the planning，scheduling and conducting of evaluations.

### 政策和程序手册；Policy and Procedure Handbook；PPH

由 ISC 编制和批准的管理文件，包括各方职责、计划安排、分析方法（MSG - 3）和表格要求等。ISC、航空器制造人、航空运营人和 MRB 成员依照此文件编写和审议 MRBR 建议书。

PPH is the management file prepared and approved by the ISC，it includes the responsibilities of all parties，scheduling，analysis method（MSG - 3）and form requirements，etc. The ISC，aircraft manufacturers，aviation operators and MRB members write and deliberate the MRBR proposal in accordance with this file.

### 政府航空器；Public Aircraft

仅为政府或下属政府机构提供服务的航空器。不包括那些政府拥有的、提供载人或载物等商业用途的航空器。

Aircraft used only in the service of a government，or a political subdivision. It does not include any government-owned aircraft engaged in carrying persons or property for commercial purposes.

### 直接进近；Straight-in Approach

用与跑道中心线延长线相交的方式进入起落航线，而不执行起落航线的任何其他部分。

Entering into the traffic pattern by interception of the extended runway centerline without executing any other portion of the traffic pattern.

### 直接目视检测；Direct Visual Testing

检测人员眼睛到被检测物体的光学路径无中断，检测时眼睛与检测面的距离不大于 60 厘米（约 25 英寸），且与检测面的角度不低于 30 度的目视检测。

The optical path from the eyes of the inspector to the object being tested is uninterrupted, and the distance between the eyes and the detection surface is not more than 60 cm (25 in), and the angle with the detection surface is not less than 30° during visual inspection.

### 直接装运授权；Direct Shipment Authorization

对产品或物品负有适航责任的生产批准持有人（PAH）颁发给供应商的书面授权，将按照 PAH 的质量/检查系统生产的物品直接运送到终端用户，不需经过 PAH 的设施处理。

The written authorization granted by a production approval holder (PAH) with responsibility for the airworthiness of a product or article, to a supplier, shipping articles produced in accordance with the PAH's quality/inspection system directly to end users without the articles being processed through the PAH's own facility.

### 值勤期-飞行；Duty Period

机组成员在接受合格证持有人安排的飞行任务后，从为了完成该次任务而到指定地点报到开始（不包括从居住地或者驻地到报到地点所用的时间），到解除任务为止的连续时间段。在高级飞行模拟机上实施训练或检查的时间应计入值勤期。在一个值勤期内，如机组成员能在有睡眠条件的场所得到休息，则该休息时间可以不计入该值勤期。

It refers to the continuous period of time from the moment when crew members report to the designated location in order to complete the flight mission arranged by the certificate holder (excluding the time from their residence to the reporting location) to the moment when the mission is terminated. The time required for training or inspection on an advanced flight simulator should be included in the duty period. During a duty period, if crew members can rest in a place with sleep conditions, the rest time may not be included in the duty period.

### 值勤期-维修；Maintenance Personnel Duty Period

维修人员在接受维修单位安排的工作任务后，从为了完成该次任务而到指定地点报到开始（不包括从居住地或驻地到报到地点所用的时间），到工作任务完成或解除为止的连续时间段。

The maintenance personnel duty period refers to the continuous time period from the time when maintenance personnel report to the designated location in order to complete the task after receiving the task arranged by the maintenance unit（excluding the time from their residence to the reporting location）to the time when the task is completed or terminated.

### 指令；Directives

书面信息或颁布的信息，这些信息有如下作用：① 描述或制订政策、组织、方法或程序；② 规定职责和权限；③ 要求采取行动；④ 包含有效地管理或操作局方活动所需的信息。

Written communications or issuances which：① prescribe or establish policy, organization, methods, or procedures；② establish responsibilities and authorities；③ require action；④ contain information needed for the effective administration or operation of agency activities.

### 指南；Guideline

有帮助的但不认为是指导的支持信息。

Supporting information that can be helpful but is not considered to be guidance.

### 指示高度；Indicated Altitude

任何高度表指示的高度。使用压力高度表或气压高度表时，它是由未修正仪表误差和未补偿标准大气条件偏差的读数所指示的高度。

Altitude as shown by any altimeter. With a pressure or barometric altimeter，it is the altitude as shown by the reading uncorrected for instrument error and uncompensated for variations from standard atmospheric conditions.

## 指示空速；Indicated Airspeed；IAS

显示在航空器总静压空速表上的航空器速度，已按海平面标准大气绝热压缩流校准，但未修正空速系统误差。

The speed of an aircraft as shown on its pitot static airspeed indicator calibrated to reflect standard atmosphere adiabatic compressible flow at sea level，but uncorrected for airspeed system errors.

## 制造；Fabrication

产品或部件制造人在按照经批准或认可的资料进行维修和改装的过程中，根据被制造部件的种类级别和适用规章生产和消耗部件/子部件的行为。

An act in which a part/subpart is made（fabricated）and consumed by the fabricator on the product，or part thereof，in the course of performing maintenance or alterations in accordance with approved or acceptable data，depending on the category（CAT）classification of the part being fabricated and the applicable regulations.

## 制造符合性；Manufacturing Conformation

民用航空产品和零部件的制造、试验、安装等符合经批准的设计。

Approved designs of manufacturing，testing and installation of civil aviation products and parts.

## 制造符合性检查；Conformity Inspection

确定试验型产品是否符合经批准的型号设计资料的检查。

A method that to determine if the prototype products conform with the approved type design.

## 制造符合性检查记录；Conformity Inspection Record

制造符合性检查代表和委任制造检查代表用以记录试验产品和试验装置制造符合性检查结果的表格之一。

One of the forms used by the Authority manufacturing inspectors or designated manufacturing inspection representatives（DMIR）to record conformity inspection results of test products and test setups.

### 制造符合性检查请求单；Request for Conformity

工程审查代表或委任工程代表请求制造符合性检查代表或委任制造检查代表进行制造符合性检查，或委托制造符合性检查代表或其他工程审查代表及委任工程代表代替其进行目击验证试验所用的文件，是制造符合性检查代表和委任制造检查代表进行制造符合性检查和目击验证试验的依据文件之一（另一依据文件为型号检查核准书）。

A form used by Authority engineers or designated engineering representatives（DER）to request Authority manufacturing inspectors or designated manufacturing inspection representatives（DMIR）to carry out conformity inspection，or to request Authority manufacturing inspector or other Authority engineers and DER instead of themselves to witness certification test. Request for conformity is one of the documents（TIA is another one）containing instructions for Authority manufacturing inspector to carry out conformity inspection and witness test.

### 制造符合性声明；Statement of Conformity；SoC

申请人正式签署的，表示改装或安装符合设计和型号设计的声明。

Official signed statement from the applicant signifying that the modification or installation conforms to the design data and type design.

### 制造国；State of Manufacture

对负责航空器最后组装的机构拥有管辖权的国家。

The state having jurisdiction over the organization responsible for the final assembly of the aircraft.

### 制造监察办公室；Manufacturing Inspection Office；MIO

监督其辖区的 MIDO 和 MISO，并为这些办公室提供组织领导和技术指导。MIO 管理辖区内所有生产工厂和委任代表。他们管理适航取证政策，办公人员配备及内部预算分配。

MIO oversees the manufacturing inspection district offices（MIDO）and manufacturing inspection satellite offices（MISO）in its geographic area and provides organizational leadership and technical guidance to these offices. The MIO manages all geographically located production facilities and designees. They administer the airworthiness certification policies，office staffing，and internal budget allocation.

### 制造检查系统；Fabrication Inspection System；FIS

保证每个已完成的部件都满足设计标准，可安装到适用的型号审定产品上的系统。

The system which ensures that each completed part conforms to its design data and is safe for installation on applicable type certificated products.

### 制造缺陷；Manufacturing Defect

在制造过程中发生异常或缺陷，可能会导致在结构强度、刚性和尺寸稳定性上有不同程度的退化。

An anomaly or flaw occurring during manufacturing that can cause varying levels of degradation in structural strength，stiffness and dimensional stability.

### 制造人；Manufacturer

按民用航空规章条例规定，从事民用航空产品及其零部件生产制造的任何人。

Any person engaged in the civil aviation products and parts manufacturing under provisions on civil aviation regulations.

### 制造人服务文件；Manufacturer's Service Documents

由 TC 持有人（或设备或部件制造商）出版的有关安全问题、产品改进、经济性、运营和维修程序的出版物。典型的出版物包括服务通告、所有运营人信函、服务简讯、服务摘要和杂志；不包括型号合格审定或批准所需要的出版物，如飞行手册和某些维修手册。

Publications by a TC holder（or appliance or component manufacturer） about safety, product improvement, economics, and operational and maintenance practices. Typical publications include service bulletins, all-operator's letters, service newsletters, service digests, and magazines, and do not include publications required for type certification or approval such as flight manuals and certain maintenance manuals.

### 制造人维修单位;Manufacturer's Maintenance Organization

由航空器或者航空器部件制造人建立的,主要维修和管理工作与生产线结合的维修机构。主要维修和管理工作与其生产线分离的视为独立的维修单位。

The maintenance organization, which is set up by the civil aircraft or aircraft component manufacturer, with its primary maintenance and management activities integrated with the production line. The maintenance organization shall be treated as an independent maintenance organization if its primary maintenance and management activities are separate from the production line.

### 质量控制资料;Quality Control Document; QCD

按民用航空规章对质量控制系统的要求所建立的、局方可以接受的资料,包括制造人及其供应商制造产品和零部件所需的方法、程序、工艺、检验、试验、规范、图表、清册和表格等。

QCD is established according to the requirements of the Civil Aviation Regulations of the quality control system which can be accepted by authority, including methods, procedures, processes, inspection, testing, specification, charts, inventory and tables from manufacture and vendor.

### 质量逃逸;Quality Escape

不符合适用的设计参数或质量系统要求,不受质量系统管理的产品或项目。

A product or article that has been released from the quality system and does not conform to the applicable design data or quality system requirements.

### 质量系统;Quality System

一个文件化的组织结构,包括职责、程序、过程和资源,通过实施管理职能来确定和执行质量原则。

A documented organizational structure containing responsibilities，procedures，processes，and resources that implement a management function to determine and enforce quality principles.

### 质量系统文件；Quality System Document

为满足民用航空规章中质量系统的要求所建立的资料，包括民用航空产品和零部件生产所需的质量手册、管理程序、作业文件等（如制造方法、检验方法、工艺/检验/试验规范、图表、清册和表格等）。

Documents established to meet the requirements of quality system in civil aviation regulations，including quality manuals，management procedures and operation documents required for the civil aviation products and articles (such as manufacturing methods，inspection methods，process/inspection/test specifications，charts，lists and forms).

### 质量系统要素；Quality System Element

对局方批准的设计或质量系统文件产生影响的特定活动或职能，如设计资料控制、制造过程控制、供应商控制等。

Specific activities or functions that affect the design or quality system documents approved by the authority，such as design data control，manufacturing process control，supplier control.

### 致命伤害；Fatal Injury

人员在受伤后 30 天内死亡的伤害。

An injury that results in death within 30 days of the occurrence that caused the injury.

### 中等光强进近灯光；Medium Intensity Approach Lights；MAL

航站区域导航系统中的一种机场灯光设施，运用辐射中等光强聚光束向驾驶员提供目视导引，驾驶员依靠此光束目视将航空器对准跑道中心线的延长线。

An airport lighting facility in the terminal area navigation system providing visual guidance to the pilot by radiating medium intensity focused

light beams by which the pilot visually aligns the aircraft with the extended
runway centerline.

### 中等总重量；Medium Gross Weight

责任方或数据提供方选择的重量，其在飞机净重和最大总重量的平均数的
10%范围之内。

A weight chosen by the sponsor or data provider that is within 10% of the
average of the numerical values of the BOW and the maximum certificated
gross weight.

### 中断飞行；Air Abort；A/A

造成航空器无法按计划执行航班的事件，包括空中中断和地面中断，空中中
断包括返航和改航（备降）等事件；地面中断包括中断起飞和滑回等事件。

An event which causes that an aircraft flight cannot be operated as
planned，including the interruption in flight as in flight turn back and diverting
and the interruption on ground as rejected takeoff，taxi-back.

### 中断进近；Missed Approach

出于以下原因未完成着陆的仪表进近：① 在批准的最低气象条件下未实现
目视接地；② 出于其他原因未完成着陆；③ 空中交通管制部门的指令。

An instrument approach not completed by landing due to：① visual
contact not established at authorized minimums；② landing not accomplished
due to other reasons；③ instructions from air traffic control.

### 中断进近程序；Missed Approach Procedure；MAP

当航空器完成仪表进近但未能着陆时所规定的飞行。

Flight prescribed when an aircraft fails to land after completing an
instrument approach.

### 中断起飞；Refused Takeoff/Rejected Takeoff；RTO

在飞机离地前被终止起飞。

A takeoff which has been stopped before the airplane lifts off the ground.

### 中国技术标准规定；Chinese Technical Standard Order；CTSO

中国针对民用航空器上指定的航空材料、零部件或机载设备的最低性能标准规定。

A minimum performance standard for specified aviation materials，parts，components，and airborne equipment used on civil aircraft.

### 中国民用航空规章；China Civil Aviation Regulation；CCAR

中国民用航空规章是由国务院负责管理民用航空活动的行政机关——中国民用航空局(CAAC)制定、发布的涉及民用航空活动的专业性规章。中国民用航空规章具有法律效力，凡从事民用航空活动的任何单位和个人都必须遵守中国民用航空规章。

China Civil Aviation Regulations are the technical regulations about civil aviation activities developed and published by the Civil Aviation Administration of China，which is the executive agency of the State Department responsible for civil aviation activities. China Civil Aviation Regulations have the force of laws，any unit or individual engaged in civil aviation activities must comply with the China Civil Aviation Regulations.

### 中间连续功率/推力；Intermediate Contingency Power and/or Thrust

通过起飞后非限制时段内一套动力设备失效或关车时的性能数据确定的功率/推力。

The power and/or thrust identified in the performance data for use after take-off when a power-unit has failed or been shut down，during periods of unrestricted duration.

### 中距指点标；Middle Marker；MM

航站区域导航系统中的一种仪表着陆系统导航设施，位于跑道中心线延伸线上，距跑道边约 3 500 英尺（约 1 067 米），发射调制在 1 300 赫兹的 75 兆赫扇形辐射方向图，用点和划交替键控，由兼容的机载设备接收，用视觉和听觉两种方式向驾驶员指示他正在飞越此设施。

An ILS navigation facility in the terminal area navigation system located approximately 3 500 ft（1 067 m）from the runway edge on the extended centerline, transmitting a 75 MHz fan-shaped radiation pattern, modulated at 1 300 Hz, keyed alternately dot and dash, and received by compatible airborne equipment, indicating to the pilot both aurally and visually, that he is passing over the facility.

## 中频;Medium Frequency; MF

从 300 至 3 000 千赫的频带。

Frequencies band from 300 to 3 000 kHz.

## 重大飞行事故;Serious Flight Accident

凡属下列情况之一者：① 人员死亡,死亡人数在 39 人及以下;② 航空器严重损坏或迫降在无法运出的地方[最大起飞重量为 5.7 吨(含)以下的航空器除外];③ 航空器失踪,机上人员在 39 人及以下。

One of the following circumstances: ① casualty with 39 and less death number; ② aircraft is severely damaged or lands in place which can not be shipped out [except aircraft with maximum take off weight less than 5. 7 t (including)]; ③ aircraft disappearance with 39 and less passengers on board.

## 重大改装;Major Modification

未列入局方批准或认可的航空器、发动机、螺旋桨技术资料中的以下改装：① 可能会显著影响重量、平衡、结构强度、性能、动力装置运行、飞行特性或适航性的其他特性;② 不是按照经认可的常规做法或基本操作的方法完成的。从设计更改的角度看,重大改装可能属于设计大改,也可能属于设计小改。

The following modifications not included in the Authority's approval or endorsement of the aircraft, engine, propeller technical data of: ① may significantly affect weight, balance, structural strength, performance, power plant operation, flight characteristics or affect the seaworthiness of the other characteristics; ② are not completed in accordance with approved conventional practices or basic operation. From the view of the change in design, the major modification may be a big change or a small change.

## 重大更改；Significant Change

对 TC 做如下一项或几项更改：一般构型、制造原理或用于项目审定的假定。这种更改不足以认为是根本性改变。

A change to the TC is significant to the extent it changes one or more of the following: general configuration, principles of construction, or the assumptions used for certification. The change is not extensive enough to be considered a substantial change.

## 重大失效状态；Major Failure Condition

此种失效状态会降低飞机性能或机组处置不利工作条件的能力，以至于会导致，例如显著降低安全裕度或功能、明显增加机组工作负荷，或在很多情况下削弱机组效能。

Major failure conditions would reduce the capability of the airplane or the ability of the crew to cope with adverse operating conditions to the extent that there would be a significant reduction in safety margins or functional capabilities. In addition, the failure condition has a significant increase in crew workload or in conditions impairing crew efficiency.

## 重大损伤；Substantial Damage

对飞机结构强度、性能或飞行特性具有不利影响的损伤和失效，且通常要求对损伤部件进行重要修理或更换。

Damage or failure that adversely affects the structural strength, performance, or flight characteristics of the aircraft, and which would normally require major repair or replacement of the affected component.

## 重大维修事故；Serious Maintenance Accident

由维修造成下列情况之一者为重大维修事故：① 航空器及部件在地面损坏，直接经济损失超过事故当时同型或同类可比新航空器（最大起飞重量小于或等于 5.7 吨的航空器除外）整机价格的 1%或直接经济损失达 100 万元（含）～500 万元，以低限为准；② 在地面发生事故死亡 3 人（含）以下，地面设备、厂房设施损坏，直接经济损失达 100 万元（含）～500 万元；③ 一般飞行事故。

Maintenance accidents which lead to the following circumstances:

① aircraft and its component are damaged on the ground, the direct economic loss is 1% more than the price of new aircraft with the same or similar model (except for the aircraft with maximum take-off weight less than or equal to 5.7 t) at that moment, or direct economic loss is between one million RMB (including) and five million RMB, whichever comes lower; ② the number of lost life is less than 3 persons (including), the damage of the ground equipment and factory facilities, direct economic loss is between one million RMB(including) and five million RMB; ③ general fight accident.

**重大修理;Major Repair**

如果不正确地实施,可能导致对重量、平衡、结构强度、性能、动力特性、飞行特性和其他适航性因素有明显影响的修理。重大修理不是按照已经被接受的方法或者通过基本的作业就能够完成的工作。

A repair that, if improperly done, might affect weight, balance, structural strength, performance, powerplant operation, flight characteristics, or other qualities affecting airworthiness, which can not be done by acceptable practices or by elementary operations.

**重量与平衡手册;Weight and Balance Manual; WBM**

WBM应包含商业飞机的管理规章中要求的所有重量和平衡相关材料,以及相关人员行使其职责时充分的补充信息。

WBM shall contain all the weight and balance materials required by government regulations for a commercial aircraft and sufficient supplementary information to allow personnel concerned to intelligently perform the duties of their position.

**重心;Center of Gravity**

物体或刚性结构的平衡点,以平均气动弦长的百分比或距某一参考点若干英寸来表示。

The balance point of an object or rigid structure expressed in terms of percent mean aerodynamic chord or in inches from a reference point.

## 重心偏移航空器的控制；Weight-shift Control Aircraft

具有可转动机翼的动力航空器，在俯仰和翻滚时，驾驶员通过机翼来调整航空器的中心。对航空器的控制取决于机翼的形变，而不是对翼面的控制。

A powered aircraft with a framed pivoting wing and a fuselage controllable only in pitch and roll by the pilot's ability to change the aircraft's center of gravity with respect to the wing. Flight control of the aircraft depends on the wing's ability to flexibly deform rather than the use of control surfaces.

## 重要安全事件；Significant Safety Event

会影响或可能影响局方飞机或机组人员安全的事件。

Events that affect or is likely to affect the safety of the Authority's aircraft or crew.

## 重要的修正/补充型号合格项目；Significant Amended/Supplemental Type Certificate Project

重要的修正/补充型号合格证申请包括：① 设计需要特殊的条件、豁免或同等安全要点，由 21 部特殊申请的审定基础；② 使用新颖、特殊的方式建造的设计；③ 运动学、动力学、任何飞行控制系统或转子驱动系统的构型设计更改；④ 改变飞行特性的设计更改；⑤ 影响主要使用困难、事件或 AD 项目的设计；⑥ 设计更改从活塞到涡喷或涡桨的发动机构型；⑦ 在批准的限制范围内持续安全飞行、着陆、运行必要的基本非承重结构；⑧ 包含从未审定过的最先进的部件系统或未发布的审定标准的相关设计；⑨ 有争议的或显著的审定；⑩ 其他重要项目或条款。

Any application for significant amended type certificate or new/amended supplement type certificate includes：① the design appears to require special conditions，exemptions，or equivalent safety findings or a certification basis derived from an unusual application of Part 21；② the design uses novel or unusual methods of construction；③ the design changes the kinematics，dynamics，or configuration of either the flight control system or rotor drive system；④ the change of design would substantially alter the aircraft's flight characteristics；⑤ the design affects an area that has been the subject of a major service difficulty，accident，or airworthiness directive action；⑥ the aircraft design changes the engine configuration from reciprocating to turbopropeller or turbojet powered; ⑦ the integrity of the basic load-bearing structure necessary for continued safe flight and landing or operation of the aircraft within approved limits is affected; ⑧ the design consists of new state-of-the-art systems of components that have not been previously certificated or for which adequate certification criteria have not been published; ⑨ the certification is likely to be controversial or highly visible; ⑩ other significant projects or amendments.

### 重要件;Important Parts

在适航部门批准的设计中,若失效会导致飞机、发动机或螺旋桨的不安全状况的任何零部件,仅为民航局的优先监控目的而设。

Design approved by airworthiness office that will undoubtedly lead to unsafety condition in any parts of the aircraft, engine or propeller insecurity, only for Authority priority monitoring purposes.

### 重要结构;Primary Structure

对支持飞行、地面和压力载荷有重要作用的结构,也叫作关键结构项目。

The structure that significantly contributes to the carrying of flight, ground, or pressure loads. It is also known as a structurally significant item (SSI).

### 重要结构项目;Structural Significant Items; SSI

飞机上承受飞行、地面、压力或控制载荷的任何结构细部、结构部件或结构组件,它们的失效会影响飞机安全飞行的结构完整性。

Any detail, element, or assembly that contributes significantly to carrying flight, ground, pressure or control loads, and whose failure could affect the structural integrity necessary for the safety of the aircraft.

### 重要零部件供应商;Important Component Vendor

向制造人提供重要零部件的供应商。

Vendors who supply important parts and components to manufacturer.

### 重要气象报告;Significant Meteorological Information; SIGMET

就对运输类(大型多发)航空器的安全特别重要的天气情况发布的天气咨询信息。SIGMET 咨询包括龙卷风、雷暴线、大冰雹、强湍流和极强湍流、严重结冰及使能见度降至小于 3 英里(约 4.8 千米)的广布的尘埃或沙暴。

A weather advisory issued concerning weather particularly significant to the safety of transport category (large multiengine) aircraft. SIGMET advisories cover tornadoes, lines of thunderstorms, large hail, severe and extreme turbulence, heavy icing, and widespread dust or sandstorms that reduce visibility to less than 3 mi (4.8 km).

### 周转件;Rotable Part

技术上可以修理并且具有厂家发布的技术文件,可以不限次数修复使用,直至无法恢复到厂家发布的技术文件要求的航空器材。

The aviation material which can be repaired technically per technical data published by OEM, and numbers of time of repair to service.

### 周转周期;Turn Around Time

从周转件、可修件拆下送修起至该航空器材恢复至合格状态、入库完毕的总时间。

The time between when rotate parts and repairable parts have been removed, repaired and when parts retrieve to standard, into the storage.

### 主差异要求;Master Difference Requirement; MDR

合格证持有人基本型别飞机与其他型别飞机之间有差异的训练、检查和近期经历的要求。MDR 是满足由航空器评审组(AEG)确定的训练、检查和近期经历要求的最低认可方式。

The different requirements for training, checking, and recent experience between the basic type of aircraft and other type of aircraft for the certificate holder. MDR is the minimum acceptable manner which meets the training, checking, and recent experience requirements determined by aircraft evaluation group(AEG).

### 主飞行控制;Primary Flight Controls

对旋翼机是驾驶员用来直接操纵航空器的俯仰、横滚、偏航和垂直运动的系统。对固定翼飞机是驾驶员用来操纵飞机副翼和襟翼、扰流板和减速板、方向舵、升降舵以及水平安定面的系统。

Primary flight controls are those used by the pilot for immediate control of pitch, roll, yaw, and vertical motion of the rotorcraft, and for control of ailerons, wing flaps, spoilers, air brake, rudder, horizontal stabilizer and elevator of the fixed-wing aircraft.

### 主飞行显示;Primary Flight Display; PFD

显示主要飞行信息的显示。

The displays used to present primary flight information.

### 主飞行仪表;Primary Flight Instrument

显示飞行机组主要参考的飞行参数数据的显示器或仪器。

The display or instrument that serves as the flightcrew's primary reference of a specific parameter of primary flight information.

### 主共用要求;Master Common Requirement; MCR

明确了合格证持有人基本型别飞机与其他型别飞机之间通用的可替代的训练、检查和近期经历的要求。

The general and replaceable requirements for training, checking, and recent experience between the basic type of aircraft and other type of aircraft for the certificate holder.

### 主管检查员;Principal Inspector; PI

经局方指派,对某一生产批准书持有人进行管理与监督的局方监察员。

Supervisor appointed by the Authority to supervise the holder of a production approval

### 主基地；Main Base

航空运营人主要的运行和维修工程管理部门所在地点（一般为航空运营人的总部）。

The location of the main operators and maintenance engineering management department of the aviation operator (usually the headquarters of the aviation operator).

### 主评估员（美国）；Principal Evaluator

一个 FAA 任命的小组组长，在特定设施进行 ACSEP 评估的唯一评估员。

An FAA-appointed team leader who acts as the sole evaluator for the performance of an ACSEP evaluation at a specific facility.

### 主起落架；Main Landing Gear；MLG

靠近飞机重心并承受大部分载荷的起落装置。

The landing gear close to the aircraft center of gravity and bears most of the load.

### 主签派系统；Primary Dispatch System

用于对飞机重量和平衡数据进行计算，签派飞机的系统。

A system that generates aircraft weight and balance data used to dispatch an aircraft for flight.

### 主任监察员；Primary Inspector；PI

局方负责对运营人进行合格审定并颁发运行规范的监察员。主任监察员包括主任运行监察员和主任维修监察员。

The inspector of Authority for certification and issuing operation specification for operator，include primary operation inspector and primary maintenance inspector.

### 主任适航监察员；Principal Airworthiness Inspector

民航局或者民航地区管理局指定的负责对某个或者某些维修单位进行监督、检查的监察员。

The airworthiness inspector, assigned by the Authority Headquarter Office or the Authority Region Office to supervise one particular or several maintenance organization(s).

### 主任维修监察员(美国);Principal Maintenance Inspector; PMI

代表 FAA AFS,主要接触运营人,负责批准和监督按照 121、125、91 部运行的飞机运营人的执行维修计划。

The representative of the FAA AFS who is the primary point of contact for an operator. This individual is responsible for the approval and surveillance of the air operator's maintenance program for operations conducted under parts 121, 125, or 129. For part 91 operations the PMI refers to the assigned maintenance inspector.

### 主视觉告警;Master Visual Alert

专门用于紧急级别(如警告和警戒)的,用来吸引机组注意力的总视觉告警。

An overall visual indication used to attract the flightcrew's attention that is specific to an alert urgency level (for example, warning or caution).

### 主视觉区域;Primary Field of View; FOV

基于最佳水平和垂直视觉区域设计眼睛参考点,只转动黄斑中心或使用中心视力就可以看到该点。

FOV is based on the optimum vertical and horizontal visual fields from the design eye reference point that can be viewed with eye rotation only using foveal or central vision.

### 主探冰系统;Primary Ice Detection System

用于确定冰保护系统是否启动的检测系统。此系统提示存在积冰或结冰条件,也能向其他飞机系统提供信息。主自动探冰系统能够自动启动防冰或除冰系统。主手动探冰系统要求机组根据主探测系统信息启动防冰或除冰。

A detection system used to determine when the IPS must be activated. This system announces the presence of ice accretion or icing conditions, and it may also provide information to other aircraft systems. A primary automatic system automatically activates the anti-icing or deicing IPS. A primary manual system requires the flightcrew to activate the anti-icing or deicing IPS upon indication from the primary ice detection system.

### 主系统；Primary System; PRIM

最有可能影响经批准的设计资料完整性和产品质量的活动或功能。

Activities of functions most likely to affect the approved design data integrity and product quality.

### 主要部件；Major Part

故障会对装置的整体运行产生不利影响的部件。

A part of which failure might adversely affect the operational integrity of the unit.

### 主要结构件；Principal Structural Elements; PSE

对飞行、着陆、负载起到重大作用的主要结构要素，这些要素的失效可能导致航空器的灾难性故障。

Principal structural elements（PSE）are those elements of primary structure which contribute significantly to carrying flight，ground，and pressurization loads，and whose failure could result in catastrophic failure of the airplane.

### 主音响告警；Master Aural Alert

专门用于紧急级别(如警告和警戒)的，用来吸引机组注意力的总音响告警。

An overall aural indication used to attract the flightcrew's attention that is specific to an alert urgency level (for example，warning or caution).

### 主最低设备清单；Master Minimum Equipment List; MMEL

由制造商制订、局方确定的在特定运行条件下可以不工作并且仍能保持可以接受的安全水平的设备清单。主最低设备清单包含这些设备不工作时航空器运行的条件、限制和程序，是运营人制订各自最低设备清单的依据。

A list of equipment，determined by the Administrator，allowed to be inoperative under particular conditions when the accepted safety level can still be maintained. MMEL includes conditions of, limitations on, and procedures of aircraft operations when the equipment is inoperative and is the basis on which operators formulate their respective MEL.

### 专项合格审定计划；Project Specific Certification Plan；PSCP

民用航空产品项目级的合格审定计划，包括申请人的审定计划（CP）信息、审查组的必要信息和审查项目特有信息。

The certification plan of civil aviation product project level, including the information from the applicant's certification plan (CP), the necessary information of authority and specific information of a specific certification project.

### 专用工具设备；Special Tools/Equipment

制造厂家技术文件中推荐的专门用于某航空器或航空器部件维修的，仅用于维修过程，而非用于确定航空器或航空器部件最终放行的工具或设备。

Tools/equipment recommended in manufacturers' technical data, only for maintaining designated aircraft or its component. They are used only in maintenance process, not used for determining the return to service of aircraft or its component.

### 专用条件；Special Condition；SC

《民用航空产品和零部件合格审定规定》（CCAR - 21）第 21.16 条规定的专用条件是针对提交进行型号合格审定的民用航空产品，出于下述原因之一使得有关的适航规章没有提供适当的或足够的安全要求，由中国民用航空局（CAAC）适航司制定并颁发的补充安全要求。① 民用航空产品具有新颖的或独特的设计特点；② 民用航空产品的预期用途是非常规的；③ 从使用中的类似民用航空产品或具有类似设计特点的民用航空产品得到的经验表明可能产生不安全状况。专用条件应具有与适用的适航规章等效的安全水平。

CAAC certification department developed and issued supplemental safety requirements to those civil aviation product which apply for type certificate. Due to the following reason that make the airworthiness regulation fail to provide applicable or enough safety requirement：① civil aviation product has new or unique design features；② civil aviation product's expected use is unconventional；③ it may lead to unsafe condition which is suggested by experience of the similar civil aviation product in use or civil aviation product with the similar design feature. Sepcial condition should have the equivalent safety level as applicable airworthiness regulation.

### 转场飞行；Ferry Flight

为使航空器返回基地或使其飞往和飞离维修基地的飞行。在某种情况下，按照特许飞行许可进行转场飞行。

A flight for the purpose of returning an aircraft to base or moving an aircraft to and from a maintenance base. Ferry flights，under certain conditions，are conducted under the terms of a special flight permit.

### 转机型训练；Transition Training

曾在相同组类、不同型别飞机的相同职位上经审定合格并服务过的机组成员和飞行签派员需要进行的改飞机型训练。

The training required for crewmembers and dispatchers who have qualified and served in the same capacity on another airplane of the same group.

### 转弯半径；Turning Radius

航空器靠自身动力转弯时画出的圆弧半径，通常给出的是最小转弯半径。

The radius of the arc described by an aircraft in making a self-powered turn，usually given as a minimum.

### 转子；Rotor

发动机和 APU 中可以高速旋转的部件。分析、测试和/或经验表明在发生非包容性故障时，转子会被抛出。对每一个发动机/APU 的型号设计，其制造商都应详细说明组成转子的所有零部件。一般地，转子至少包括轮盘、轮毂、鼓筒、密封件、叶轮、叶片和隔叶块。

The rotating components of the engine and APU that can be released during uncontained failure，which is shown by analysis, test，and/or experience. The engine or APU manufacturer should define those components that constitute the rotor for each engine and APU type design. Typically rotors have included, as a minimum, disks, hubs, drums, seals, impellers, blades and spacers.

### 状态信息;Status Information

飞机系统及其周边的当前状态信息。

Information about the current condition of an airplane system and its surroundings.

### 撞击区;Impact Area

飞机很可能被转子失效过程中产生的非包容性碎片碰撞的区域。

The area of the airplane likely to be impacted by uncontained fragments generated during a rotor failure.

### 准合格审定状态;Pre-certification

飞机适航证颁发前的飞机状态。

An aircraft's state of condition before the issuance of the airworthiness certificate.

### 着陆方向指示器;Landing Direction Indicator; LDI

以视觉方式指示即时起降方向的一种装置。

A device which visually indicates the active takeoff and landing direction.

### 着陆航向信标台;Localizer; LOC

航站区电子导航系统中的一种仪表着陆系统导航设施,在进近和着陆过程中向航空器提供对跑道中心线的水平导引,方法是辐射两个信号调制的甚高频无线电波方向图,这两个信号在以等强度接收时,由兼容的机载设备显示为"在航道上"指示;在以不等强度接收时,显示为"偏离航向"指示。

An ILS navigation facility in the terminal area electronic navigation system, providing horizontal guidance to the runway centerline for aircraft during approach and landing by radiating a directional pattern of VHF radio waves modulated by two signals which, when received with equal intensity, are displayed by compatible airborne equipment as an "on-course" indication; and when received in unequal intensity are displayed as an "off course" indication.

### 着陆基准速度;Landing Reference Speed; $V_{REF}$

飞机正常着陆时,高度 50 英尺(约 15 米)时的基准速度。通常是飞机在着陆形态时的失速速度的 1.3 倍。

The reference speed at the 50 ft (15 m) height in a normal landing. This speed is generally equal to 1.3 times the stall speed in the landing configuration.

### 着陆决断点;Landing Decision Point; LDP

着陆决断点是在进场与着陆航迹上可以按第 29.85 条完成中断着陆的最后一点。

Landing decision point in the approaching and landing track is the last point to interrupt landing which can be completed by 29.85.

### 咨询;Advisory

针对要求机组注意并可能要求机组随后响应的情况的那一级或一类告警。

The level or category of alert for conditions that requires flightcrew awareness and may require subsequent flightcrew response.

### 咨询通告;Advisory Circular; AC

局方一系列的对外出版物,由各种政策、指南和信息性的非规章材料组成。

A series of external Authority publications consisting of all non-regulatory material of a policy, guidance, and informational nature.

### 姿态;Attitude; ATT

运动或静止航空器的位置或方位,它取决于在某种固定基准轴坐标系中航空器轴线与某基准线(或面)的相对关系。

The position or orientation of an aircraft either in motion or at rest as determined by the relationship between its axes and some reference line or plane on some fixed system of reference axes.

### 姿态管理；Attitude Management

识别危险姿态并通过应用适当的纠正方法修正该姿态的能力。

The ability to recognize hazardous attitudes in oneself and the willingness to modify them as necessary through the application of an appropriate antidote thought.

### 姿态指示器；Attitude Indicator

用于即时并直接给出飞行器真实飞行姿态的唯一设备,是最基本的姿态参考。

The only instrument that portrays both instant and direct actual flight attitude，which is the basic attitude reference.

### 子供应商；Sub Supplier

向供应商提供零件或服务的任何人。

Anyone provides parts or services to suppliers.

### 自动定向仪；Automatic Direction Finder; ADF

由无线电接收机、辨向和定向(环形)天线及方位指示器组成的无线电装置,利用地面站发射的无线电信号自动指示航空器相对地面发射机的方位。

A radio device composed of a radio receiver，sense and directional (loop) antennas，and a bearing indicator，using radio transmissions from ground stations to automatically indicate the bearing of an aircraft relative to the ground transmitter.

### 自动飞行航迹；Automatic Flight Path

一种稳定的飞行,它获得能针对预定飞行航迹提供航线指引或高度指引(或两者兼而有之)的导航系统发出的控制信息。

The type of stabilized flight which obtains control information from a navigation system capable of providing course line or altitude guidance，or both，aiming at a desired flight path.

### 自动飞行控制系统；Automatic Flight Control System；AFCS

由所有能引导飞机飞行的子系统所组成的一个系统。

A system composed of all necessary subsystems needed to guide the flight of an airplane.

### 自动高度控制系统；Automatic Altitude-keeping System

用于自动控制航空器，使其保持在某个气压高度的系统。

Any system which is designed to automatically control the aircraft to a referenced pressure-altitude.

### 自动跟踪；Automatic Tracking

利用雷达输入信号自动跟踪选定的目标，并以模拟或数字方式提供连续位置数据的装置。

A device which automatically follows a selected target using radar input signals and provides continuous positional data in analog or digital form.

### 自动记录带中继；Automatic Tape Relay

一种通信方法，其信息靠电传打字机的记录带接收和传送，没有人工参与。

A method of communication whereby messages are received and transmitted in teletypewriter tape form without manual intervention.

### 自动驾驶；Autopilot

提供航空器自动控制手段的装置和组件。

Units and components which furnish a means of automatically controlling the aircraft.

### 自动检测系统；Automatic Inspection System

通过自动旋转或移动检测探头或装置来完成过程检测，且能够通过配置的报警或记录系统自动记录缺陷显示的系统。

A system which can perform process test by automatically rotating or moving the detection probe or device，and can automatically record the defect display by the configured alarm or record system.

### 自动进近限制高度；Automatic Approach Limit

沿进近航迹的某一点，在此点上自动进近设备应与航空器的操纵系统脱离开，因为航向结构的变化莫测会因自动进近设备引起航空器不稳定或不安全的机动动作。

That point along the approach path at which automatic approach equipment should be released from control of the aircraft because of vagaries in the course structure，causing erratic or unsafe maneuvering of the aircraft by the automatic approach equipment.

### 自动数据处理；Automatic Data Processing；ADP

用机械、电气或电子手段自动记录、处理、储存、传送、提供数据或对数据进行其他形式的处理。

The automatic recording，processing，storing，transmitting，presenting，or other manipulation of data by mechanical，electrical or electronic means.

### 自动数据交换系统；Automatic Data Interchange System；ADIS

天气资料的高速传送/接收设施。

High speed transmitting/receiving facility for weather data.

### 自动油门系统；Auto Throttle System；ATS

由机组选择的提供自动发动机推力控制的系统。

A system selected by the crew to provide automatic engine thrust control.

### 自动中继装置；Automatic Relay Installation

一种电传打字机装置，其中的自动设备用来传递电路之间的信息。此术语包括自动与半自动两种装置。

A teletypewriter installation where automatic equipment is used to transfer messages between circuits. This term includes both automatic and semiautomatic installations.

## 自动着陆系统；Automatic Landing System

设计用于引导航空器在没有事先援助的情况下安全着陆的一种系统。

A system designed to guide an aircraft to a safe landing without prior assistance.

## 自检设备；Built-in Test Equipment；BITE

被动的故障指示系统。如果功能信号流中断或超出最大可接受水平，将出现视觉或听觉警告表明已发生故障。

System as a passive fault indicator. If the functional signal flow stops or increases beyond a maximum acceptance level，a visual/aural warning is displayed to indicate that a malfunction has occurred.

## 自燃温度；Auto-ignition Temperature

易燃蒸气和空气混合物在正常的没有外部点火源（如火焰或火花）的大气中均匀加热时会自发点燃的最低温度。

The minimum temperature at which an optimized flammable vapor and air mixture will spontaneously ignite when heated to a uniform temperature in a normal atmosphere without an external source of ignition，such as a flame or spark.

## 自修正自动导航；Self-correcting Automatic Navigation；SCAN

一种推测导航装置，当它确认来自全向指向标或其他以地面为基地的无线电辅助设施或雷达的信号精确时，会自动、连续地按上述设施修正自身位置。将推测导航与外部无线电信号相组合，所得到的精度高于单独使用任一种可得到的精度。

A dead reckoning device which automatically and continuously corrects itself from omnirange or other ground-based radio aids or radar when it determines that these signals are accurate. Combining dead reckoning and external radio signals to give greater accuracy than that available from either type alone.

## 自愿性共识标准；Voluntary Consensus Standards

国内和国际自愿性共识标准组织制定或采用的标准。

Standards developed or adopted by voluntary consensus standards bodies, both domestic and international.

### 自愿性共识标准组织；Voluntary Consensus Standards Bodies

通过协商对自愿性共识标准进行规划、制定、实施和协调的国内或国际组织。

Domestic or international organizations which plan, develop, establish, or coordinate voluntary consensus standards using agreed-upon procedures.

### 自制件；In-house Fabricated Part

不是依据航空器或航空器部件的制造厂家公开发布的持续适航性文件中给定的设计数据、材料或加工方法制造的航空器部件。

An aircraft component which is designed and manufactured pursuant to the design data, material or process other than those specified in the continual airworthiness documents published by civil aircraft or aircraft component manufacturer.

### 总垂直误差；Total Vertical Error；TVE

航空器飞行的实际气压高度与其指定的气压高度（飞行高度层）之间的垂直几何差。

The vertical geometric difference between the actual pressure altitude flown by an aircraft and its assigned pressure altitude (flight level).

### 总管/歧管压力；Manifold Pressure

在进气系统适当位置测得的绝对压力，通常用英寸汞柱表示。

The absolute pressure measured at the appropriate location in the intake system, usually expressed by inches of mercury high.

### 总体环境；Total Environment

围绕设备并对其产生影响的情形或条件，包括制造、操作、存储、运输、任务、维护及修理。

The circumstances and conditions which surround and influence the equipment. The total environment includes manufacturing, handling, storage, shipping, mission, maintenance, and repair.

### 租出飞机(美国);Loaned Aircraft

由联邦直管机构拥有、但由非直管机构按不含补偿条款的协议进行保管的飞机。

A federal aircraft owned by an executive agency, but in the custody of a non-executive agency under an agreement that does not include compensation.

### 租赁航空器;Rental Aircraft

在商业协议下由承租人在协议规定的时间里专属运行的航空器。该航空器由承租人运行但不负责维护。在局方规章管辖下的租赁民用航空器可以通过开发市场或合同协议获得,并在局方租赁方案内使用。

An aircraft hired commercially under an agreement in which the executive agency has exclusive use of the aircraft for an agreed upon time. The executive agency operates, but does not maintain, a rental aircraft. Authority-operated civil aircraft obtained through open market or contract agreements and used within the Authority rental program.

### 租赁时间;Rental Time

租赁时间的起始根据租赁合同条款或基于转速记录器确定。当使用机组记录时间作为租赁时间时,要注意精确记录以便符合约定的付费义务。

This time begins and ends according to the terms of the contract or based on a recording tachometer. When using crew-recorded times, recording them accurately to ensure an equitable payment obligation.

### 阻燃;Flame Resistant

移去火源后不会使火焰蔓延到超出安全限度的程度。

After removing the ignition source, it will not burn to the extend that the flame spreads beyond the safety limit.

### 阻止级抖振;Deterrent Level of Buffet

严重的抖振,明确地构成了对进一步减小空速或增加攻角的阻止。

A severe level of buffet that constitutes a clear deterrent to further decrease in airspeed or increase in angle of attack.

### 组合视觉系统；Combined Vision System; CVS

在单一集成显示器上整合的增强视觉系统和合成视觉系统信息的系统。

A system which combines information from an enhanced vision system and a synthetic vision system in a single integrated display.

### 最大动力涡轮机超速；Maximum Power-turbine Overspeed

自由动力涡轮机最大转速，若不慎在该转速下持续工作长达 20 秒，刚认为使用或维修操作中的发动机不需要废弃。

The maximum rotational speed of the free power-turbine，inadvertent occurrence of which for periods of up to 20 seconds，has been agreed not to require rejection of the engine from service or maintenance action.

### 最大动力涡轮机自转速度；Maximum Power-turbine Speed for Autorotation

在任何持续时间的自转过程中都允许的最大动力涡轮机转速。

The maximum rotational speed of the power-turbine permitted during autorotation for periods of unrestricted duration.

### 最大发动机超速；Maximum Engine Overspeed

发动机最大转速，若不慎在该转速下持续工作长达 20 秒，则认为使用或维修操作中的发动机不需要废弃。

The maximum engine rotational speed，inadvertent occurrence of which for periods of up to 20 seconds，has been agreed not to require rejection of the engine from service or maintenance action.

### 最大改航时间；Maximum Diversion Time

运营人的延程运行当局针对一个航班批准的最长改航时间，用于延程运行航路计划。在计算最大改航时间时，假设飞机在标准条件下的静止大气中，以一台发动机不工作的巡航速度飞行。

For the purposes of ETOPS route planning，the longest diversion time authorized for a flight under the operator's ETOPS authority. It is calculated under standard conditions in still air at a one-engine-inoperative cruise speed.

### 最大可用高度；Maximum Usable Altitude；MUA

由技术因素（例如干扰或信号衰减）决定的高度。

The altitude which is determined by technical factors such as interference or signal attenuation.

### 最大连续额定功率和/或推力；Maximum Continuous Power and/or Thrust Rating

系列和新翻修发动机在指定条件和相应可接受限制内运行时试车台接受的最小功率和/或推力，见发动机型号合格证数据单中的规定。

The minimum test bed acceptance power and/or thrust，as stated in the engine type certificate data sheet，of series and newly overhauled engines when running at the specified conditions and within the appropriate acceptance limitations.

### 最大连续功率和/或推力；Maximum Continuous Power and/or Thrust

性能资料里规定的用于不受限制的持续时间内的功率和/或推力。

The power and/or thrust identified in the performance data for use during periods of unrestricted duration.

### 最大旅客座位布局（最大载客量）；Maximum Passenger Seating Configuration（or Maximum Passenger Capacity）

基于出口布局和撤离率，飞机上可安装的乘客座椅的理论最大数量。不是所有的飞机都被批准可以按照理论最大旅客座位布局安装乘客座椅。

The theoretical maximum number of passenger seats that can be installed in an airplane based upon the exit configuration and the exit ratings. Not all airplanes are approved for installation of the theoretical maximum passenger seating configuration.

### 最大排气超温；Maximum Exhaust Gas Overtemperature

发动机最大排气温度，若不慎在该温度下使用长达 20 秒，则认为使用或维修操作中的发动机不需要废弃。

The maximum engine exhaust gas temperature, inadvertent use of which for periods of up to 20 seconds, has been agreed not to require rejection of the engine from service or maintenance action.

### 最大起飞重量;Maximum Take-off Weight

因设计或运行限制,航空器能够起飞时所容许的最大重量。

The maximum weight allowed for an aircraft to take off due to design or operational limitations.

### 最大容许转动速度;Maximum Permissible Rotational Speed

在正常或可能的紧急的运行中允许的最大螺旋桨转动速度。

The maximum propeller rotational speed permitted in normal or possible emergency operation.

### 最大商载;Maximum Commercial Load

① 对在局方批准的技术文件中列出的最大无燃油重量的航空器,用最大无燃油重量减去空重、适用的航空器携带设备的重量和运行载重(包括最少机组、餐饮及与餐饮有关的补给和设备,不包括可用燃油和滑油)。② 对其他航空器,用审定的最大起飞重量减去空重、适用的航空器携带设备的重量和运行载重(包括最少燃油、滑油和机组重量)。机组、滑油和燃油的重量计算如下。a) 对于每名机组成员(含随身携带行李):(a) 男性飞行机组成员为 82 千克;(b) 女性飞行机组成员为 64 千克;(c) 男性客舱乘务员为 82 千克;(d) 女性客舱乘务员为 59 千克;(e) 对于未指明性别的客舱乘务员为 64 千克。b) 滑油为 157 千克或者根据航空器型别证书中列出的滑油容量计算出的数据。c) 燃油按照中国民用航空规章要求实施飞行所需的最低燃油重量。

Maximum Commercial Load equals to: ① The maximum zero fuel weight minus the empty weight, the weight of applicable equipment on board, and operation load (including the weight of minimum crew, catering, and supplies and equipment related to catering, excluding available fuel and oil) for aircraft with its maximum zero fuel weight listed in technical documents authorized by the Administrator. ② The certificated maximum takeoff weight minus the empty weight, the weight of applicable equipment on board, and operation load (including the weight of minimum fuel, oil, and crew) for other aircraft. Weight of crew, oil, and fuel is calculated as follows. a) For every crewmember (including hand baggage): (a) Male flight crewmember is 82 kilograms; (b) Female flight crewmember is 64 kilograms; (c) Male cabin attendant is 82 kilograms; (d) Female cabin attendant is 59 kilograms; (e) Cabin attendant with unidentified sex is 64 kilograms. b) Oil is 157 kilograms or calculated in accordance with the oil quantity listed in the aircraft type certificate. c) Fuel minimum fuel weight required to perform flights under the Chinese Civil Aviation Regulations.

### 最大审定乘客容量；Maximum Certificated Occupant Capacity

由有管辖权的局方审定的每个特定型号的飞机能运载的最大人数。

Maximum certificated occupant capacity is the maximum number of persons that can be carried for each specific aircraft model as certified by the Authority having jurisdiction.

### 最大调节转速；Maximum Governed Rotational Speed

由设定螺旋桨调速器或控制机构来确定的最大转速。

The maximum rotational speed as determined by setting the propeller governor or control mechanism.

### 最大推荐巡航功率条件；Maximum Recommended Cruising Power Conditions

发动机手册中推荐的适用于巡航运行的机轴转动速度、发动机歧管压力和任何其他参数。

The crankshaft rotational speed, engine manifold pressure and any other parameters recommended in the engine manuals as appropriate for cruising operation.

### 最大稳定性特性速度 $V_{FC}/M_{FC}$；Maximum Speed for Stability Characteristics；$M_{FC}$

一个不得小于最大使用限制速度（$V_{MO}/M_{MO}$）与经演示验证飞行俯冲速度

($V_{DF}/M_{DF}$)之间平均值的速度,但是在马赫数作为限制因素的高度上,$M_{FC}$ 不必超过发生有效速度警告的马赫数。

The average speed between no less than the maximum operating limit speed($V_{MO}/M_{MO}$) and the flight diving speed($V_{DF}/M_{DF}$) by the demonstration, however, $M_{FC}$ should not exceed the Mach number when the effective speed warning occurs in the height as a limiting factor of the Mach number.

## 最大应急额定功率和/或推力;Maximum Contingency Power and/or Thrust Rating

系列和新翻修发动机在指定条件和相应可接受限制范围内运行时,试车台接受的最小功率和/或推力,见发动机型号合格证数据单中的规定。

The minimum test bed acceptance power and/or thrust, as stated in the engine type certificate data sheet, of series and newly overhauled engines when running at the specified conditions and within the appropriate acceptance limitations.

## 最大应急功率和/或推力;Maximum Contingency Power and/or Thrust

性能资料中规定的最大应急功率和/或推力,在动力装置起飞、受阻碍着陆过程中或终止进近之前出现故障或关车时使用,限制持续使用不超过 2.5 分钟。

The power and/or thrust identified in the performance data for use when a power-unit has failed or been shut down during take-off, baulked landing or prior to a discontinued approach and limited in use for a continuous period of not more than 2.5 min.

## 最大允许表面温度;Maximum Allowable Surface Temperatures

燃油箱(油箱壁、挡板或任何部件)内的表面温度,该温度在所有正常或故障的情况下为燃油箱提供一个安全裕度,即批准的燃油最低预期自动点火温度以下至少 50°F(约 10℃)。燃油的自动点火温度会因各种因素(环境压力、停留时间、燃油类型等)而有所不同。例如在静态海平面条件下,该数值对喷气飞机 A 是 450°F(232.2℃)。这导致在一个受影响的部件表面,最大允许表面温度为 400°F(204.4℃)。

A surface temperature within the fuel tank (the tank walls, baffles, or any components) that provides a safe margin under all normal or failure conditions, which is at least 50℉ (10℃) below the lowest expected auto-ignition temperature of the approved fuels. The auto-ignition temperature of fuels will vary because of a variety of factors (ambient pressure, dwell time, fuel type). For example, the value for Jet A, under static sea level conditions, is 450℉ (232.2℃). This results in a maximum allowable surface temperature of 400℉ (204.4℃) for an affected component surface.

### 最低穿越高度；Minimum Crossing Altitude；MCA

某些无线电定位点处的最低高度，当航空器朝较高的最小航线仪表飞行规则高度方向飞行时必须横越此最低高度。

The lowest altitude at certain radio fixes at which an aircraft must cross when proceeding in the direction of a higher minimum en route IFR altitude.

### 最低接收高度；Minimum Reception Altitude；MRA

接收足够信号以确定具体 VOR/VORTAC/TACAN 位置所需的最低高度。

The lowest altitude required to receive adequate signals to determine specific VOR/VORTAC/TACAN fixes.

### 最低接通高度；Minimum Engage Height

飞机起飞后，允许机组接通自动驾驶仪的最低高度。

The minimum height after takeoff at which the flightcrew is permitted to engage the autopilot.

### 最低设备清单；Minimum Equipment List；MEL

运营人依据主最低设备清单并考虑各航空器的构型、运行程序和条件，为其运行所编制的设备清单。最低设备清单经局方批准后，允许航空器在规定条件下，所列设备不工作时继续运行。最低设备清单应等于或严格于相应机型的主最低设备清单。

An equipment list prepared by an operator in accordance with MMEL for its operations, considering types, operations procedures, and conditions of its aircraft. After being authorized by the Administrator, MEL allows aircraft with inoperable equipment listed in MEL to operate under specified conditions. MEL requirements shall be equal to or stricter than those in MMEL of the applicable aircraft type.

### 最低使用高度；Minimum Use Height；MUH

在适航验证或审查过程中规定的高度，在标准或规定条件下，在此高度以上一个可能的系统故障不会导致从指定基准面（如机场标高）或指定障碍物净空面不可接受地减少飞行路径间隙的显著路径位移。

A height specified during airworthiness demonstration or review above which, under standard or specified conditions, a probable failure of a system is not likely to cause a significant path displacement unacceptably reducing flight path clearance from specified reference surfaces (for example, airport elevation) or specified obstacle clearance surfaces.

### 最低下降高度；Minimum Descent Altitude/Minimum Descent Height；MDA/ MDH

在非精密进近或者盘旋进近中，如果不能建立必需的目视参考，则不能继续下降的特定高度（海拔高度）或高（相对高度）。

The lowest altitude or height to which descent is not authorized during nonprecision approach or circle-to-land maneuvering unless the necessary visual reference can be established.

### 最低油量；Minimum Fuel Quantity

飞行过程中应当向空中交通管制报告并采取应急措施的最小燃油量，该特定的最小燃油量能够使飞机在 450 米（1 500 英尺）以上的高度加上等待空速的机场标高上飞行 30 分钟，其中应当考虑到规定的燃油油量指示系统误差。

A particular minimum fuel quantity in flight that shall be reported to air traffic control to take emergency measures and is able to enable aircraft to fly to the landing airport and flight for 30 min at the altitude of more than 450 m (1 500 ft) plus the airport elevation at the holding airspeed, considering the specified fuel quantity indication system margin.

### 最低越障高度；Minimum Obstruction Clearance Altitude；MOCA

VOR(甚高频全向无线电指向标)航路、偏离航路航线或航线段上两个无线电定位点之间的规定高度,它在整个航段上均符合越障要求,并且仅在 VOR 的 22 海里范围内具有可接受的导航信号覆盖能力。

The specified altitude in effect between radio fixes on VOR airways, off airway routes, or route segments which meets obstruction clearance requirements for the entire route segment and which assumes acceptable navigational signal coverage only within 22 n mile of a VOR.

### 最高表面温度；Maximum Surface Temperature

电气设备在允许的最不利条件下运行时,其表面或任一部分可能达到的并有可能引燃周围爆炸性气体环境的最高温度。

The maximum temperature which may be reached the surface or any part of the electrical equipment and ignite the surrounding explosive gas atmospheres when it is running under the most adverse permissible conditions.

### 最小飞行重量；Minimum Flying Weight；MFW

带有最少机组和燃油的飞机空重。

The airplane empty weight with minimum crew and fuel.

### 最小风险爆炸位置；Least Risk Bomb Location；LRBL

在飞机上安置易爆或易燃设备的位置,该位置可以将爆炸时的后果降至最小。

The location on the airplane where an explosive or incendiary device should be placed to minimize the effects to the airplane in case of detonation.

### 最小航线仪表飞行规则高度；Minimum En Route IFR Altitude；MEA

两个无线电定位点之间的规定高度,它保证了这两点之间有可接受的导航信号覆盖,并符合越障要求。为航路或航段、区域导航低航线或高航线或其他直航线所规定的最小航线仪表飞行规则高度,适用于划定航路、航段或航线的两个无线电定位点之间的航路、航段或航线的整个宽度。

The altitude in effect between radio fixes which assures acceptable navigational signal coverage and meets obstruction clearance requirements between those fixes. The MEA prescribed for a airway or segment thereof，area navigation low or high route，or other direct route；applies to the entire width of the airway，segment or route between the radio fixes defining the airway，segment，or route.

### 最小离地速度；Minimum Unstick Speed；$V_{MU}$

根据校准的空气速度，最小离地速度 $V_{MU}$ 应为飞机能够离开地面并继续起飞而不显示任何危险特性的速度。

The minimum unstick speed $V_{MU}$, in terms of calibrated air speed, shall be the speed at and above which the airplane can be made to lift off the ground and to continue the takeoff without displaying any hazardous characteristics.

### 最小弯曲半径；Minimum Bend Radius

为了避免损坏导线绝缘层，单个导线或线束的最小弯曲半径应该与导线制造商的规范要求相一致。在飞机制造商的标准线路施工手册中能找到最小弯曲半径的指导。其他行业标准也包含有关最小弯曲半径的指导，如欧洲航空航天工业协会的文件 AECMA EN3197 或 SAE AS50881。

To avoid damage to wire insulation，the minimum radius of bends in single wires or bundles should be in accordance with the wire manufacturer's specifications. Guidance on the minimum bend radius can be found in the airplane manufacturer's standard wiring practices manual. Other industry standards such as the European Association of Aerospace Industries' document AECMA EN3197 or SAE AS50881 also contain guidance on minimum bend radius.

### 最小引导高度；Minimum Vectoring Altitude；MVA

由雷达操纵员引导仪表飞行规则的航空器飞行的最低高度，以平均海平面以上的英尺数表示。此高度保证通信、雷达覆盖并符合越障要求。

The lowest altitude，expressed in feet above MSL，that an IFR aircraft will be vectored by a radar controller. The altitude assures communications, radar coverage and meets obstacle clearance criteria.

### 最终起飞速度；Final Takeoff Speed；$V_{FTO}$

飞机在航路形态单发不工作条件下，在起飞航迹终点处所达到的速度。

The speed of the airplane that exists at the end of the takeoff path in the en route configuration with one engine inoperative.

### 作业任务航空器；Job Task Aircraft

由运营人提供的用于审定、研究或局方机组成员要求的特种程序的飞机。

An aircraft provided by an operator in which certification, research, or special procedures require an on-board Authority crewmember.

### 座舱压力高度；Cabin Pressure Altitude

与飞机座舱内压力相对应的压力高度。对无增压座舱的飞机，"座舱气压高度"和"飞行高度"是相同的。

The pressure altitude corresponding with the pressure in the cabin of the airplane. For airplanes without pressurized cabins, "cabin pressure altitude" and "flight altitude" is the same.

### 座椅布局；Seating Configuration

飞机内部平面图，定义了机组人员和乘客在滑行、起飞、着陆和飞行条件下可用的座位。

The airplane interior floor plan, which defines the seating positions available to crew and passengers during taxi, takeoff, landing, and in-flight conditions.

### 座椅/约束系统；Seating/Restraint System

包括座椅结构、靠垫、坐垫、安全带、肩带和连接装置的系统。

A system that includes the seat structure, cushions, upholstery, the safety belt, the shoulder harness, and the attachment devices.

**座椅主载荷通路;Seat Primary Load Path**

座椅内从载荷作用点开始到对座椅系统或子系统载荷产生反作用的结构为止的承载构件。主载荷通路随评估参数不同可分为以下几种：① 结构主载荷通路从座椅安全带/肩带到座椅系统与航空器结构的连接接头；② 腰椎主载荷通路从底部坐垫到座椅系统与航空器结构的连接接头；③ 排与排之间头部伤害准则的主载荷通路从拟人试验装置头部接触点到座椅主结构的连接部位；④ 头部路径，如前排座椅或大倾斜角座椅，其至载荷通路与结构的主载荷通路相同。

The components within the seat that carry the load from the point of load application to the structure that reacts the load from the seat system or sub-system. The primary load path varies depending on the parameter being evaluated, as follows：① structural — from seat belt/harness to fittings attaching seat system to aircraft structure；② lumbar — from bottom cushion to fittings attaching seat system to aircraft structure；③ row-to-row head injury criterion — from point of ATD head contact to the attachment of seat primary structure；④ head path（front row or large pitch seats）— same as structural.

**座椅组;Family of Seats**

不考虑座椅位置数量，依据等效部件在主传力路径建立的一组座椅组件。

A group of seat assemblies，regardless of the number of seat places，built from equivalent components in the primary load path.

# 其 他

**Ⅰ类产品;Type Ⅰ**

已具有型号合格证的航空器、发动机或螺旋桨。

Aircrafts、engines、propellers which have already obtained Type Certification.

**(维修)Ⅰ类问题;Level Ⅰ Finding**

不符合 CCAR‐145 在厂房设施、工具设备、器材、人员及适航性资料等硬件方面的要求,并且是在日常维修工作中不可缺少而短期内不能解决的问题;或严重不符合 CCAR‐147 在培训教室、培训场地、设施设备、人员及培训资料、管理系统等方面的要求,并且是在日常培训工作中不可缺少而短期内不能解决的问题。

The finding that does not meet the requirements of CCAR‐145 on the hardware aspects of the facilities、tools、equipments、materials、personnel、and airworthiness data、which is indispensable in the daily maintenance work、and can not be solved in the short term; or the finding that does not meet the requirements of CCAR‐147 on the aspects of the training room、training venues、facilities、equipment、personnel and training information management system seriously、which is indispensable in the daily training、and can not be solved in the short term.

**Ⅰ类仪表着陆系统;ILS Category Ⅰ**

一种仪表着陆系统,它通过仪表着陆系统的覆盖范围限制提供可接受的导引信息,直到航向信标航道线与下滑航迹在包含跑道入口的水平面上方100英尺高度相交。Ⅰ类仪表着陆系统能支持的着陆最低条件为200英尺离跑道入口高度(HAT)和 1 800 英尺跑道能见度距离(RVR)。

An ILS which provides acceptable guidance information from the coverage limits of the ILS to the point at which the localizes course line intersects the glide path at a height of 100 ft above the horizontal plane containing the runway threshold. A Category Ⅰ ILS supports landing minima as low as 200 ft (HAT) and 1 800 ft (RVR).

## Ⅱ类产品；Type Ⅱ

Ⅱ类产品指其破损会危及Ⅰ类产品安全的主要部件,如机翼、机身、起落架、动力传动装置、操纵面等,以及 CTSOA 规定的材料、零部件和机载设备。

The main component of which damage will endanger the safety of Type Ⅰ product，such as wing, fuselage, leading gear, power plant, control surface and so on，and the material, component, airborne equipment obtained by CTSOA.

## (维修)Ⅱ类问题；Level Ⅱ Finding

在管理系统及其工作程序上不符合 CCAR - 145 的要求,以及实际维修工作与管理程序不符合的情况普遍存在或重复出现的问题;或培训程序不符合CCAR - 147 要求的情况,此种情况有可能降低培训标准,并且长时间不予以纠正将影响被培训学员今后的维修水平。

The finding that management system and its work procedure does not meet the requirements of CCAR - 145，as well as the actual repair work and management of the procedure does not meet widespread or recurring; or training procedure does not meet the requirements of CCAR - 147, in which case it will be possible to reduce the training standards，and will affect the maintenance level of the trainees in the future if it is not corrected at long periods.

## Ⅱ类仪表运行许可；Category Ⅱ Pilot Authorization

持有人仪表等级或航线运输驾驶员证书(需分别签署)的一部分,授权持有人作为指定机型机长实施Ⅱ类运行。

A part of the holder's instrument rating or airline transport pilot certificate ( but separately issued ) that authorizes the holder to conduct Category Ⅱ operations as pilot in command of specified types of airplanes.

### Ⅱ类仪表着陆系统;ILS Category Ⅱ

一种仪表着陆系统,它通过仪表着陆系统的覆盖范围限制提供可接受的导引信息,直到航向信标航道线与下滑航迹在包含跑道入口的水平面上方 50 英尺高度相交。Ⅱ类仪表着陆系统能支持的着陆最低条件为 100 英尺(HAT)和 1 200 英尺(RVR)。

An ILS which provides acceptable guidance information from the coverage limits of the ILS to the point at which the localizes course line intersects the glide path at a height of 50 ft above the horizontal plane containing the runway threshold. A Category Ⅱ ILS supports landing minima as low as 100 ft (HAT) and 1 200 ft (RVR).

### Ⅱ类仪表着陆系统训练;ILS Category Ⅱ Training

性能在Ⅱ类的标准之内[至 100 英尺(HAT)],进行Ⅰ类飞行的设施,并被宣称适用于Ⅱ类资格鉴定训练。

A Category Ⅰ operational use facility with performance within the standards for Category Ⅱ (to 100 ft HAT), which is advertised as acceptable for Category Ⅱ qualifications training.

### Ⅱ类运行;Category Ⅱ Operation

决断高低于 60 米(约 200 英尺)但不低于 30 米(约 100 英尺),跑道视程不低于 350 米的精密仪表进近和着陆。

A precision instrument approach and landing with a decision height lower than 60 m (200 ft), but not lower than 30 m (100 ft), and a runway visual range not less than 350 m.

### Ⅲ类产品;Type Ⅲ

Ⅲ类产品指Ⅰ、Ⅱ类产品以外的产品,包括按民航局认为适用的技术标准制造的标准零件。

Any product other than Type Ⅰ and Type Ⅱ, including the standard component which is manufactured according to the CTSO.

## (维修)Ⅲ类问题;Level Ⅲ Finding

除Ⅰ、Ⅱ类问题以外的任何问题。

The finding not included in the finding level Ⅰ or Ⅱ.

## Ⅲ类运行;Category Ⅲ Operation

就航空器运行而言,指使用局方或其他相关部门颁发的Ⅲ类仪表着陆系统仪表进近程序向机场跑道做仪表着陆系统进近,并进行着陆。

With respect to the operation of aircraft, Category Ⅲ operation means an ILS approach to, and landing on, the runway of an airport using a Category Ⅲ ILS instrument approach procedure issued by the Administrator or other appropriate authority.

## ⅢA类仪表着陆系统飞行;ILS‐CAT ⅢA Operation

朝向和沿着跑道表面的飞行,无决断高限制,跑道能见度不小于700英尺。

Operation, with no decision height limitation, that is to and along the surface of the runway with a runway visual range not less than 700 ft.

## ⅢA类运行;Category ⅢA Operation

在下列情况下的精密仪表进近和着陆:① 决断高低于30米(约100英尺)或无决断高;② 跑道视程不低于200米。

A precision instrument approach and landing with: ① a decision height lower than 30 m (100 ft) or no decision height; ② a runway visual range not less than 200 m.

## ⅢB类仪表着陆系统飞行;ILS‐CAT ⅢB Operation

朝向和沿着跑道表面的飞行,无决断高限制,不依靠外部目视参照物,但随后利用外部目视参照物进行滑行,跑道能见度不小于150英尺。

Operation, with no decision height limitation, that is to and along the surface of the runway without reliance on external visual reference; and, subsequently, taxiing with external visual reference with a runway visual range not less than 150 ft.

### Ⅲ B 类运行;Category Ⅲ B Operation

在下列情况下的精密仪表进近和着陆：① 决断高低于 15 米（约 50 英尺）或无决断高；② 跑道视程低于 200 米但不低于 50 米。

A precision instrument approach and landing with：① a decision height lower than 15 m（50 ft）or no decision height；② a runway visual range less than 200 m but not less than 50 m.

### Ⅲ C 类运行;Category Ⅲ C Operation

无决断高和跑道视程限制的精密仪表进近和着陆。

A precision instrument approach and landing with no decision height and no runway visual range limitations.

# 附录  适航术语范畴文件清单

## 一、中国民航适航管理文件

### （一）民航法规

| 1 | 中华人民共和国主席令第二十四号 | 中华人民共和国民用航空法 |

### （二）管理条例

| 1 | NA | 中华人民共和国民用航空器适航管理条例 |
| 2 | 中华人民共和国国务院令第 232 号 | 中华人民共和国民用航空器国籍登记条例 |

### （三）适航规章

| 1 | CCAR - 21 - R4 | 民用航空产品和零部件合格审定的规定 |
| 2 | CCAR - 25 - R4 | 运输类飞机适航标准 |
| 3 | CCAR - 26 | 运输类飞机的持续适航和安全改进规定 |
| 4 | CCAR - 33 - R2 | 航空发动机适航规定 |
| 5 | CCAR - 34 - R1 | 涡轮发动机飞机燃油排泄和排气排出物规定 |
| 6 | CCAR - 36 - R3 | 交通运输部关于修改《航空器型号和适航合格审定噪声规定》的决定 |
| 7 | CCAR - 37AA | 民用航空材料、零部件和机载设备技术标准规定 |
| 8 | CCAR - 39AA | 民用航空器适航指令规定 |
| 9 | CCAR - 43 - R1 | 维修和改装一般规则 |
| 10 | CCAR - 45 - R3 | 民用航空器国籍登记规定 |
| 11 | CCAR - 60 - R1 | 飞行模拟训练设备管理和运行规则 |
| 12 | CCAR - 61 - R4 | 民用航空器驾驶员合格审定规则 |
| 13 | CCAR - 91 - R4 | 一般运行和飞行规则 |

| 14 | CCAR - 121 - R7 | 交通运输部关于修改《大型飞机公共航空运输承运人运行合格审定规则》的决定 |
| 15 | CCAR - 135R3 | 小型商业运输和空中游览运营人运行合格审定规则 |
| 16 | CCAR - 145 - R4 | 民用航空器维修单位合格审定规定 |
| 17 | CCAR - 142 | 飞行训练中心合格审定规则 |
| 18 | CCAR - 147R1 | 民用航空器维修培训机构合格审定规定 |
| 19 | CCAR - 183AA - R1 | 民用航空器适航委任代表和委任单位代表管理规定 |
| 20 | CCAR - 183FS | 民用航空飞行标准委任代表和委任单位代表管理规定 |
| 21 | CCAR - 183SE | 中国民用航空计量技术委任代表和委任单位代表规定 |
| 22 | CCAR - 395 - R2 | 民用航空器事件调查规定 |
| 23 | CCAR - 398 | 民用航空安全管理规定 |

（四）技术标准规定

| 1 | CTSO - C8e | 垂直速度（爬升率）表 |
| 2 | CTSO - C10b | 气压高度表（敏感型） |
| 3 | CTSO - C13f | 救生衣 |
| 4 | CTSO - C22g | 安全带 |
| 5 | CTSO - C26d | CCAR23、27、29 部航空器机轮、刹车和机轮刹车组件 |
| 6 | CTSO - C30c | 航空器航行灯 |
| 7 | CTSO - C37d | VHF 无线电通讯发射设备（工作频带 117.975～137.000 MHz） |
| 8 | CTSO - C39b | 航空器座椅和卧铺 |
| 9 | CTSO - C43c | 温度表 |
| 10 | CTSO - C44c | 燃油流量表 |
| 11 | CTSO - C49b | 磁滞式电动转速表（指示器和传感器） |
| 12 | CTSO - C51a | 航空器飞行数据记录器 |
| 13 | CTSO - C52b | 飞行指引仪 |
| 14 | CTSO - C62e | 航空轮胎 |

| 15 | CTSO‐C63c | 机载脉冲气象和地形雷达 |
|---|---|---|
| 16 | CTSO‐C64a | 旅客连续供氧面罩组件 |
| 17 | CTSO‐C72c | 单独漂浮装置 |
| 18 | CTSO‐C87a | 机载低空无线电高度表 |
| 19 | CTSO‐C88b | 自动气压高度编码发生设备 |
| 20 | CTSO‐C96a | 防撞灯系统 |
| 21 | CTSO‐C106 | 大气数据计算机 |
| 22 | CTSO‐C123c | 驾驶舱音频记录器 |
| 23 | CTSO‐C124c | 飞行数据记录器 |
| 24 | CTSO‐C127b | 旋翼航空器、运输类飞机和小飞机座椅系统 |
| 25 | CTSO‐135a | 运输类飞机机轮和机轮刹车组件 |
| 26 | CTSO‐148 | 航空器机械紧固件 |
| 27 | CTSO‐C166b | 基于1 090兆赫扩展电文的广播式自动相关监视(ADSB)和广播式交通情报服务(TIS‐B)设备 |
| 28 | CTSO‐C175 | 飞机厨房手推车、物品箱及相关组件 |

（五）适航管理程序

| 1 | AP‐00‐AA‐2018‐01R5 | 适航审定培训管理程序 |
|---|---|---|
| 2 | AP‐11‐AA‐2010‐01 | 适航规章和环境保护要求制定和修订程序 |
| 3 | AP‐21‐01R2 | 进口民用航空产品和零部件认可审定程序 |
| 4 | AP‐21‐2 | 关于国产民用航空产品服务通告管理规定 |
| 5 | AP‐21‐7 | 民用航空产品和零部件适航证件的编号规则 |
| 6 | AP‐21‐8 | 仅依据型号合格证生产的审定和监督程序 |
| 7 | AP‐21‐10 | 批准放行证书/适航批准标签的使用程序 |
| 8 | AP‐21‐13 | 代表外国适航当局进行生产监督的工作程序 |
| 9 | AP‐21‐15 | 进口民用航空器重要改装设计合格审定程序 |
| 10 | AP‐21‐AA‐2008‐05R2 | 民用航空器及其相关产品适航审定程序 |
| 11 | AP‐21‐AA‐2009‐18 | 认可审查资料归档管理程序 |

| 12 | AP - 21 - AA - 2009 - 19 | 美国民用航空产品和 TSO 件认可审定程序 |
| 13 | AP - 21 - AA - 2014 - 31R1 | 航空器型号合格审定试飞安全计划 |
| 14 | AP - 21 - AA - 2019 - 31 | 生产批准和监督程序 |
| 15 | AP - 21 - AA - 2020 - 12 | 技术标准规定项目批准书合格审定程序 |
| 16 | AP - 21 - AA - 2020 - 13 | 零部件制造人批准书合格审定程序 |
| 17 | AP - 21 - AA - 2022 - 11 | 型号合格审定程序 |
| 18 | AP - 21 - AA - 2022 - 16R1 | 民用航空产品和零部件故障、失效和缺陷报告处理程序 |
| 19 | AP - 21 - AA - 2022 - 31R1 | 生产批准和监督程序 |
| 20 | AP - 39 - 01R1 | 适航指令的颁发和管理 |
| 21 | AP - 45 - AA - 2022 - 01R4 | 民用航空器国籍登记管理程序 |
| 22 | AP - 183 - AA - 2018 - 01R1 | 适航委任代表管理程序 |
| 23 | AP - 183 - AA - 2018 - 11 | 适航委任单位代表管理程序 |

（六）咨询通告

| 1 | AC - 21 - 2 | 机载系统和设备合格审定中的软件审查方法 |
| 2 | AC - 21 - 13 | 在 RVSM 空域实施 300 米（1 000 英尺）垂直间隔标准运行的航空器适航批准 |
| 3 | AC - 21 - 1317 | 航空器高强辐射场（HIRF）保护要求 |
| 4 | AC - 21 - AA - 2008 - 15 | 运营人航空器适航检查单 |
| 5 | AC - 21 - AA - 2008 - 213 | 研发试飞和验证试飞特许飞行证颁发程序 |

| 6 | AC-21-AA-2010-16 | 民用航空器监造工作要求 |
| 7 | AC-21-AA-2013-14R6 | 航空器内、外部标记和标牌 |
| 8 | AC-25.1191-1 | "1301"固定式灭火瓶的批准 |
| 9 | AC-25.1301-1 | 获得 CTSOA 的航空集装单元在运输类飞机上的安装批准 |
| 10 | AC-25.1529-1 | 审定维修要求 |
| 11 | AC-25.733-1 | 国产轮胎在进口运输类飞机上装机批准技术要求 |
| 12 | AC-25.735-1 | 刹车磨损限制要求 |
| 13 | AC-25-AA-2008-02 | 运输类飞机炭刹车盘替换件合格审定符合性方法 |
| 14 | AC-36-AA-2008-04 | 航空器型号与适航合格审定噪声规定 |
| 15 | AC-61-023 | 驾驶员机型资格规范评审及评审结论的应用 |
| 16 | AC-61-FS-2014-12R3 | 航空器型别等级和训练要求 |
| 17 | AC-66-FS-001R5 | 航空器维修人员执照申请指南 |
| 18 | AC-66-FS-002R1 | 航空器维修基础知识和实作培训规范 |
| 19 | AC-66-FS-010 | 航空维修技术英语等级测试指南 |
| 20 | AC-91-010R2 | 国产航空器的运行符合性评审 |
| 21 | AC-91-037 | 航空器主最低设备清单的制定和批准 |
| 22 | AC-91-FS/AA-2010-14 | 在无雷达区使用1 090兆赫扩展电文广播式自动相关监视的适航和运行批准指南 |
| 23 | AC-121/135-2008-26 | 关于航空运营人安全管理体系的要求 |
| 24 | AC-121/135-53R1 | 民用航空器维修方案 |
| 25 | AC-121/135-FS-2012-45R1 | 飞行品质监控(FOQA)实施与管理 |
| 26 | AC-121-102R1 | 大型飞机公共航空运输机载应急医疗设备配备和训练 |

| 27 | AC - 121/135 - FS - 2008 - 28 | 驾驶舱观察员座椅和相关设备 |
|----|------------------------------|----------------------------|
| 28 | AC - 121 - 50R1 | 地面结冰条件下的运行 |
| 29 | AC - 121 - 51 | 维修工程管理手册编写指南 |
| 30 | AC - 121 - 52R1 | 航空器投入运行的申请和批准 |
| 31 | AC - 121 - 55R1 | 航空器的修理和改装 |
| 32 | AC - 121 - 64R1 | 质量管理系统 |
| 33 | AC - 121 - 66 | 维修计划和控制 |
| 34 | AC - 121 - 68 | 航空器空重和重心控制 |
| 35 | AC - 121 - 7 | 航空人员的维修差错管理 |
| 36 | AC - 121 - FS - 057R1 | 飞机地面勤务 |
| 37 | AC - 121 - FS - 075 | 公共运输航空运营人维修系统的设置 |
| 38 | AC - 121 - FS - 2015 - 21R1 | 高原机场运行 |
| 39 | AC - 121 - FS - 2018 - 031R1 | 电子飞行包(EFB)运行批准指南 |
| 40 | AC - 121 - FS - 2018 - 59 - R1 | 飞机维修记录和档案 |
| 41 | AC - 121 - FS - 2019 - 25R1 | 缩减时间的转机型课程和混合机队飞行 |
| 42 | AC - 145 - 04 | 维修记录与报告表格填写指南(中英文) |
| 43 | AC - 145 - 05 | 维修单位手册编写指南(中英文) |
| 44 | AC - 145 - 09 | 国家标准和行业标准的采用(中英文) |
| 45 | AC - 145 - 10 | 维修单位自制工具设备(中英文) |
| 46 | AC - 145 - 14 | 维修工时管理 |
| 47 | AC - 145 - 2R1 | 国外/地区维修单位申请指南(中英文) |
| 48 | AC - 145 - 3R1 | 民用航空器维修单位批准清单 |
| 49 | AC - 145 - 08 | 航空器及航空器部件维修技术文件(中英文) |
| 50 | AC - 145 - FS - 001R1 | 国内维修单位的申请和批准 |
| 51 | AC - 145 - FS - 006R3 | 航空器航线维修 |
| 52 | AC - 145 - FS - 013R2 | 维修单位培训大纲的制定 |
| 53 | AC - 145 - FS - 015R1 | 维修单位的质量安全管理体系 |

| 54 | AC - 147 - 5 | 民用航空器维修培训机构管理手册编写指南（中英文） |
|---|---|---|
| 55 | AC - 147 - 6 | 民用航空器维修培训机构年度报告填写指南 |
| 56 | AC - 147 - FS - 001R1 | 维修培训机构申请指南 |
| 57 | AC - 147 - FS - 004R3 | 机型、发动机型号维修培训实施规范 |
| 58 | AC - 183 - 01 | 生产检验委任代表工作程序手册编写指南 |
| 59 | AC - FS - 121 - 2015 - 125 | 航空器承运人控制风险管控系统实施指南 |

（七）行业标准

| 1 | MH/T 0001—1997 | 民航标准编写规定 |
|---|---|---|
| 2 | MH/T 0003—1997 | 民航标准编写规定 |
| 3 | MH/T 0013—1998 | 民用航空专用计量器具检定规程体系表 |
| 4 | MH/T 0020—2012 | 航空器材管理术语 |
| 5 | MH/T 0021—2000 | 航空公司计量工作导则 |
| 6 | MH/T 0027—2006 | 航空货运设备集装箱操作系统图形符号 |
| 7 | MH/T 3001—2012 | 航空器无损检测人员资格鉴定与认证 |
| 8 | MH/T 3002—2006 | 航空器无损检测　超声检测 |
| 9 | MH/T 3003—2000 | 民用航空器维修工程爆炸危险场所安全规程 |
| 10 | MH/T 3004—2002 | 航空器电缆识别标记 |
| 11 | MH/T 3005—2002 | 航空管路识别 |
| 12 | MH/T 3006—2011 | 民用航空维修用吊具检测技术规范 |
| 13 | MH/T 3007—2004 | 航空器无损检测　渗透检验 |
| 14 | MH/T 3008—2012 | 航空器无损检测　磁粉检测 |
| 15 | MH/T 3009—2012 | 航空器无损检测　射线照相检测 |
| 16 | MH/T 3010.1—2006 | 民用航空器维修　管理规范　第1部分：民用航空器试飞 |
| 17 | MH/T 3010.2—2006 | 民用航空器维修　管理规范　第2部分：民用航空器在经停站发生故障的处理 |
| 18 | MH/T 3010.3—2006 | 民用航空器维修　管理规范　第3部分：民用航空器维修事故与差错 |
| 19 | MH/T 3010.4—2006 | 民用航空器维修　管理规范　第4部分：民用航空器维修工作单（卡）的编制 |

| 20 | MH/T 3010.5—2006 | 民用航空器维修　管理规范　第 5 部分：民用航空器冬季的维修 |
| 21 | MH/T 3010.6—2006 | 民用航空器维修　管理规范　第 6 部分：民用航空器维修人员的技术档案 |
| 22 | MH/T 3010.7—2006 | 民用航空器维修　管理规范　第 7 部分：民用航空器维修记录的填写 |
| 23 | MH/T 3010.8—2006 | 民用航空器维修　管理规范　第 8 部分：民用航空器维修人员的行为规范 |
| 24 | MH/T 3010.9—2012 | 民用航空器维修　管理规范　第 9 部分：地面指挥民用航空器的信号 |
| 25 | MH/T 3010.10—2006 | 民用航空器维修　管理规范　第 10 部分：维修人员与机组联络的语言 |
| 26 | MH/T 3010.11—2006 | 民用航空器维修　管理规范　第 11 部分：民用航空器地面维修设备和工具 |
| 27 | MH/T 3010.12—2006 | 民用航空器维修　管理规范　第 12 部分：民用航空器的清洁 |
| 28 | MH/T 3010.13—2006 | 民用航空器维修　管理规范　第 13 部分：民用航空器发动机的清洗 |
| 29 | MH/T 3010.14—2006 | 民用航空器维修　管理规范　第 14 部分：民用航空器航线维修规则 |
| 30 | MH/T 3010.15—2006 | 民用航空器维修　管理规范　第 15 部分：民用航空器一般勤务规则 |
| 31 | MH/T 3010.16—2009 | 民用航空器维修　管理规范　第 16 部分：民用航空器线路维护 |
| 32 | MH/T 3010.17—2009 | 民用航空器维修　管理规范　第 17 部分：民用航空器防静电维护 |
| 33 | MH/T 3010.18—2010 | 民用航空器维修　管理规范　第 18 部分：维修人为因素方案指南 |
| 34 | MH/T 3010.19—2011 | 民用航空器维修　管理规范　第 19 部分：发动机状态监控地面站设计指南 |
| 35 | MH/T 3010.20—2011 | 民用航空器维修　管理规范　第 20 部分：航空器结构维修记录 |

| 36 | MH/T 3011.1—2006 | 民用航空器维修　地面安全　第 1 部分：民用航空器轮挡 |
| 37 | MH/T 3011.2—2006 | 民用航空器维修　地面安全　第 2 部分：民用航空器的停放与系留 |
| 38 | MH/T 3011.3—2006 | 民用航空器维修　地面安全　第 3 部分：民用航空器的牵引 |
| 39 | MH/T 3011.4—2006 | 民用航空器维修　地面安全　第 4 部分：民用航空器的顶升 |
| 40 | MH/T 3011.5—2010 | 民用航空器维修　地面安全　第 5 部分：民用航空器的地面试车 |
| 41 | MH/T 3011.6—2006 | 民用航空器维修　地面安全　第 6 部分：民用航空器的操纵面试验 |
| 42 | MH/T 3011.7—2006 | 民用航空器维修　地面安全　第 7 部分：民用航空器的加油和放(抽)油 |
| 43 | MH/T 3011.8—2006 | 民用航空器维修　地面安全　第 8 部分：民用航空器部件的吊装 |
| 44 | MH/T 3011.9—2006 | 民用航空器维修　地面安全　第 9 部分：民用航空器地面溢油的预防和处理 |
| 45 | MH/T 3011.10—2006 | 民用航空器维修　地面安全　第 10 部分：机坪防火 |
| 46 | MH/T 3011.11—2006 | 民用航空器维修　地面安全　第 11 部分：民用航空器局部喷漆、客舱整新和焊接 |
| 47 | MH/T 3011.12—2006 | 民用航空器维修　地面安全　第 12 部分：地面消防设施维修、使用和管理 |
| 48 | MH/T 3011.13—2006 | 民用航空器维修　地面安全　第 13 部分：红色警告标记的使用 |
| 49 | MH/T 3011.14—2006 | 民用航空器维修　地面安全　第 14 部分：民用航空器地面紧急救援 |
| 50 | MH/T 3011.15—2006 | 民用航空器维修　地面安全　第 15 部分：民用航空器燃油箱的维修 |
| 51 | MH/T 3011.16—2006 | 民用航空器维修　地面安全　第 16 部分：民用航空器座舱地面增压试验 |

| 52 | MH/T 3011.17—2006 | 民用航空器维修　地面安全　第 17 部分：民用航空器燃油沉淀物的检查 |
| 53 | MH/T 3011.18—2006 | 民用航空器维修　地面安全　第 18 部分：民用航空器的风害防护 |
| 54 | MH/T 3011.19—2006 | 民用航空器维修　地面安全　第 19 部分：民用航空器除冰、防冰液的使用 |
| 55 | MH/T 3011.21—2006 | 民用航空器维修　地面安全　第 21 部分：民用航空器地面加温 |
| 56 | MH/T 3011.22—2006 | 民用航空器维修　地面安全　第 22 部分：地面高压瓶的充装和使用 |
| 57 | MH/T 3011.23—2006 | 民用航空器维修　地面安全　第 23 部分：民用航空器地面设备的安全技术规范 |
| 58 | MH/T 3011.24—2006 | 民用航空器维修　地面安全　第 24 部分：勤务车辆停靠民用航空器的规则 |
| 59 | MH/T 3011.25—2006 | 民用航空器维修　地面安全　第 25 部分：民用航空器充氧 |
| 60 | MH/T 3011.26—2008 | 民用航空器维修　地面安全　第 26 部分：民用航空器地面用电安全 |
| 61 | MH/T 3011.27—2011 | 民用航空器维修　地面安全　第 27 部分：氧气系统维护 |
| 62 | MH/T 3012.1—2008 | 民用航空器维修　地面维修设施　第 1 部分：维修机库 |
| 63 | MH/T 3012.2—2008 | 民用航空器维修　地面维修设施　第 2 部分：喷漆机库 |
| 64 | MH/T 3012.3—2008 | 民用航空器维修　地面维修设施　第 3 部分：发动机修理作业场所 |
| 65 | MH/T 3012.4—2008 | 民用航空器维修　地面维修设施　第 4 部分：机械附件修理作业场所 |
| 66 | MH/T 3012.5—2008 | 民用航空器维修　地面维修设施　第 5 部分：电子附件修理作业场所 |
| 67 | MH/T 3012.6—2008 | 民用航空器维修　地面维修设施　第 6 部分：电器附件修理作业场所 |

| 68 | MH/T 3012.7—2008 | 民用航空器维修　地面维修设施　第 7 部分：电瓶充电修理作业场所 |
| 69 | MH/T 3012.8—2008 | 民用航空器维修　地面维修设施　第 8 部分：高压气瓶修理作业场所 |
| 70 | MH/T 3012.9—2008 | 民用航空器维修　地面维修设施　第 9 部分：氧气附件修理作业场所 |
| 71 | MH/T 3012.10—2008 | 民用航空器维修　地面维修设施　第 10 部分：紧急救生设备修理作业场所 |
| 72 | MH/T 3012.11—2008 | 民用航空器维修　地面维修设施　第 11 部分：机械加工修理作业场所 |
| 73 | MH/T 3012.12—2008 | 民用航空器维修　地面维修设施　第 12 部分：电镀作业场所 |
| 74 | MH/T 3012.13—2008 | 民用航空器维修　地面维修设施　第 13 部分：热处理作业场所 |
| 75 | MH/T 3012.14—2008 | 民用航空器维修　地面维修设施　第 14 部分：喷砂、喷丸作业场所 |
| 76 | MH/T 3012.15—2008 | 民用航空器维修　地面维修设施　第 15 部分：计量检测实验室 |
| 77 | MH/T 3012.16—2008 | 民用航空器维修　地面维修设施　第 16 部分：灭火瓶维修作业场所 |
| 78 | MH/T 3012.17—2010 | 民用航空器维修　地面维修设施　第 17 部分：维修机坞 |
| 79 | MH/T 3013.1—2008 | 民用航空器维修　职业安全健康　第 1 部分：地面设备的安全管理规则 |
| 80 | MH/T 3013.2—2008 | 民用航空器维修　职业安全健康　第 2 部分：用电安全管理规则 |
| 81 | MH/T 3013.3—2008 | 民用航空器维修　职业安全健康　第 3 部分：压力容器安全管理规则 |
| 82 | MH/T 3013.4—2008 | 民用航空器维修　职业安全健康　第 4 部分：地面气瓶安全管理规则 |
| 83 | MH/T 3013.5—2008 | 民用航空器维修　职业安全健康　第 5 部分：起重设备管理规则 |

| 84 | MH/T 3013.6—2008 | 民用航空器维修　职业安全健康　第 6 部分：焊接与切割管理规则 |
| --- | --- | --- |
| 85 | MH/T 3013.10—2012 | 民用航空器维修　职业安全健康　第 10 部分：管理体系实施指南 |
| 86 | MH/T 3014.1—2007 | 民用航空器维修　航空器材　第 1 部分：航空器材仓库 |
| 87 | MH/T 3014.2—2007 | 民用航空器维修　航空器材　第 2 部分：危险品的存储 |
| 88 | MH/T 3014.3—2007 | 民用航空器维修　航空器材　第 3 部分：航空器材包装 |
| 89 | MH/T 3014.4—2007 | 民用航空器维修　航空器材　第 4 部分：站坪航空器材 |
| 90 | MH/T 3014.5—2007 | 民用航空器维修　航空器材　第 5 部分：航空器材检验、储存、发付 |
| 91 | MH/T 3014.6—2007 | 民用航空器维修　航空器材　第 6 部分：航空器材的地面装运 |
| 92 | MH/T 3015—2006 | 航空器无损检测　涡流检测 |
| 93 | MH/T 3016—2007 | 航空器渗漏检测 |
| 94 | MH/T 3017—2008 | 民用航空金属零件旋片喷丸强化工艺 |
| 95 | MH/T 3018—2009 | 民用航空器地面维修用通讯设备性能要求 |
| 96 | MH/T 3019—2009 | 航空器无损检测　目视检测 |
| 97 | MH/T 3020—2011 | 金属硬度标尺的转换 |
| 98 | MH/T 3021—2011 | 燃气涡轮发动机燃油喷嘴测试 |
| 99 | MH/T 3022—2011 | 航空器复合材料构件红外热像检测 |
| 100 | MH/T 3023—2011 | 推进系统中保险钢索、保险丝、止动垫片和开口销的一般应用方法 |
| 101 | MH/T 6002—2008 | 民用航空油料设备完好技术规范 |
| 102 | MH/T 6003—1996 | 客舱内部装饰用塑料板材标准 |
| 103 | MH/T 6004—2015 | 民用航空油料计量管理 |
| 104 | MH/T 6005—2009 | 民用航空器加油规范 |
| 105 | MH/T 6007—1998 | 飞机清洗剂 |
| 106 | MH/T 6020—2012 | 民用航空燃料质量控制和操作程序 |

| 107 | MH/T 6039—2015 | 电镀工艺和飞机用化学品的机械氢脆评估试验方法 |
| 108 | MH/T 6040—2017 | 航空材料烟密度试验方法 |
| 109 | MH/T 6043—2015 | 溶剂型航空器零部件清洗剂 |
| 110 | MH/T 6046—2008 | 含摩擦剂的飞机外表面清洗剂 |
| 111 | MH/T 6047—2008 | 航空毛毯四层水平燃烧试验方法 |
| 112 | MH/T 6058—2017 | 航空地毯清洗剂 |
| 113 | MH/T 6059—2017 | 飞机维护用化学品全浸泡腐蚀试验方法 |
| 114 | MH/T 6060—2015 | 航空地毯 |
| 115 | MH/T 6061—2010 | 飞机厨房手推车、物品箱及相关组件的最低设计和性能 |
| 116 | MH/T 6062—2010 | 增稠型飞机除冰防冰液的粘度测定 |
| 117 | MH/T 6063—2010 | 飞机烤箱在位清洗剂 |
| 118 | MH/T 6064—2010 | 飞机舱内硬表面清洗剂 |
| 119 | MH/T 6065—2010 | 飞机清洗及化学维护用品对飞机涂漆表面影响的试验方法 |
| 120 | MH/T 6066—2010 | 液体油料中能活细菌和真菌的计数 过滤和培养程序 |
| 121 | MH/T 6067—2010 | 接触液体或半液体化合物的丙烯酸类塑料应力银纹化试验方法 |
| 122 | MH/T 6068—2017 | 航空燃料中游离水、固体颗粒物和其他污染物现场检测方法 |
| 123 | MH/T 6069—2015 | 机场道面固体除冰防冰剂 |
| 124 | MH/T 6070—2010 | 民用航空器橡胶和金属轮挡 |
| 125 | MH/T 6071—2011 | 夹层腐蚀试验方法 |
| 126 | MH/T 6072—2011 | 飞机发动机清洗用品对钛合金应力腐蚀的试验方法 |
| 127 | MH/T 6073—2011 | 飞机粘胶去除剂 |
| 128 | MH/T 6074—2011 | 环氧和聚氨酯漆无氯溶剂脱漆剂 |
| 129 | MH/T 6075—2011 | 飞机驾驶舱玻璃清洗剂 |
| 130 | MH/T 9001—2008 | 亚音速喷气飞机噪声合格审定飞行试验等效程序 |

| 131 | MH/T 9002—2008 | 温度和湿度函数的大气噪声吸收标准值 |
| 132 | MH/T 9003—2008 | 电声学　航空噪声测量仪器　在运输类飞机噪声合格审定中测量 1/3 宽带倍频声压级装置的性能要求 |

（八）管理手册

| 1 | AEG - H | 航空器评审组工作手册 |
| 2 | CASM R2 | 持续适航监察员手册 |
| 3 | MD - FS - AEG002 | MMEL 建议项目政策指南 |
| 4 | MD - FS - AEG004 | 运行符合性清单的编制及应用 |

（九）其他资料

| 1 | NA | 民用航空术语汇编（FAA Order 1000. 15A 中国民用航空总局航空器适航审定司　中国民用航空总局航空安全技术中心编译） |
| 2 | NA | 适航指令手册（翻译自 FAA - AIR - M - 8040. 1） |

## 二、国外民航局及组织的相关文件

（一）FAA Regulation

| 1 | CFR Part 1 | Definitions and Abbreviations |
| 2 | CFR Part 21 | Certification Procedures for Products and Parts |
| 3 | CFR Part 25 | Airworthiness Standards: Transport Category Airplanes |
| 4 | CFR Part 26 | Continued Airworthiness and safety Improvements for Transport Category Airplanes |
| 5 | CFR Part 33 | Airworthiness Standards: Aircraft Engines |
| 6 | CFR Part 34 | Fuel Venting and Exhaust Emission Requirements for Turbine Engine Powered Airplanes |
| 7 | CFR Part 36 | Noise Standards: Aircraft Type and airworthiness Certification |
| 8 | CFR Part 39 | Airworthiness Directives |
| 9 | CFR Part 43 | Maintenance, Preventive Maintenance, Rebuilding, and Alteration |

| 10 | CFR Part 60 | Flight Simulation Training Device Initial and Continuing Qualification and Use |
|----|-------------|-------------------------------------------------------------------------------|
| 11 | CFR Part 65 | Certification: Airmen Other Than Flight Crewmembers |
| 12 | CFR Part 91 | General Operating and Flight Rules |
| 13 | CFR Part 121 | Operating Requirements: Domestic, Flag, and Supplemental Operations |
| 14 | CFR Part 135 | Operating Requirements: Commuter and On Demand Operations and Rules Governing Persons on Board Such Aircraft |
| 15 | CFR Part 142 | Training Centers |
| 16 | CFR Part 145 | Repair Stations |
| 17 | CFR Part 147 | Aviation Maintenance Technician Schools |
| 18 | CFR Part 183 | Representatives of the Administrator |

（二）FAA AC

| 1 | 20 - 18B | Qualification Testing of Turbo-jet Engine Thrust Reversers |
|----|----------|------------------------------------------------------------|
| 2 | 20 - 24D | Approval of Propulsion Fuels and Lubricating Oils |
| 3 | 20 - 29B | Use of Aircraft Fuel Anti-icing Additives |
| 4 | 20 - 30B | Aircraft Position Light and Anticollision Light Installations |
| 5 | 20 - 33B | Technical Information Regarding Civil Aeronautics Manuals 1, 3, 4a, 4b, 5, 6, 7, 8, 9, 13, and 14 |
| 6 | 20 - 40 | Placards for Battery-excited Alternators Installed in Light Aircraft |
| 7 | 20 - 41A | Substitute Technical Standard Order (TSO) Aircraft Equipment |
| 8 | 20 - 42D | Hand Fire Extinguishers for Use in Aircraft |
| 9 | 20 - 43C | Aircraft Fuel Control |
| 10 | 20 - 44 | Glass Fiber Fabric for Aircraft Covering |

| 11 | 20 – 45 | Safetying of Turnbuckles on Civil Aircraft |
| 12 | 20 – 47 | Exterior Colored Band Around Exits on Transport Airplanes |
| 13 | 20 – 53C | Protection of Aircraft Fuel Systems Against Fuel Vapor Ignition Caused by Lightning |
| 14 | 20 – 56A | Marking of TSO – C72b Individual Flotation Devices |
| 15 | 20 – 62E | Eligibility, Quality, & Identification of Aeronautical Replacement Parts |
| 16 | 20 – 65A | U. S. Airworthiness Certificates and Authorizations for Operation of Domestic and Foreign Aircraft |
| 17 | 20 – 67B | Airborne VHF Communication Equipment Installations |
| 18 | 20 – 68B | Recommended Radiation Safety Precautions for Ground Operation of Airborne Weather Radar |
| 19 | 20 – 69 | Conspicuity of Aircraft Instrument Malfunction Indicators |
| 20 | 20 – 71 | Dual Locking Devices on Fasteners |
| 21 | 20 – 73A | Aircraft Ice Protection |
| 22 | 20 – 74 | Aircraft Position and Anticollision Light Measurements |
| 23 | 20 – 77B | Use of Manufacturers' Maintenance Manuals |
| 24 | 20 – 88A | Guidelines on the Marking of Aircraft |
| 25 | 20 – 94A | Digital Clock Installation in Aircraft |
| 26 | 20 – 100 | General Guidelines for Measuring Fire-extinguishing Agent Concentrations in Powerplant Compartments |
| 27 | 20 – 104 | Revised Powerplant Engineering Report No. 3A Standard Fire Test Apparatus and Procedure (for Flexible Hose Assemblies) |

| 28 | 20 - 106 | Aircraft Inspection for the General Aviation Aircraft Owner |
| 29 | 20 - 107B | Composite Aircraft Structure |
| 30 | 20 - 115D | Radio Technical Commission for Aeronautic, Inc. Document RTCA/DO - 178B |
| 31 | 20 - 116 | Marking Aircraft Fuel Filler Openings with Color Coded Decals |
| 32 | 20 - 117 | Hazards Following Ground Deicing and Ground Operations in Conditions Conducive to Aircraft Icing |
| 33 | 20 - 118A | Emergency Evacuation Demonstration |
| 34 | 20 - 119 | Fuel Drain Valves |
| 35 | 20 - 120 | Nondirectional Beacon Frequency Congestion |
| 36 | 20 - 122A | Anti-misfueling Devices: Their Availability and Use |
| 37 | 20 - 123 | Avoiding or Minimizing Encounters with Aircraft Equipped with Depleted Uranium Balance Weights during Accident Investigations |
| 38 | 20 - 124 | Water Ingestion Testing for Turbine Powered Airplanes |
| 39 | 20 - 125 | Water in Aviation Fuels |
| 40 | 20 - 127 | Use of Society of Automotive Engineers (SAE) Class H11 Bolts |
| 41 | 20 - 128A | Design Considerations for Minimizing Hazards Caused by Uncontained Turbine Engine and Auxiliary Power Unit Rotor Failure |
| 42 | 20 - 129 | Airworthiness Approval of Vertical Navigation (VNAV) Systems for Use in the U. S. National Airspace System (NAS) and Alaska |
| 43 | 20 - 131A | Airworthiness Approval of Traffic Alert and Collision Avoidance Systems (TCAS II) and Mode S Transponders |

| 44 | 20 – 133 | Cockpit Noise and Speech Interference Between Crewmember |
| 45 | 20 – 134 | Test Procedures for Maximum Allowable Airspeed Indicators |
| 46 | 20 – 135 | Powerplant Installation and Propulsion System Component Fire Protection Test Methods, Standards and Criteria |
| 47 | 20 – 138D | Airworthiness Approval of Positioning and Navigation Systems |
| 48 | 20 – 140C | Guidelines for Design Approval of Aircraft Data Communications Systems |
| 49 | 20 – 141B | Airworthiness and Operational Approval of Digital Flight Data Recorder Systems |
| 50 | 20 – 143 | Installation, Inspection, and Maintenance of Controls for General Aviation Reciprocating Aircraft Engines |
| 51 | 20 – 144A | Recommended Method for FAA Approval of Aircraft Fire Extinguishing System Components |
| 52 | 20 – 146A | Methodology for Dynamic Seat Certification by Analysis for Use in Part 23, 25, 27, and 29 Airplanes and Rotorcraft |
| 53 | 20 – 147A | Turbojet, Turboprop, and Turbofan Engine Induction System Icing and Ice Ingestion |
| 54 | 20 – 150B | Airworthiness Approval of Satellite Voice Equipment Supporting Air Traffic Service (ATS) Communication |
| 55 | 20 – 151C | Airworthiness Approval of Traffic Alert and Collision Avoidance Systems (TCAS Ⅱ), Versions 7.0 & 7.1 and Associated Mode S Transponders |
| 56 | 20 – 152A | RTCA, Inc., Document RTCA/DO – 254, Design Assurance Guidance for Airborne Electronic Hardware |

| 57 | 20 - 154 | Guide for Developing a Receiving Inspection System for Aircraft Parts and Material |
| 58 | 20 - 155A | SAE Documents to Support Aircraft Lightning Protection Certification |
| 59 | 20 - 156 | Aviation Databus Assurance |
| 60 | 20 - 157 | How to Prepare Reliability Assessment Plans for Aircraft Systems and Equipment |
| 61 | 20 - 158A | The Certification of Aircraft Electrical and Electronic Systems for Operation in the High-Intensity Radiated Fields ( HIRF ) Environment |
| 62 | 20 - 160A | Onboard Recording of Controller Pilot Data Link Communication in Crash Survivable Memory |
| 63 | 20 - 161 | Aircraft Onboard Weight and Balance Systems |
| 64 | 20 - 162B | Airworthiness Approval and Operational Allowance of RFID Systems |
| 65 | 20 - 163 | Displaying Geometric Altitude Relative to Mean Sea Level |
| 66 | 20 - 164A | Designing and Demonstrating Aircraft Tolerance to Portable Electronic Devices |
| 67 | 20 - 165B | Airworthiness Approval of Automatic Dependent Surveillance Broadcast ( ADS B) Out Equipment for Operation in the National Airspace System (NAS) |
| 68 | 20 - 166A | Issue Paper Process |
| 69 | 20 - 167A | Airworthiness Approval of Enhanced Vision System，Synthetic Vision System，Combined Vision System，and Enhanced Flight Vision System Equipment |
| 70 | 20 - 168 | Certification Guidance for Installation of Non-Essential， Non-Required Aircraft Cabin Systems & Equipment (CS&E) |

| 71 | 20 - 170 | Integrated Modular Avionics Development. Verification, Integration and Approval Using RTCA/DO - 297 and Technical Standard Order C153 |
| 72 | 20 - 171 | Alternatives to RTCA/DO - 178B for Software in Airborne Systems and Equipment |
| 73 | 20 - 173 | Installation of Electronic Flight Bag Components |
| 74 | 20 - 174 | Development of Civil Aircraft and Systems |
| 75 | 20 - 175 | Controls for Flight Deck Systems |
| 76 | 20 - 178 | Flammability Testing of Aircraft Cabin Interior Panels After Alterations |
| 77 | 21. 17 - 2A | Type Certification-Fixed Wing Gliders (Sailplanes), Including Powered Gliders |
| 78 | 21. 25 - 1 | Issuance of Type Certificate: Restricted Category Agricultural Airplanes |
| 79 | 21 - 12C | Application for U. S. Airworthiness Certificate, FAA Form 8130 - 6 |
| 80 | 21 - 16G | RTCA Document DO - 160 Versions D, E and F, Environmental Conditions and Test Procedures for Airborne Equipment |
| 81 | 21 - 19B | Installation of Used Aircraft Engines in New Production Aircraft |
| 82 | 21 - 22 | Injury Criteria for Human Exposure to Impact |
| 83 | 21 - 23B | Airworthiness Certification of Civil Aircraft, Engine, Propellers, and Related Products Imported to the United States |
| 84 | 21 - 25B | Approval of Modified Seats and Berths |
| 85 | 21 - 26A | Quality Control for the Manufacture of Composite Structures (Consolidated Reprint Includes Change 1) |

| 86 | 21 - 28 | Airworthiness Certification of U. S. - Produced Aircraft and Engine Kits Assembled Outside the United States |
| 87 | 21 - 29D | Detecting and Reporting Suspected Unapproved Parts Including Changes 1 and 2 |
| 88 | 21 - 31A | Quality Control for the Manufacture of Non-Metallic Compartment Interior Components |
| 89 | 21 - 32B | Control of Products and Parts Shipped Prior to Type Certificate Issuance |
| 90 | 21 - 34 | Shoulder Harness-safety Belt Installations |
| 91 | 21 - 37 | Primary Category Aircraft |
| 92 | 21 - 40A | Guide for Obtaining a Supplemental Type Certificate |
| 93 | 21 - 41A | Replacing MIL - S - 8879C with SAE AS8879 |
| 94 | 21 - 43A | Issuance of Production Approvals Under Subparts G, K, & O |
| 95 | 21 - 44A | Issuance of Export Airworthiness Approvals Under 14 CFR Part 21 Subpart L |
| 96 | 21 - 47 | Submittal of Data to an ACO, a DER or an ODA for a Major Repair or Major Alteration |
| 97 | 21 - 48 | Using Electronic Modeling Systems as Primary Type Design Data |
| 98 | 21 - 51 | Applicants Showing of Compliance and Certifying Statement of Compliance |
| 99 | 25. 335 - 1A | Design Dive Speed |
| 100 | 25. 491 - 1 | Taxi, Takeoff and Landing Roll Design Loads |
| 101 | 25. 562 - 1B | Dynamic Evaluation of Seat Restraint Systems and Occupant Protection on Transport Airplanes |
| 102 | 25. 571 - 1D | Damage Tolerance and Fatigue Evaluation of Structure |

| 103 | 25.613 - 1 | Material Strength Properties and Material Design Values |
| 104 | 25.629 - 1B | Aeroelastic Stability Substantiation of Transport Category Airplanes |
| 105 | 25.703 - 1 | Takeoff Configuration Warning Systems |
| 106 | 25.723 - 1 | Shock Absorption Tests |
| 107 | 25.735 - 1 | Brakes and Braking Systems Certification Tests and Analysis |
| 108 | 25.773 - 1 | Pilot Compartment View Design Considerations |
| 109 | 25.775 - 1 | Windows and Windshields |
| 110 | 25.783 - 1A | Fuselage Doors and Hatches |
| 111 | 25.795 - 1A | FlightDeck Intrusion Resistance |
| 112 | 25.795 - 2A | Flight Deck Penetration Resistance |
| 113 | 25.795 - 3 | Flightdeck Protection (Smoke and Fumes) |
| 114 | 25.795 - 4 | Passenger Cabin Smoke Protection |
| 115 | 25.795 - 5 | Cargo Compartment Fire Suppression |
| 116 | 25.795 - 6 | Least Risk Bomb Location |
| 117 | 25.795 - 7 | Survivability of Systems |
| 118 | 25.795 - 8 | Interior Design to Facillitate Searches |
| 119 | 25.803 - 1A | Emergency Evacuation Demonstrations |
| 120 | 25.807 - 1 | Uniform Distribution of Exits |
| 121 | 25.812 - 1A | Floor Proximity Emergency Escape Path Marking |
| 122 | 25.812 - 2 | Floor Proximity Emergency Escape Path Marking Systems Incorporating Photoluminescent Elements |
| 123 | 25.853 - 1 | Flammability Requirements for Aircraft Seat Cushions |
| 124 | 25.856 - 1 | Thermal/Acoustic Insulation Flame Propagation Test Method Details |
| 125 | 25.856 - 2A | Installaiton of Thermal/Acoustic Insulation for Burnthrough Proctection |

| 126 | 25.869 – 1A | Fire Protection Systems |
| 127 | 25.899 – 1 | Electrical Bonding and Protection Against Static Electricity |
| 128 | 25.939 – 1 | Evaluating Turbine Engine Operating Characteristics |
| 129 | 25.963 – 1 | Fuel Tank Access Covers |
| 130 | 25.981 – 1D | Fuel Tank Ignition Source Prevention Guidelines |
| 131 | 25.981 – 2A | Fuel Tank Flammability |
| 132 | 25.994 – 1 | Design Considerations to Protect Fuel Systems during a Wheels-up Landing |
| 133 | 25.1309 – 1A | System Design and Analysis |
| 134 | 25.1322 – 1 | Flightcrew Alerting |
| 135 | 25.1329 – 1C | Approval of Flight Guidance Systems |
| 136 | 25.1353 – 1A | Electrical Equipment and Installations |
| 137 | 25.1357 – 1A | Circuit Protective Devices |
| 138 | 25.1360 – 1 | Protection Against Injury |
| 139 | 25.1362 – 1 | Electrical Supplies for Emergency Conditions |
| 140 | 25.1365 – 1 | Electrical Appliances, Motors, and Transformers |
| 141 | 25.1435 – 1 | Hydraulic System Certification Tests and Analysis |
| 142 | 25.1455 – 1 | Waste Water/Potable Water Drain System Certification Testing |
| 143 | 25.1523 – 1 | Minimum Flightcrew |
| 144 | 25.1529 – 1A | Instructions for Continued Airworthiness of Structural Repairs on Transport Airplanes |
| 145 | 25.1581 – 1 | Airplane Flight Manual |
| 146 | 25.1701 – 1 | Certification of Electrical Wiring Interconnection Systems on Transport Category Airplanes |
| 147 | 25 – 7D | Flight Test Guide for Certification of Transport Category Airplanes |
| 148 | 25 – 8 | Auxiliary Fuel Systems Installations |

| 164 | 25.869.1A | Fire Protection Systems |
|---|---|---|
| 165 | 39 - 7D | Airworthiness Directives |
| 166 | 39 - 8 | Continued Airworthiness Assessments of Powerplant and Auxiliary Power Unit Installations of Transport Category Airplanes |
| 167 | 43.9 - 1G | Instructions for Completion of FAA Form 337 |
| 168 | 43.13 - 1B | Acceptable Methods, Techniques, and Practices-aircraft Inspection and Repair [Large AC. This includes Change 1.] |
| 169 | 43.13 - 2B | Acceptable Methods, Techniques, and Practices-aircraft Alterations |
| 170 | 43 - 2B | Minimum Barometry for Calibration and Test of Atmospheric Pressure Instruments |
| 171 | 43 - 4B | [Large AC] Corrosion Control for Aircraft |
| 172 | 43 - 6D | Altitude Reporting Equipment and Transponder System Maintenance and Inspection Practices |
| 173 | 43 - 9C | Maintenance Records |
| 174 | 43 - 10C | United States — Canadian Bilateral Aviation Safety Agreement Maintenance Implementation Procedures |
| 175 | 43 - 12A | Preventive Maintenance |
| 176 | 43 - 14 | Maintenance of Weather Radar Radomes |
| 177 | 43 - 15 | Recommended Guidelines for Instrument Shops |
| 178 | 43 - 18 | Fabrication of Aircraft parts by Maintenance Personnel |
| 179 | 43 - 206 | Inspection, Prevention, Control, and Repair of Corrosion on Avionics Equipment |
| 180 | 43 - 207 | Correlation, Operation, Design, and Modification of Turbofan/Jet Engine Test Cells |

| 181 | 43 – 208 | Maintenance of Emergency Evacuations Systems for Aircraft Operating Under Part 121 |
| 182 | 43 – 210A | Standardized Procedures for Requesting Field Approval of Data, Major Alterations, and Repairs |
| 183 | 43 – 213A | Parts Marking Identification |
| 184 | 60 – 6B | Airplane Flight Manuals (AFM), Approved Manual Materials, Markings, and Placards Airplanes |
| 185 | 60 – 22 | Aeronautical Decision Making |
| 186 | 91. 21 – 1D | Use of Portable Electronic Devices Aboard Aircraft |
| 187 | 91 – 21D | Use of Portable Electronic Devices Aboard Aircraft |
| 188 | 91 – 26 | Maintenance and Handling of Airdriven Gyroscopic Instruments |
| 189 | 91 – 36D | Visual Flight Rules (VFR) Flight Near Noise-Sensitive Areas |
| 190 | 91 – 50 | Importance of Transponder Operation and Altitude Reporting |
| 191 | 91 – 53A | Noise Abatement Departure Profile |
| 192 | 91 – 55 | Reduction of Electrical System Failures Following Aircraft Engine Starting |
| 193 | 91 – 56B | Continuing Structural Integrity Program for Airplanes |
| 194 | 91 – 57C | Model Aircraft Operating Standards |
| 195 | 91 – 59A | Inspection and Care of General Aviation Aircraft Exhaust Systems |
| 196 | 91 – 62A | Use of Child Seats in Aircraft |
| 197 | 91 – 63D | Temporary Flight Restrictions (TFRs/TFR) |
| 198 | 91 – 65 | Use of Shoulder Harness in Passenger Seats |

| 199 | 91 - 66 | Noise Abatement for Helicopters |
| 200 | 91 - 70B | Oceanic and Remote Continental Airspace Operations |
| 201 | 91 - 72 | Waivers of Provisions of Title 14 of the Code of Federal Regulations Part 91 |
| 202 | 91 - 74B | Pilot Guide: Flight in Icing Conditions |
| 203 | 91 - 75 | Attitude Indicator |
| 204 | 91 - 76A | Hazard Associated with Sublimation of Solid Carbon Dioxide (Dry Ice) Aboard Aircraft |
| 205 | 91 - 77 | General Aviation, Coded Departure Routes (CDR) |
| 206 | 91 - 78 | Use of Class 1 or Class 2 Electronic Flight Bag (EFB) |
| 207 | 91 - 79A | Runway Overrun Prevention |
| 208 | 91 - 80 | Air Traffic Control System Command Center (ATCSCC) Hotlines |
| 209 | 91 - 81 | Dual J80 Route Procedure |
| 210 | 91 - 82A | Fatigue Management Programs for Airplanes with Demonstrated Risk of Catastrophic Failure Due to Fatigue |
| 211 | 91 - 84 | Fractional Ownership Programs |
| 212 | 91 - 85B | Approval of Aircraft and Operators for flight in Reduced Vertical |
| 213 | 120 - 104 | Establishing and Implementing Limit of Validity to Prevent Widespread Fatigue Damage |
| 214 | 121. 195 - 1A | Operational Landing Distances for Wet Runways: Transport Category Airplanes |
| 215 | 121. 321 - 1 | Compliance with Requirements of § 121. 321, Operations in Icing |
| 216 | 121 - 6 | Portable Battery-powered Megaphones |

| 217 | 121 - 22C | Maintenance Review Board Report Maintenance Type Board，and OEM/TCH Inspection Program Procedures |
| 218 | 121 - 24D | Passenger Safety Information Briefing and Briefing Cards |
| 219 | 121 - 25 | Additional Weather Information：Domestic and Flag Air Carriers |
| 220 | 121 - 28 | Preparation and Loading of Magnetron Magnetic Materials for Air Transportation |
| 221 | 121 - 31 | Flight Crew Sleeping Quarters and Rest Facilities |
| 222 | 121 - 32A | Dispatch Resource Management Training |
| 223 | 121 - 33B | Emergency Medical Equipment |
| 224 | 121 - 34B | Emergency Medical Equipment Training |
| 225 | 121 - 35 | Management of Passengers During Ground Operations without Cabin Ventilation |
| 226 | 121 - 36 | Management of Passengers Who May Be Sensitive to Allergens |
| 227 | 121 - 37A | Voluntary Disclosure Reporting Program-Hazardous Materials |
| 228 | 121 - 38 | Reporting Hazardous Materials Discrepancies to the Federal Aviation Administration |
| 229 | 135 - 4A | Aviation Security：Air Taxi Commercial Operators（ATCO） |
| 230 | 135 - 7B | Part 135：Additional Maintenance Requirements for Aircraft Type Certificated for Nine or Less Passenger Seats |
| 231 | 135 - 9 | FAR Part 135 Icing Limitations |
| 232 | 135 - 10C | Approved Aircraft Inspection Program |
| 233 | 135 - 16 | Ground Deicing & Anti-icing Training & Checking |

| 234 | 145 - 2 | Repair Station Limited Ratings Beech 18 Series Aircraft |
| 235 | 145 - 4A | Inspection, Retread, Repair, and Alterations of Aircraft Tires |
| 236 | 145 - 5 | Repair Station Internal Evaluation Programs |
| 237 | 145 - 9A | Guide for Developing and Evaluating Repair Station and Quality Control Manuals |
| 238 | 145 - 10 | Repair Station Training Program |
| 239 | 145 - 11A | Repair Station Guidance for Compliance with the Safety Agreement between the United States and the European Union |
| 240 | 147 - 3C | Certification and Operation of Aviation Maintenance Technician Schools |

## (三) FAA Order

| 1 | 1100. 128 | Implementation of Noise Type Certification Standards |
| 2 | 2150. 3C | FAA Compliance and Enforcement Program |
| 3 | 4040. 26C | Aircraft Certification Service Flight Test Risk Management Program |
| 4 | 4040. 9E | FAA Aircraft Management Program |
| 5 | VS 8000. 370C | Aviation Safety (AVS) Safety Policy |
| 6 | 8000. 51C | Aircraft Certification Directorates' Delegation of Authority |
| 7 | 8000. 79 | Use of Electronic Technology and Storage of Data |
| 8 | 8020. 11D | Aircraft Accident and Incident Notificaiton, Investigation, and Reporting |
| 9 | 8040. 1C | Airworthiness Directives |
| 10 | 8100. 11D | Decision Paper Criteria for Undue Burden and No Undue Burden Determinations Under 14 CFR Part 21 |

| 11 | 8100. 15B | Organization Designation Authorization Procedures |
|----|-----------|---|
| 12 | 8100. 16 | Aircraft Certification Service Policy Statement, Policy Memorandum, and Deviation Memorandum Systems |
| 13 | 8100. 8D | Designee Management Handbook |
| 14 | 8110. 100B | Special Airworthiness Information Bulletin |
| 15 | 8110. 103B | Alternative Methods of Compliance (AMOC) |
| 16 | 8110. 105A | Simple and Complex Electronic Hardware Approval Guidance |
| 17 | 8110. 106 | Assessing the Relative Risk of Title 14 CFR Parts |
| 18 | 8110. 107A | Monitor Safety/Analyze Data |
| 19 | 8110. 112A | Standardized Procedures for Usage of Issue Papers and Development of Equivalent Levels of Safety Memorandums |
| 20 | 8110. 113 | Approval of Flammability Test Data in Support of Major Repairs or Major Alterations |
| 21 | 8110. 115 | Certification Project Initiation and Certification Project Notification |
| 22 | 8110. 23 | FAA Approval of Recognition (Logo) Lights Installations |
| 23 | 8110. 35B | Aircraft Noise Certification Historical Database (RIS 8110. 1) |
| 24 | 8110. 4C | Type Certification |
| 25 | 8110. 49A | Software Approval Guidelines |
| 26 | 8110. 51 | Acceptability of Previously Approved Certification Compliance Data from Foreign Sources |
| 27 | 8110. 52B | Type Validation and Post-type Validation Procedures |

| 28 | 8110.53 | Reciprocal Acceptance of Repair Design Data Approvals Between FAA and TCCA |
| 29 | 8110.54A | Instructions for Continued Airworthiness Responsibilities, Requirements, and Contents |
| 30 | 8110.55B | How to Evaluate and Accept Processes for Aeronautical Database Suppliers |
| 31 | 8110.56B | Restricted Category Type Certification |
| 32 | 8120.23A | Certificate Management of Production Approval Holders |
| 33 | 8130.21H | Procedures for Completion and Use of the Authorized Release Certificate, FAA Form 8130-3, Airworthiness |
| 34 | 8130.2J | Airworthiness Certification of Aircraft and Related Products |
| 35 | 8150.1D | Technical Standard Order Program |
| 36 | 8260.3E | Calculation of Radio Altimeter Height |
| 37 | 8260.43C | Flight Procedures Management Program |
| 38 | 8300.13A | Repair Assessment Program |
| 39 | 8400.13F | Procedures for the Evaluation and Approval of Facilities for Special Authorization Category Ⅰ Operations and All Category Ⅱ and Ⅲ Operations |
| 40 | 8620.2B | Applicability and Enforcement of Manufacturer's Data |
| 41 | 9550.8 | Human Factors Policy |

（四）其他 FAA 文件

| 1 | CPI_guide_Ⅱ | The FAA and Industry Guide to Product Certification |
| 2 | FAA-IR-M-8040.1A | AD Handbook |
| 3 | NA | SAFOs-safety Alerts for Operators |
| 4 | NA | INFOs-information for Operators |

（五）欧洲航空安全局（EASA）文件

| | | |
|---|---|---|
| 1 | AMC‑20 | General Acceptable Means of Compliance for Airworthiness of Products，Parts and Appliances |
| 2 | CS‑25 | Certification Specifications and Acceptable Means of Compliance for Large Aeroplanes |
| 3 | CS‑34 | Aircraft Engine Emissions and Fuel Venting |
| 4 | CS‑36 | Certification Specifications for Aircraft Noise |
| 5 | CS‑Definitions | Definitions and Abbrebiations |
| 6 | CS‑E | Certification Specifications for Engines |
| 7 | CS‑ETSO | European Technical Standard Orders |
| 8 | CS‑FSTD(A) | Aeroplane Flight Simulation Training Devices |
| 9 | (EC)No. 2042/2003 | Continuing Airworthiness of Aircraft and Aeronautical Products，Parts and Appliances，and on the Approval of Organizations and Personnel |
| 10 | （EU）No. 748/2012 (Part 21) | Certification of Aircraft and Related Products，Parts and Appliances，and of Design and Production Organizations |

（六）国际民航组织（ICAO）文件

| | | |
|---|---|---|
| 1 | ICAO Annex 6 | Operation of Aircraft |
| 2 | ICAO Annex 8 | Airworthiness of Aircraft |
| 3 | ICAO Annex 16 Volume I | ICAO Annex 16‑Environmental Protection：Volume I：Aircraft Noise, Aircraft Noise Certification |
| 4 | ICAO Annex 16 Volume II | ICAO Annex 16‑Environmental Protection：Volume II：Aircraft Engine Emissions |

（七）厂家公司文件

| | | |
|---|---|---|
| 1 | NA | 波音公司中英文民用航空字典 |
| 2 | NA | 波音787缩略语手册 |
| 3 | NA | 空客缩写字典 |
| 4 | NA | A380缩略语手册 |

| 5 | NA | 适航技术咨询专用术语定义与缩略语 |
| 6 | NA | 工程常用缩略语手册 |
| 7 | NA | 东航技术部词典 |

（八）其他组织文件

| 1 | ATA2200 | Air Transport Association of America Spec 2200 |
| 2 | RTCA/DO - 160G | Environmental Conditions and Test Procedures for Airborne Equipment |
| 3 | RTCA/DO - 178C | Software Considerations in Airborne Systems and Equipment Certification |
| 4 | RTCA/DO - 254 | Design Assurance Guidance for Airborne Electronic Hardware |
| 5 | RTCA/DO - 297 | Integrated Modular Avionics (IMA) Development Guidance and Certification Considerations |
| 6 | RTCA/DO - 330 | Certification Evidence for its VxWorks Product Line |
| 7 | SAE ARP 4754 | Guidelines for Development of Civil Aircraft and System |
| 8 | SAE ARP 4761 | Guidelines and Methods for Conducting the Safety Assessment Process on Civil Airborne Systems and Equipment |

# 参 考 文 献

[ 1 ] 《航空工业科技词典》编辑委员会.航空工业科技词典[M].北京：国防工业出版社,1982.

[ 2 ] 曹鹤荪.英汉词典与空间技术词典[M].北京：国防工业出版社,1976.

[ 3 ] 华人杰.英汉航空航天新词典[M].上海：上海科学普及出版社,1999.

[ 4 ] 郭博智,陈迎春.商用飞机专业术语[M].北京：航空工业出版社,2011.

[ 5 ] 全国语言与术语标准化技术委员会.GB/T 20001.1—2001 标准编写规则 第1部分：术语[S].北京：中华人民共和国国家质量监督检验检疫总局,2001.

[ 6 ] 中国民用航空局.CCAR-21-R4 民用航空产品和零部件合格审定规定[S].北京：中国民用航空局,2017.

[ 7 ] 中国民用航空局.CCAR-61-R5 民用航空器驾驶员合格审定规则[S].北京：中国民用航空局,2016.

[ 8 ] 中国民用航空局.CCAR-121-R7 大型飞机公共航空运输承运人运行合格审定规则[S].北京：中国民用航空局,2021.

[ 9 ] 中国民用航空局.CCAR-183AA-R1 民用航空适航委任代表和委任单位代表管理规定[S].北京：中国民用航空局,2017.

[10] 中国民用航空局.CCAR-395-R2 民用航空器事件调查规定[S].北京：中国民用航空局,2022.

[11] 中国民用航空局.CCAR-398 民用航空安全管理规定[S].北京：中国民用航空局,2018.

[12] FAA. 14 CFR Part 1 Definitions and Abbreviations[S]. Washington：FAA, 2021.

[13] FAA. AC 21-43A Production Under 14 CFR Part 21,Subparts F,G, K,and O Document Information[S]. Washington：FAA,2015.

[14] 中国民用航空局.AC-183-01 生产检验委任代表工作程序手册编写指南[S].北京：中国民用航空局,1996.

[15] 中国民用航空局.AC-61-023 驾驶员机型资格规范评审及评审结论的应用[S].北京：中国民用航空局,2018.

[16] 中国民用航空局.AP-183-AA-2018-11 适航委任单位代表管理程序[S].北京：中国民用航空局,2018.

[17] 中国民用航空局.AP-21-2 关于国产民用航空产品服务通告管理规定[S].北京：中国民用航空局,1988.

[18] 中国民用航空局.AP-21-AA-2014-31R1 航空器型号合格审定试飞安全计划[S].北京：中国民用航空局,2014.

[19] 中国民用航空局.AP-21-AA-2019-31 生产批准和监督程序[S].北京：中国民用航空局,2019.

[20] 中国民用航空局.AP-21-AA-2020-12 技术标准规定项目批准书合格审定程序[S].北京：中国民用航空局,2020.

[21] 中国民用航空局.AP-21-AA-2020-13 零部件制造人批准书合格审定程序[S].北京：中国民用航空局,2020.

[22] 中国民用航空局.AP-21-AA-2022-11 型号合格审定程序[S].北京：中国民用航空局,2022.

[23] 中国民用航空局.AP-21-AA-2022-16R1 民用航空产品和零部件故障、失效和缺陷报告处理程序[S].北京：中国民用航空局,2022.

[24] 中国民用航空局.AP-21-AA-2022-31R1 生产批准和监督程序[S].北京：中国民用航空局,2022.

[25] 中国南方航空股份有限公司.A380 缩略语手册[G].中国南方航空股份有限公司,2022.

[26] 中国南方航空股份有限公司.波音 B787 缩略语手册[G].中国南方航空股份有限公司,2022.

[27] Airbus.空客缩写字典[S].Toulouse：Airbus,2007.

[28] 中航商用飞机有限责任公司.工程常用缩略语手册[G].中航商用飞机有限责任公司,2007.

[29] 中国东方航空集团有限公司.东航技术部词典[G].中国东方航空集团有限公司,2011.

[30] 中国民用航空局.MH/T0020-2012 民用航空器材管理术语[S].北京：

中国民用航空局,2012.

[31] 中华人民共和国政策法规司.中华人民共和国民用航空法[S].北京：中国法制出版社,2021.

[32] 中国民用航空局.民用航空术语汇编[S].北京：中国民用航空局,2001.

[33] EASA. CS-Definitions and abbreviations used in Certification Specifications for products, parts and appliances[S]. Toulouse：EASA, 2010.

# 英文名称索引